A NOTE FROM THE AUTHOR

Congratulations on your decision to take the AP Statistics exam! Whether or not you're completing a year-long AP Statistics course, this book can help you prepare for the exam. In it you'll find information about the exam as well as Kaplan's test-taking strategies, a targeted review that highlights important concepts on the exam, and practice tests. Take the diagnostic test to see which subject you should review most, and use the full-length exams to get comfortable with the testing experience. The review chapters include summaries of the most important AP exam topics, so that even if you haven't completed them in class, you won't be surprised on test day. Don't miss the strategies for answering the free-response questions: You'll learn how to cover the key points AP graders will want to see.

By studying college-level statistics in high school, you've placed yourself a step ahead of other students. You've developed your critical thinking and time-management skills, as well as your understanding of the practice of statistical research. Now it's time for you to show off what you've learned on the exam.

Best of luck,

Bruce Simmons

Mary Jean Bland

Barbara Wojciechowski

RELATED TITLES

AP Biology

AP Calculus AB & BC

AP Chemistry

AP English Language & Composition

AP English Literature & Composition

AP Environmental Science

AP European History

AP Human Geography

AP Macroeconomics/Microeconomics

AP Physics B & C

AP Psychology

AP U.S. Government & Politics

AP U.S. History

AP World History

SAT Premier with CD-ROM

SAT: Strategies, Practice, and Review

SAT Subject Test: Biology E/M

SAT Subject Test: Chemistry

SAT Subject Test: Literature

SAT Subject Test: Mathematics Level 1

SAT Subject Test: Mathematics Level 2

SAT Subject Test: Physics

SAT Subject Test: Spanish

SAT Subject Test: U.S. History

SAT Subject Test: World History

AP® STATISTICS

2011

Bruce Simmons
Mary Jean Bland
Barbara Wojciechowski

KAPLAN
PUBLISHING
New York

Published by Kaplan Publishing, a division of Kaplan, Inc.
1 Liberty Plaza, 24th Floor
New York, NY 10006

Printed in the United States of America

10 9 8 7 6 5 4 3 2 1

ISBN-13: 978-1-60714-537-0

TABLE OF CONTENTS

PART THREE: AP STATISTICS REVIEW

PART FOUR: PRACTICE TESTS

ABOUT THE AUTHORS

Bruce Simmons has taught AP Statistics and other math courses at St. Stephen's Episcopal School in Austin, Texas, since 1994.

Mary Jean Bland has taught at numerous schools and universities, and has served as a statistician for the Internal Revenue Service for 21 years.

Barbara Wojciechowski has taught mathematics at various institutions, including Johnson & Wales University, the Citadel, and the College of Charleston.

KAPLAN PANEL OF AP EXPERTS

Congratulations—you have chosen Kaplan to help you get a top score on your AP exam.

Kaplan understands your goals, and what you're up against—achieving college credit and conquering a tough test—while participating in everything else that high school has to offer.

You expect realistic practice, authoritative advice, and accurate, up-to-the-minute information on the test. And that's exactly what you'll find in this book, as well as in every other book in the AP series. To help you (and us!) reach these goals, we have sought out leaders in the AP community. Allow us to introduce our experts:

MICHAEL KAPLAN teaches at Fair Lawn High School in New Jersey, where he has been teaching AP Statistics for ten years.

LEE KUCERA has been teaching Statistics at Capistrano Valley High School in Mission Viejo, California, for 12 years, where she has been a teacher for 26 years. She has also been a reader for the AP Statistics exam since its inception in 1997.

JODIE MILLER has been a mathematics teacher at the secondary and college levels for the past ten years. As a participant in the AP Statistics program, she has served as an exam reader, table leader, workshop consultant, and multiple-choice item writer for the College Board since 1999.

MARY MORTLOCK has been teaching Statistics since 1996, beginning at Thomas Jefferson High School for Science & Technology in Alexandria, Virginia, and continuing on to her current position at Cal Poly State University in California, where she has taught since 2001. Mary has also been a reader for the AP Statistics exam since 2000.

DR. MURRAY SIEGEL has been teaching Statistics at the high school level since 1976 and at the college level since 1977. He has been a College Board AP Statistics consultant since 1995, served as an exam reader in 1998 and 2002, and served as an exam table leader from 2003–2005. He currently teaches AP Statistics at the South Carolina Governor's School for Science & Mathematics.

Part One

THE BASICS

CHAPTER 1: INSIDE THE AP STATISTICS EXAM

Advanced Placement (AP) exams have been around for a half century. While the format and content have changed over the years, the basic goal of the AP program remains the same: to give high school students a chance to earn college credit or advanced placement. To do this, a student needs to do two things:

1. Find a college that accepts AP scores
2. Do well enough on the exam

The first part is easy, since a majority of colleges accept AP scores in some form or another. The second part requires a little more effort. If you have worked diligently all year in your coursework, you've laid the groundwork. The next step is familiarizing yourself with the test.

INTRODUCTION TO THE AP STATISTICS EXAM

If you are reading this book, chances are that you are already in (or thinking about being in) an AP Statistics course. You can spend $86 and take the AP Statistics exam, and if you score high enough on it, many universities will give you three hours of credit (one class) in a related introductory statistics course. Or you can spend $1,500 (a rough estimate for three credit hours) and take the introductory statistics course in college—usually in an auditorium filled with at least 100 students where you are unable to meet the professor let alone ask any questions or get extra help—where you will learn the EXACT same material that is taught in your AP Statistics course.

You don't have to be a statistical genius to know that 3 hours is a relatively small part of 36 hours, a little more than 8%. Likewise, $86 is a relatively small percentage of $1,500, roughly 5%. By doing well on a 3-hour exam, you can forgo the headache-inducing chaos of an introductory course—for a fraction of the price!

Depending on the college, a score of 4 or 5 on the AP Statistics test will allow you to leap over the freshman intro course and jump right into more advanced classes. These classes are usually smaller in size, better focused, more intellectually stimulating, and simply put, just more interesting than a basic course. If you are solely concerned about fulfilling your math requirement so you can get on with your study of Elizabethan music or political science or some such non-statistics-related area, the AP test can help you there, too. Ace the AP Statistics exam and, depending on the requirements of the college you choose, you may never have to take a mathematics class again.

OVERVIEW OF THE TEST STRUCTURE

The College Board—the company that creates the AP exams—releases a list of the topics covered on the exam. The College Board even provides the percentage amount that each topic appears on the exam. This information is useful to give you an idea of not only what is covered on the exam, but also how much of the exam covers each topic. Anyone considering taking the test should check out the following breakdown.

TOPICS COVERED ON THE AP STATISTICS EXAMINATION

I. Exploring Data: Describing Patterns and Departures from Patterns (25%)

A. Constructing and interpreting graphical displays of distributions of univariate data

B. Summarizing distribution of univariate data

C. Comparing distributions of univariate data

D. Exploring bivariate data

E. Exploring categorical data

II. Sampling and Experimentation: Planning and Conducting a Study (15%)

A. Overview of methods of data collection

B. Planning and conducting surveys

C. Planning and conducting experiments

D. Generalizability of results and types of conclusions that can be drawn from observational studies, experiments, and surveys

III. Anticipating Patterns: Exploring Random Phenomena Using Probability and Simulation (25%)

A. Probability

B. Combining independent random variables

C. The normal distribution

D. Sampling distributions

IV. Statistical Inference: Estimating Population Parameters and Testing Hypotheses (35%)

A. Estimation (point estimators and confidence intervals)

B. Tests of significance

The AP Statistics exam is three hours long, and consists of two parts: a multiple-choice section and a free-response section. In Section I, you have 90 minutes to answer 40 multiple-choice questions with five answers each. This section is worth 50% of your total score.

AP EXPERT TIP

The free-response problems will almost always involve real-life scenarios. It is important to understand how the statistical concepts relate to the context of the problem.

Section II consists of six "free-response" questions that are worth the other 50% of your total score. The term "free-response" means roughly the same thing as "large, multistep, and involved," since you will spend 90 minutes answering only six questions. Although these free-response questions are long and often broken down into multiple parts, they usually do not cover obscure topics. Instead, they take a fairly basic statistical concept and ask you a bunch of questions about it.

Section II is broken into two subparts. One part consists of five questions that require 12 minutes apiece to answer. These questions usually relate to one topic or category. The other part consists of an "investigative task." This means the question will require about 25 minutes of the allotted time and will be a broad-ranging question involving numerous topics and concepts. Whatever free-response questions you are given will require a lot of statistical work, but it will be fundamental statistical work.

CALCULATORS

Not only are calculators allowed on the AP Statistics exam, but you are EXPECTED to bring a graphing calculator with statistical capabilities. Minicomputers, pocket organizers, electronic writing pads (e.g., Newton), and calculators with QWERTY will not be allowed. Your calculator's memory can be used only to store programs that bring your calculator's abilities/functions up to the level of other acceptable calculators. Accessing any notes on or copying and storing any part of the exam into your calculator will be considered cheating on the exam. If you forget your calculator, you will be up the proverbial creek without a paddle since calculators CANNOT be shared. If you sometimes think you are a walking embodiment of Murphy's Law, you may bring up to TWO calculators (in case one blows up in the middle of the exam).

Formulas and Tables

Formulas and tables are an intricate part of understanding and studying statistics. However, the AP Statistics course is more concerned with you developing and understanding fundamental concepts than memorizing and regurgitating formulas. You will be given an appropriate list of formulas and tables to help in answering the questions. You are expected to have a *working familiarity* with the formulas and graphs, so this information is intended to be needed only as a reference.

A list of the formulas and tables that will be provided appears on page 431.

HOW THE EXAM IS SCORED

When your three hours of testing are up, your exam is sent away for grading. The multiple-choice part is handled by a machine, while qualified graders—current high school and college statistics teachers—grade your responses to Section II. After a seemingly interminable wait, your composite score will arrive. Your results will be placed into one of the following categories, reported on a 5-point scale.

5. Extremely well qualified (to receive college credit or advanced placement)
4. Well qualified
3. Qualified
2. Possibly qualified
1. No recommendation

Some colleges will give you college credit for a score of 3 or higher, but it's much safer to score a 4 or a 5. If you have an idea of which colleges you want to go to, check out their Websites or call the admissions office to find out what their particular rules regarding AP scores are. If you don't get the grade required by your college, it is always a good idea to take a copy of the syllabus and a copy of your AP exam results to show the department when you arrive. In some cases, you *may* be able to obtain credit even though you have not met the official requirement.

REGISTRATION AND FEES

You can register for the exam by contacting your guidance counselor or AP Coordinator. If your school doesn't administer the exam, contact AP Services for a listing of schools in your area that do. The fee for each AP exam is $86. For students with financial need, a $22 reduction is available. To learn about other sources of financial aid, contact your AP Coordinator.

For more information on all things AP, visit collegeboard.com or contact AP Services:

AP Services
P.O. Box 6671
Princeton, NJ 08541-6671
Phone: 609-771-7300 or 888-225-5427 (toll-free in the United States and Canada)
E-mail: apexams@info.collegeboard.org

ADDITIONAL RESOURCES

The Cartoon Guide to Statistics
Larry Gonick and Woollcott Smith, 1994
HarperCollins Publishers, Inc., New York, NY
ISBN: 0-062-73102-5

Elementary Statistics (9th Edition)
Robert Johnson and Patricia Kuby, 2003
Duxbury Press
ISBN: 0-534-39915-0

Introduction to the Practice of Statistics (4th Edition)
David S. Moore and George P. McCabe, 2002
W. H. Freeman and Company, New York, NY
ISBN: 0-716-79657-0

Introduction to Statistics and Data Analysis (2nd Edition)
Roxy Peck, Chris Olsen, and Jay Devore, 2005
Brooks/Cole-Thomson Learning, Belmont, CA
ISBN: 0-534-46710-5

Statistics (3rd Edition)
David Freedman, Robert Pisani, and Roger Purves, 1997
W. W. Norton & Company
ISBN: 0-393-97083-3

Statistical Methods (8th Edition)
George W. Snedecor and William G. Cochran, 1989
Iowa State Press, Ames, IA
ISBN: 0-813-81561-4

Stats: Modeling the World (1st Edition)
David E. Bock, Paul F. Velleman, and Richard D. De Veaux, 2004
Pearson—Addison Wesley, Boston, MA
ISBN: 0-201-73735-3

Workshop Statistics: Discovery with Data and the Graphing Calculator (2nd Edition)
Allan J. Rossman, Beth L. Chance, and J. Barr Von Oehsen, 2001
Key College Publishing, Emeryville, CA
ISBN: 1-930-19004-2

CHAPTER 2: STRATEGIES FOR SUCCESS: PREPARING FOR THE EXAM

Sixty years ago, there was only one standardized test, administered by the U.S. Army to determine which enlistees were qualified for officer training. The idea behind the U.S. Army officer's exam has adapted and flourished in both the public and private sectors. Nowadays, you can't go a semester of school without taking some letter-jumble exam like the PSAT, SAT, ACT, or ASVAB. As you may already know, developing certain test-taking strategies is a good way to help prepare yourself for these exams. Since everyone reading this has taken a standardized test of one kind or another, you are probably familiar with some of the general strategies that help students increase their scores on a standardized exam.

GENERAL TEST-TAKING STRATEGIES

1. **Pacing.** Since many tests are timed, proper pacing is essential to answering every question in the time allotted. Too often students spend too much time on questions in the beginning and they run out of time before getting a chance at every problem.
2. **Process of Elimination.** On every multiple-choice test you ever take, the answer is given to you. The only difficulty resides in the fact that the correct answer is hidden among incorrect choices. Even so, the multiple-choice format means you don't have to pluck the answer out of the air. If you can eliminate answer choices you know are incorrect, it will increase your chances of identifying the correct answer.
3. **Patterns and Trends.** The key word here is the *standardized* in "standardized testing." Being standardized means that tests don't change greatly from year to year. Sure, each question won't be the same, and different topics may be covered in one administration to the next, but there will also be a lot of overlap from year to year. That's the nature of *standardized* testing: if the test changed wildly each time it came out, it would be useless as a tool for

comparison. Because of this, certain patterns can be uncovered about any standardized test. Learning about these trends and patterns can help students taking the test for the first time.

4. **The Right Approach.** Having the right mindset plays a large part in how well people do on a test. Those students who are nervous about the exam and hesitant to make guesses often fare much worse than students with an aggressive, confident attitude. Students who start with question 1 and plod on from there don't score as well as students who pick and choose the easy questions first before tackling the harder ones. People who take a test cold have more problems than those who take the time learning about the test beforehand. In the end, simple factors like these distinguish people who are good test-takers from those who struggle even if they know the material.

These points are valid for every standardized test, but they are quite broad in scope. The rest of this chapter will discuss how these general ideas can be modified to apply specifically to the AP Statistics exam. These test-specific strategies—combined with the factual information covered in your course and in this book's review—will help you succeed on this test.

HOW THIS BOOK CAN HELP, PLUS SPECIFIC AP STATISTICS EXAM STRATEGIES

If you're holding this book, chances are you are already gearing up for the AP Statistics exam. Your teacher has spent the year cramming your head full of the statistical know-how you will need to have at your disposal. But there is more to the AP Statistics exam than statistical know-how. You have to be able to work around the challenges and pitfalls of the test—and there are many—if you want your score to reflect your abilities. You see, studying statistics and preparing for the AP Statistics exam is not the same thing. Rereading your textbook is helpful, but it is not enough.

That's where this book comes in. We'll show you how to marshal your knowledge of statistics and put it to brilliant use on test day. We'll explain the ins and outs of the test structure and question format so you won't experience any nasty surprises. We'll even give you answering strategies designed specifically for the AP Statistics exam.

Effectively preparing yourself for the AP Statistics exam means doing some extra work. You need to review your text *and* master the material in this book. Is the extra push worth it? If you have any doubts, think of all the interesting things you could be doing in college instead of taking an intro course filled with facts you already know.

We have scrutinized and analyzed the AP Statistics exam more times than the results of a regression analysis on winning lotto numbers. We did this to learn everything about the test, so we could then pass on this information to you. *Kaplan's AP Statistics* contains precisely the information you will need to ace the test. There's nothing extra in here to waste your time—no pointless review material you won't be tested on, no rah-rah speeches. Just the most potent test-preparation tools available:

1. **Test Strategies Geared Specifically to the AP Statistics Exam.** Many books give the same talk about Process of Elimination that's been used for every standardized test given in the past 20 years. We're going to talk about Process of Elimination as it applies to the AP Statistics exam, and only the AP Statistics exam. There are several skills and general strategies that work for this particular test and these will be covered in the next chapter.
2. **A Well-Crafted Review of All the Relevant Subjects.** The best test-taking strategies in the world won't get you a good score if you can't tell the difference between the Central Limit Theorem and a dot plot of univariate data. At its core, this AP exam covers a wide range of statistical topics, and learning these topics is *absolutely* necessary. However, chances are good you're already familiar with these subjects, so an exhaustive review is not needed. In fact, it would be a waste of your time. No one wants that, so we've tailored our review section to focus on how the relevant topics typically appear on the exam, and what you need to know in order to answer the questions correctly. If a topic doesn't come up on the AP Statistics exam, we don't cover it. If it does appear on the exam, we'll provide the facts you need to navigate the problem safely.
3. **Full-Length Practice Tests.** Few things are better than experience when it comes to standardized testing. Taking a practice AP exam gives you an idea of what it is like to answer statistics questions for three hours. It's definitely not a fun experience, but it is a helpful one. Practice exams give you the opportunity to find out what areas are your strongest, and what topics you should spend some additional time studying. And the best part is that it does not count! Mistakes you make on the practice test teach you how to avoid those same mistakes on the real test.

HOW TO APPROACH THE MULTIPLE-CHOICE QUESTIONS

The worst thing that can be said about the AP Statistics multiple-choice questions is that they count for 50% of your total score. Although you might not like multiple-choice questions, there's no denying the fact that it's easier to guess on a multiple-choice question than on an open-ended question. On a multiple-choice problem, the correct answer is always there in front of you; the trick is to find it among incorrect answers.

Remember, there are a total of 40 multiple-choice questions on the AP Statistics exam. There are two distinct question types.

1. **Content Questions.** These are the questions that DO NOT deal with math. As their name implies, these questions are concerned with statistical concepts expressed through words, not numbers. Therefore, you can easily identify these questions by looking at the answer choices, because all the answers will be expressed verbally, not numerically. Since very little or no calculation is required to answer these questions, they tend to take very little time to answer. Pay attention to the fact that more of these questions can be answered more quickly than questions that call for calculations. These questions account for less than half of the

questions on the Statistics multiple-choice exam, usually 10–15 questions per exam. A typical Content Question looks like this:

> A volunteer for a city council candidate's campaign conducts polls to estimate the proportion of people in the city who are planning to vote for this candidate in the upcoming city council elections. Fifteen days before the election, the volunteer plans to double the size of the sample. The main purpose of this is to:
>
> (A) decrease the standard deviation of the sampling distribution of the sample proportion
> (B) increase the variability in the population
> (C) increase bias due to the interviewer effect
> (D) reduce the effects of confounding variables
> (E) avoid nonresponse bias

Don't worry about the answer yet; we'll come back to this question later. It is important to note that although there is not a set order of difficulty for the questions on the AP Statistics exam, you may find that there are more "easy" questions near the beginning of the test than near the end. Therefore—to use statistical terms—question 35 has a higher probability of being a "difficult" question than question 3.

2. **Calculation Questions.** As their name implies, Calculation Questions are the questions that involve math. On the AP Statistics exam, these questions make up the majority of multiple-choice questions. Since these questions require both conceptual knowledge and a familiarity with formulas, they tend to take more time to answer than the Content Questions. Even when you know how to solve a question, you may still have to flip back and forth between the provided tables and formulas and the question, which is not only inconvenient, but time-consuming. You can easily identify these questions by the equations or numerical values in the answer choices. A typical Calculation Question looks like this:

> A least-squares regression line was fitted to the weights (in pounds) versus age (in years) of a group of young adults. The equation of the line is $\hat{y} = 34.3 + 5.5t$, where $\hat{y}$ is the predicted weight and t is the age of the young adult. A 20-year-old has an actual weight of 120 pounds. Which of the following is the residual weight, in pounds, for this person?
>
> (A) 30.7
> (B) 24.3
> (C) 20.0
> (D) –24.3
> (E) –30.7

WHAT TO KNOW FOR THE AP EXAM

Remember that the AP Statistics exam is mainly concerned with your understanding of statistical concepts and applying those concepts effectively to "real-life" scenarios. The test writers do not want you to waste your time memorizing formulas, so they provide you with a list of formulas and tables on the exam. **These formulas are only helpful if you are familiar with their applications before you take the exam.**

A list of the formulas and tables that will be provided to you on the exam is found on page 431.

You may be able to answer calculation questions quickly using your calculator, so you should be very familiar with all of the statistical functions on your calculator. See chapter 19 for a review of the statistics commands on the graphing calculator.

INCREASE THE PROBABILITY OF SCORING WELL ON THE MULTIPLE-CHOICE SECTION

There are two factors that you can use on the multiple-choice section of the AP Statistics exam to help increase your scoring output:

1. **Time.** Most standardized test takers start with question 1 and work consecutively through the exam until they get to the end or run out of time, whichever comes first. Students run out of time because they get stuck on some problems that are more difficult, and thus are more time-consuming. When this happens, the student not only misses the difficult questions, but by running out of time, does not get a chance to answer the easier questions at the end. The result is a lower score.

   Since you know that you have 90 minutes to answer 40 multiple-choice questions, make sure you don't spend too much time on any one question. *Answer the questions that take the least amount of time first;* this way, you won't get bogged down on one question and miss the opportunity to answer 3 or 4 other questions.

2. **Knowledge.** What students fail to realize is that they don't need to get every multiple-choice question right to score well on the AP Statistics exam. Don't try for perfection. All you need is a score of 4 or 5, and that means you need to get a large portion, but not all, of the multiple-choice questions right. On an AP exam, nothing is gained by a perfect score.

   If you are not going to have enough time to get to all the questions, make sure that you get to all the questions that you KNOW. This means you should not answer the questions in consecutive order, but rather, in order of your ability. Take a minute and look over the list of topics covered on the AP Statistics exam on pages 4 and 5 of chapter 1. Divide these topics into two separate lists: "Statistics I know" and "Statistics I have trouble with." Keep this list in mind when you begin the multiple-choice section of the exam. On your first pass through the questions, answer all the questions that deal with the concepts in the "Statistics I know" list. Save all the questions dealing with "Statistics I have trouble with" list for the second pass. *Remember: You want to correctly answer as many questions as possible, so be sure you are able to answer all the ones you KNOW in the given amount of time.*

On the AP Statistics exam:

- There is no order of difficulty. Easy questions are the ones you know, medium questions are the ones you "sort of" know, and harder questions are the ones you do not know.
- No two questions are connected to each other in any way.
- There's no system to when certain statistics concepts appear and in what question.

There is no overall pattern as to how the multiple-choice questions are presented to you. The questions are random and disconnected, and therefore your approach can be the same way, so long as you follow the three basic steps to answering AP Statistics questions:

1. **First pass**—Read through and answer all of the Content Questions you know first. Since these questions require little or no math, they will take less time to answer. Remember that you can identify Content Questions by looking at the answer choices, so read through the exam and pick out the Content Questions.
2. **Second pass**—After you have answered all the Content Questions you know, read through the Calculation Questions. If a Calculation Question falls into the "Statistics I know" category, answer it and move on to the next Calculation Question.
3. **Third pass**—This is for all the questions that fall into the "Statistics I have trouble with" category. Saving these for last helps you maximize your score output because you have already answered all the questions you know. For these remaining questions, try to answer Content Questions first, since they generally require less time and effort. Be aware of the time limit and focus on eliminating incorrect answers rather than finding the one correct answer.

For the difficult questions, you should always try to eliminate some answer choices and then make an educated guess. Admittedly, the AP Statistics exam is a test of specific knowledge, so picking the right answer from the bad answer choices is harder to do than it is in other standardized tests. Still, it can be done. The following two keys to the AP Statistics exam are easy to remember and will come in handy on the tougher multiple-choice problems.

1. Comprehensive, Not Sneaky

Some tests are sneakier than others. They have convoluted writing, questions designed to trip you up mentally, and a host of other little tricks. Students taking a sneaky test often have the proper facts, but get the question wrong because of a trap in the question itself.

The AP Statistics exam is NOT a sneaky test. Its objective is to see how much statistical knowledge you have. To do this, it asks a wide range of questions from an even wider range of statistics topics. The exam tries to cover as many different facts in statistics as it can, which is why the problems jump around from chi-square distribution to least-squares regression lines. The test works hard to be as comprehensive as it can be, so that students who only know one or two statistical topics will soon find themselves struggling.

Understanding these facts about how the test is designed can help answer questions on it. The AP Statistics exam is comprehensive, not sneaky; it makes questions hard by asking about hard subjects, not by crafting hard questions. And you have taken an AP Statistics course, so...

You don't have time to ponder every tough question, so trusting your instincts can help you from getting bogged down and wasting time on a problem. You might not get every educated guess correct, but again, the point isn't about getting a perfect score. It's about getting a good score, and surviving hard questions by going with your gut feeling is a good way to achieve this.

On other problems, though, you might have no inkling of what the correct answer should be. In that case, turn to the following key idea.

AP EXPERT TIP

Trust your instincts when guessing. If you think you know the right answer, chances are you dimly remember the topic being discussed in your AP course. The test is about knowledge, not traps, so trusting your instincts will help more often than not.

2. Think "Good Statistics!"

The AP Statistics exam rewards good statisticians. The test wants to foster future statisticians by covering fundamental topics. What the test does not want is bad statistics. It *does not* want answers that are *factually incorrect, too extreme to be true,* or *irrelevant to the topic at hand.*

Yet these bad statistics answers invariably appear, because it's a multiple-choice test and you have to have four incorrect answer choices around the one right answer. So if you don't know how to answer a problem, look at the answer choices and think "Good Statistics." This will lead you to find some poor answer choices that can be eliminated.

Take a look at the following sample question:

> A volunteer for a city council candidate's campaign conducts polls to estimate the proportion of people in the city who are planning to vote for this candidate in the upcoming city council elections. Fifteen days before the election, the volunteer plans to double the size of the sample. The main purpose of this is to
>
> (A) decrease the standard deviation of the sampling distribution of the sample proportion.
> (B) reduce the variability in the population.
> (C) increase bias due to the interviewer effect.
> (D) reduce the effects of confounding variables.
> (E) increase nonresponse bias.

Even if you forget exactly what happens when the sample size is doubled, you have still taken a year of statistics and have a good idea of what the general goals of statistical analysis are. Looking at your answer choices, has it ever been a goal of statisticians to *increase* bias? Absolutely not. Thinking "Good Statistics," you know that the whole foundation of statistics is to calculate *reasoned* and *objective* predictions. Therefore, answers (C) and (E) must be wrong. You now have a one out of three shot, so take a guess.

You would be surprised how many times the correct answer on a multiple-choice question is a simple, blandly-worded fact like "Probability is used to anticipate the distribution of data." No breaking news there, but it is Good Statistics: a carefully worded statement that is factually accurate.

Thinking "Good Statistics" can help you in two ways.

1. It can help you cross out extreme answer choices or choices that are untrue or out of place.
2. It can occasionally point you toward the correct answer, since the correct answer will be a factual piece of information sensibly worded.

No single strategy is 100% effective every time, but on a tough multiple-choice problem, these techniques can sometimes make the difference between a score of 3 and a score of 4.

Of course, the multiple-choice questions only account for 50% of your total score. To get the other 50%, you have to tackle Section II: the free-response questions.

HOW TO APPROACH THE FREE-RESPONSE QUESTIONS

On the AP Statistics exam, you will be required to answer six free-response (essay) questions in 90 minutes. These questions are divided up into two parts: Part A and Part B. Part A consists of five questions and makes up 75% of the Section II grade. You should spend about 65 minutes on Part A. Part B consists of one investigative task and makes up 25% of the Section II grade. You should save 25 minutes to answer Part B.

This means that each question in Part A is worth 15% of your Section II score and should take roughly 12 minutes apiece to answer. The one question in Part B is worth twice as much as one question in Part A and should take about twice the time to answer.

AP EXPERT TIP

It's possible that Part B could include a statistical topic that you have never even seen before! However, you will be able to understand this problem and earn points with a strong understanding of more basic statistical concepts.

Before answering each question, make sure you understand what is being asked and jot down any thoughts you have about the answer. Write down any keywords you want to mention.

- **Part A** will require you to answer relatively short statistical problems, usually broken up into smaller parts.
- **Part B**, the longer investigative task, will require you to use different statistical strategies and techniques related to the investigative task. This question is usually broken up into three or four smaller subparts.

Free-response questions can come in any shape or size, but there are some things you can know about them beforehand.

Two Points to Remember about Free-Response Questions

AP EXPERT TIP

Each part of a free-response question is worth points, so don't give up on a free-response question just because one part seems too difficult.

1. **Most Questions Are Stuffed with Smaller Questions.** You usually won't get one broad question like, "Discuss the impact of statistical analysis in our world." Instead, you'll get an initial setup followed by questions labeled (a), (b), (c), and so on. Each of these smaller questions often requires as much work as any individual multiple-choice question, so be sure to allow enough time to completely answer each one.
2. **Expressing Smart Statistics Earns You Points!** For each subquestion on a free-response question, points are given for solving correctly. The AP Statistics graders use a scoring rubric, which is a blueprint for what a good answer should look like. Every subsection of a question has two to five key ideas/solutions attached to it. If you write about one of those ideas or solve part of the problem, you earn yourself a point. There's a limit to how many points you can earn on a single subquestion, but it basically boils down to this: Writing and expressing smart things about each question will earn you points toward that question. *Even if you don't know how to solve for the exact answer, if you know certain equations are involved in solving for the answer, write them out—you may get a few points that can help you out in your overall Section II score.*

Again, you have 12 minutes for each of the first five questions and 25 minutes for the remaining investigative task, so don't rush unnecessarily; try to focus on good problem-solving methods. Use your time to be as precise as you can for each subquestion. Sometimes doing well on one subquestion earns you enough points to cover up for another subquestion you are not strong in. When all the points are tallied for that free-response problem, you come out ahead on total points, even though you didn't ace every subquestion.

On the other hand, do not ramble on beyond what is necessary to explain the answer for each question. It IS possible to lose points by discussing things that are not relevant to the question at hand.

Every free-response question will involve a real-life scenario, and it is important to understand how the statistical concepts will relate to this scenario. Just showing the statistics in your free-response solution will not earn full credit. It is important that you clearly communicate the meaning of your solution in the context of the problem.

Beyond these points, there's a bit of a risk in the free-response questions since there are only six questions. If you get a question on a topic you're weak in, things might look pretty grim for that problem. Still, take heart. Quite often, you'll earn some points on every question since there will be some subquestions or segment that you are familiar with. Remember, the goal is not perfection. If you can ace the longer question and slug your way to partial credit on the shorter ones, or vice versa, you will put yourself in position to get a good score on the entire test. The total test score is the Big Picture, so don't lose sight of it just because you don't know the answer to one subquestion on Section II.

Be sure to use all the strategies discussed in this chapter when taking the practice exams. Trying out these strategies here will get you comfortable with them, and you should be able to put them to good use on the real exam.

Of course, all the strategies in the world can't save you if you don't know anything about statistics. The next part of the book will help you review the primary concepts and facts that you can expect to encounter on the AP Statistics exam, but first take the short Diagnostic test to determine which areas to focus on.

TIME MANAGEMENT

You have learned a great deal about statistics throughout the school year, and it is important that you begin your review early. Do not save it all until the last week! Schedule your time so that you can review a manageable amount of material each week for at least a few weeks prior to the exam. Set weekly goals for yourself. It is also important to make time to complete the practice tests in this book.

If you don't start early, there is no reason to panic. Take the Diagnostic Test in Part Two of this book so that you can then focus your studies on the specific subjects in which you need help. Also, plan to spend some time working on the practice tests in this book before the AP exam.

It is also a good idea to work together with friends to prepare for the AP test. Choose friends who are serious about studying. Set specific goals for the review sessions, such as completing and reviewing a practice test. You can work together to make sure you all understand the concepts completely. It also seems more fun to study together and it is a good opportunity to relieve some stress while still making progress towards the AP exam. Of course, studying in groups should not replace your individual studying.

STRESS MANAGEMENT

You can beat anxiety the same way you can beat the AP Statistics exam—by knowing what to expect beforehand and developing strategies to deal with it. Following are some time-tested and simple techniques for dealing with stress.

Visualize

Sit in a comfortable chair in a quiet setting. If you wear glasses, take them off. Close your eyes and breathe in a deep, satisfying breath of air. Really fill your lungs until your rib cage is fully expanded and you can't take in any more. Then, exhale the air completely. Imagine you're blowing out a candle with your last little puff of air. Do this two or three more times, filling your lungs to their maximum and emptying them totally. Keep your eyes closed, comfortably but not tightly. Let your body sink deeper into the chair as you become even more comfortable.

Close your eyes and start remembering a real-life situation in which you did well on a test. If you can't come up with one, remember a situation in which you did something that you were really proud of—a genuine accomplishment. Make the memory as detailed as possible. Think about the sights, the sounds, the smells, even the tastes associated with this remembered experience. Remember how confident you felt as you accomplished your goal. Now start thinking about the AP Statistics exam. Keep your thoughts and feelings in line with that prior, successful experience. Don't make comparisons between them. Just imagine taking the upcoming test with the same feelings of confidence and relaxed control.

This exercise is a great way to bring the test down to earth. You should practice this exercise often, especially when you feel burned out on exam preparation. The more you practice it, the more effective the exercise will be for you.

Exercise

Whether it's jogging, walking, biking, mild aerobics, push-ups, or a pickup basketball game, physical exercise is a very effective way to stimulate both your mind and body and to improve your ability to think and concentrate. Lots of students get out of the habit of regular exercise when they're prepping for the exam. Also, sedentary people get less oxygen to the blood, and hence to the brain, than active people. You can watch TV fine with a little less oxygen; you just can't think as well.

Any big test is a bit like a race. Finishing the race strong is just as important as being quick early on. If you can't sustain your energy level in the last sections of the exam, you could blow it. Along with a good diet and adequate sleep, exercise is an important part of keeping yourself in fighting shape and thinking clearly for the long haul.

There's another thing that happens when students don't make exercise an integral part of their test preparation. Like any organism in nature, you operate best if all your "energy systems" are in balance. Studying uses a lot of energy, but it's all mental. When you take a study break, do something active. Take a 5-to 10-minute exercise break for every 50 or 60 minutes that you study. The physical exertion helps keep your mind and body in sync. This way, when you finish studying for the night and go to bed, you won't lie there unable to sleep because your head is wasted while your body wants to run a marathon.

One warning about exercise: It's not a good idea to exercise vigorously right before you go to bed. This could easily cause sleep-onset problems. For the same reason, it's also not a good idea to study right up to bedtime. Make time for a "buffer period" before you go to bed. Take 30 to 60 minutes to take a long hot shower, to meditate, to read a relaxing book, or even to watch TV.

Stay Drug-Free and Eat Healthy

Using drugs to prepare for or take a big test is not a good idea. Don't take uppers to stay alert. Amphetamines make it hard to retain information. Mild stimulants, such as coffee, cola, or over-the-counter caffeine pills can help you study longer since they keep you awake, but they can also lead to agitation, restlessness, and insomnia. Some people can drink a pot of coffee sludge and sleep like a baby. Others have one cup and start to vibrate. It all depends on your tolerance for caffeine. Remember, a little anxiety is a good thing. The adrenaline that gets pumped into your bloodstream helps you stay alert and think more clearly.

You can also rely on your brain's own endorphins. Endorphins have no side effects and they're free. It just takes some exercise to release them. Running, bicycling, swimming, aerobics, and power walking all cause endorphins to occupy the happy spots in your brain's neural synapses. In addition, exercise develops your mental stamina and increases the oxygen transfer to your brain.

To reduce stress you should eat as healthily as possible. Fruits and vegetables can help reduce stress. Low-fat protein such as fish, skinless poultry, beans, and legumes (like lentils), or whole grains such as brown rice, whole wheat bread, and pastas (no bleached flour) are good for your brain chemistry. Avoid sweet, high-fat snacks. Simple carbohydrates like sugar can make stress worse, and fatty foods may lower your immunity. Don't eat excessively salty foods either. They can deplete potassium, which you need for nerve functions.

Isometrics

Here's another natural route to relaxation and invigoration. You can do it whenever you get stressed out, including during the test. Close your eyes. Start with your eyes and—without holding your breath—gradually tighten every muscle in your body (but not to the point of pain) in the following sequence:

- Close your eyes tightly.
- Squeeze your nose and mouth together so that your whole face is scrunched up. (If it makes you self-conscious to do this in the test room, skip the face-scrunching part.)
- Pull your chin into your chest, and pull your shoulders together.
- Tighten your arms to your body, then clench your fists.
- Pull in your stomach.
- Squeeze your thighs and buttocks together, and tighten your calves.
- Stretch your feet, then curl your toes (watch out for cramping in this part).

At this point, every muscle should be tightened. Now, relax your body, one part at a time, in reverse order, starting with your toes. Let the tension drop out of each muscle. The entire process might take five minutes from start to finish (maybe a couple of minutes during the test). This clenching and unclenching exercise will feel silly at first, but if you get good at it, it can help you feel very relaxed.

COUNTDOWN TO THE TEST

It's almost over. Eat a power snack, drink some carrot juice—or whatever makes healthy sense to keep yourself going. Here are Kaplan's strategies for the three days leading up to the test.

Three Days Before the Test

Take a full-length practice test under timed conditions. Use the techniques and strategies you've learned in this book. Approach the test strategically, actively, and confidently.

Two Days Before the Test

Go over the results of your practice test. Don't worry too much about your score, or about whether you got a specific question right or wrong. The practice test doesn't count. But do examine your performance on specific questions with an eye to how you might get through each one faster and better on the test to come.

WARNING: DO NOT take a full practice exam if you have fewer than 48 hours left before the test. Doing so will probably exhaust you and hurt your score on the actual test. Maybe it will help to think of the AP Statistics exam as a marathon. Racers don't run a marathon the day before the real thing—they rest and conserve their energy!

The Night Before the Test

DO NOT STUDY. Get together an "AP Statistics exam kit" containing the following items:

- A calculator with fresh batteries
- A watch
- A few No. 2 pencils
- Erasers
- Photo ID card
- A snack—there are breaks, and you'll probably get hungry

Know exactly where you're going, exactly how you're getting there, and exactly how long it takes to get there. It's probably a good idea to visit your test center sometime before the day of the test, so that you know what to expect—what the rooms are like, how the desks are set up, and so on.

Relax the night before the test. Read a good book, take a long hot shower, watch a movie or something on TV. Get a good night's sleep. Go to bed early and leave yourself extra time in the morning.

The Morning of the Test

- Eat breakfast. Make it something substantial, but not anything too heavy or greasy.
- Don't drink a lot of coffee if you're not used to it, bathroom breaks cut into your time, and too much caffeine is a bad idea.
- Dress in layers so that you can adjust to the temperature of the test room.
- Read something non-test related. Warm up your brain with a newspaper or a magazine.
- Be sure to get there early. Allow yourself extra time for traffic, mass transit delays, and/or detours.
- Be confident! Remind yourself that you have put the time in to prepare for this exam, and you are ready to do well!

During the Test

Don't be shaken. If you find your confidence slipping, remind yourself how well you've prepared. You know the structure of the test; you know the instructions; you've had practice with—and have learned strategies for—every question type.

Remember that parts of this test will be challenging, and you are not expected to score 100% on the exam. In fact, earning about 75% of the possible points should result in a 5 on the exam. If you encounter a challenging question, remember that it is probably just as challenging for everyone. Use the strategies you've learned to earn the maximum number of points on the entire exam.

If something goes really wrong, don't panic. If the test booklet is defective—two pages are stuck together or the ink has run—raise your hand and tell the proctor you need a new book. If you accidentally mis-grid your answer page or put the answers in the wrong section, raise your hand and tell the proctor. He or she might be able to arrange for you to re-grid your test after it's over, when it won't cost you any time.

After the Test

You might walk out of the exam thinking that you blew it. This is a normal reaction. Lots of people—even the highest scorers—feel that way. You tend to remember the questions that stumped you, not the ones that you knew.

We're positive that you will have performed well and scored your best on the exam because you followed the Kaplan strategies outlined in this section. Be confident in your preparation, and celebrate the fact that the AP Statistics exam is soon to be a distant memory.

| Part Two |

DIAGNOSTIC TEST

AP STATISTICS DIAGNOSTIC TEST

The questions in this diagnostic are designed to cover most of the topics you will encounter on the AP Statistics exam. After you take it, you can use the results to give yourself a broad idea of what subjects you are strong in and what topics you need to review more. You can use this information to tailor your approach to the following review chapters. Ideally, you'll want to read all the chapters, but if pressed, you can start with the chapters and subjects you know you need to work on.

Give yourself 45 minutes for the 20 multiple-choice questions and 25 minutes for the free-response question. Time yourself, and take the entire test without interruption—you can always text your friend back *after* you finish. Also, no TV or music! You won't get to watch TV or listen to music while taking the real AP Statistics exam, so you may as well get used to it now.

Be sure to read the explanations for all questions, even those you answered correctly. Even if you got the problem right, reading another person's answer can give you insight that will prove helpful on the real exam.

Good luck on the diagnostic!

Diagnostic Test Answer Grid

To compute your score for the diagnostic test, calculate the number of questions you got wrong, then deduct $\frac{1}{4}$ of that number from the number of right answers. So, if you got 5 questions wrong out of 20, you did *not* get a score of 75. Deduct $\frac{1}{4}$ of 5(1.25) from the number of questions you got right (15). The final score out of 20 is 13.75. To set this equal to a score out of 100, set up a proportion:

$$\frac{13.75}{20} = \frac{n}{100}$$

$$20n = 1375$$

$$n = 68.75.$$

A score of 69 is a 2, so you can definitely do better. If your score is low, keep on studying to improve your chances of getting credit for the AP Exam.

1.	Ⓐ Ⓑ Ⓒ Ⓓ Ⓔ	11.	Ⓐ Ⓑ Ⓒ Ⓓ Ⓔ
2.	Ⓐ Ⓑ Ⓒ Ⓓ Ⓔ	12.	Ⓐ Ⓑ Ⓒ Ⓓ Ⓔ
3.	Ⓐ Ⓑ Ⓒ Ⓓ Ⓔ	13.	Ⓐ Ⓑ Ⓒ Ⓓ Ⓔ
4.	Ⓐ Ⓑ Ⓒ Ⓓ Ⓔ	14.	Ⓐ Ⓑ Ⓒ Ⓓ Ⓔ
5.	Ⓐ Ⓑ Ⓒ Ⓓ Ⓔ	15.	Ⓐ Ⓑ Ⓒ Ⓓ Ⓔ
6.	Ⓐ Ⓑ Ⓒ Ⓓ Ⓔ	16.	Ⓐ Ⓑ Ⓒ Ⓓ Ⓔ
7.	Ⓐ Ⓑ Ⓒ Ⓓ Ⓔ	17.	Ⓐ Ⓑ Ⓒ Ⓓ Ⓔ
8.	Ⓐ Ⓑ Ⓒ Ⓓ Ⓔ	18.	Ⓐ Ⓑ Ⓒ Ⓓ Ⓔ
9.	Ⓐ Ⓑ Ⓒ Ⓓ Ⓔ	19.	Ⓐ Ⓑ Ⓒ Ⓓ Ⓔ
10.	Ⓐ Ⓑ Ⓒ Ⓓ Ⓔ	20.	Ⓐ Ⓑ Ⓒ Ⓓ Ⓔ

DIAGNOSTIC TEST

SECTION I: MULTIPLE-CHOICE QUESTIONS

Time—45 Minutes 20 Questions

Directions: Section 1 of this examination contains all multiple-choice questions. Decide which of the suggested answers best suits the question. You may use the formulas and tables found in the Appendix on page 431 on the test.

1. The following is a display of census report data of marital status (never married, married, divorced) against age for a sample of residents of Portland, Oregon.

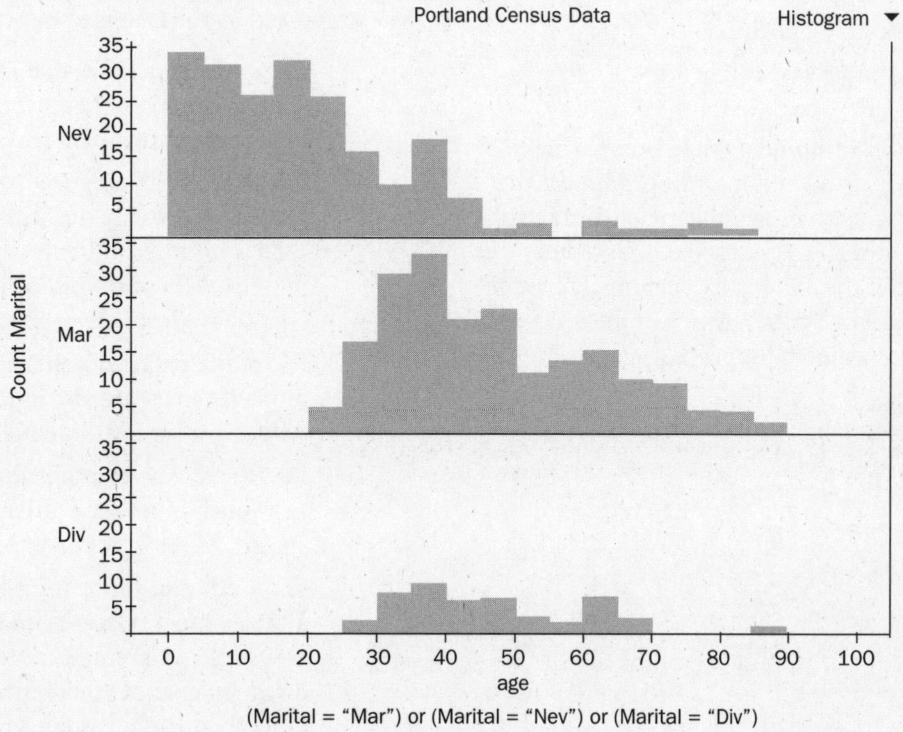

Which choice correctly interprets information about the sample represented in the graph?

(A) The median age of those never married is greater than the mean age of those never married.

(B) The range of ages of those who are divorced is greater than the range of ages of those never married.

(C) There are more people age 25 to 35 years old who were never married than married people in the same age range.

(D) At least one of the three distributions has an outlier on the low end of the distribution.

(E) Ignoring outliers, the distribution of those who are divorced is the least skewed.

GO ON TO THE NEXT PAGE

2. A confidence interval for a population mean is computed from a random sample with mean $\bar{x}$, standard deviation s_x, and sample size n. Which of the following will decrease the margin of error by a factor of $\frac{1}{4}$, assuming all assumptions for computation of the confidence interval are satisfied?

   (A) Use sample size $\frac{n}{2}$.
   (B) Use sample size $2n$.
   (C) Use sample size $4n$.
   (D) Use sample size $8n$.
   (E) Use sample size $16n$.

3. A classroom equipment manufacturer makes meter sticks. Students at a school who use the meter sticks suspect that the meter sticks tend to be too long. If these students wanted to test whether the meter sticks genuinely have a mean length of 1 meter, which of the following hypotheses would be appropriate?

   (A) H_0: $\bar{x} = 1$, H_a: $\bar{x} > 1$
   (B) H_0: $\bar{x} = 1$, H_a: $\bar{x} < 1$
   (C) H_0: $\mu = 1$, H_a: $\mu > 1$
   (D) H_0: $\mu = 1$, H_a: $\mu < 1$
   (E) H_0: $\mu > 1$, H_a: $\mu \leq 1$

4. A teacher wants to investigate the effect of class absences on grades for students at her school. She selects a simple random sample of students and records the following data for each student: the number of days absent in the fall term, the number of days absent in the spring term, the average grade for the fall term, and the average grade for the spring term. If these sample statistics show the teacher that students with more absences in a term also have lower average grades, which of the following generalizations would be best supported by the results of her study?

   (A) The study concludes that students can, in general, improve their average grades by being absent from fewer classes.
   (B) The study concludes that students with more absences are dealing with illness, injury, or other difficult circumstances that force them to miss school, and as a result have lower average grades.
   (C) The study concludes that students with more absences are less interested in school and as a result have lower grades.
   (D) The study cannot establish a cause-effect relationship between absences and average grades. It is not an experiment.
   (E) The study cannot by itself establish a cause-effect relationship between absences and average grades without being independently replicated first. Only then could students conclude that being absent fewer times will tend to result in higher average grades.

GO ON TO THE NEXT PAGE

5. One thousand subjects take a standardized test on English grammar. The scores, which are approximately normally distributed with a mean of 550 and standard deviation of 90, are ordered from lowest to highest. Which of the following intervals shows the approximate range of scores in places 300 to 600 on the ordered list?

(A) 503 to 573

(B) 515 to 585

(C) 523 to 586

(D) 527 to 597

(E) 532 to 559

6. An "unfair" coin is flipped numerous times, and it is determined that heads appears 1.5 times for every time tails appears. Under the same conditions, the coin is flipped 4 times. What is the probability that heads will appear exactly twice?

(A) 65.4%

(B) 34.6%

(C) 37.5%

(D) 14.4%

(E) 62.5%

7. A direct mail advertiser mails a product offer to a simple random sample of 1,000 addresses in a large city. Sixty of the recipients of the offer actually order the product. A 90% confidence interval for the corresponding population proportion is 0.060 ± 0.012. Which of the following statements must be true?

(A) There is a 90% probability that the proportion of all the city's residents who will order the product in response to receiving the offer is between 4.8% and 7.2%.

(B) The proportion of all the city's residents who will order the product in response to receiving the offer is within 90% of the interval (0.048, 0.072).

(C) If 1,000 sample proportions were found by sampling this city's population as described in the question, then 900 of the sample proportions would have a standard error between 4.8% and 7.2%.

(D) We can be 90% confident that the proportion of all the city's residents who will order the product in response to being mailed the offer is between 4.8% and 7.2%.

(E) If $p = 0.065$, then this sample result would be unlikely to occur.

8. When a fair, six-sided die is tossed, the mean toss is 3.5 with a standard deviation of 1.7. What are the mean and standard deviation for the sum of ten independent tosses of this die?

(A) Mean 3.5, standard deviation 0.5

(B) Mean 35.0, standard deviation 17.0

(C) Mean 35.0, standard deviation 5.4

(D) Mean 35.0, standard deviation 1.7

(E) Mean 35.0, standard deviation 0.5

GO ON TO THE NEXT PAGE

9. A drug company wants to test a new medication designed to relieve chronic pain from arthritis. The managers of the company believe the new medication will do a better job than the company's current medication and they want to design an experiment to show this. Which of the following experimental designs will best allow the experimenters to determine whether the new medication is, in fact, superior?

(A) Randomly assign either the new medication or a placebo that offers no pain relief to each subject for a period of time. Compare the results of the treatments.

(B) Randomly assign either the current medication or the new medication to each subject for a period of time. Compare the results of the treatments.

(C) Randomly assign the subjects to two groups. Group 1 will take the current medication for a period of time, then take the new medication for a period of time. Group 2 will do the same but in the opposite order. Compare the results of the two treatments for each test subject.

(D) Give the current medication to all test subjects for a period of time, then give the new medication to all test subjects for a period of time. Compare the results of the treatments for each test subject.

(E) Have the subjects take either the current medication or the new medication for a period of time with the subjects themselves choosing the treatment. Compare the results of the treatments.

10. Assume the duration of human pregnancies is roughly normal with mean 267 days and standard deviation 16 days. If a random sample of 400 pregnant women is selected from a large city, which of these is the approximate probability that the mean duration of their pregnancies will be less than 270 days?

(A) $P\left(z < \dfrac{270-267}{\dfrac{16}{\sqrt{400}}}\right)$

(B) $P\left(z < \dfrac{267-270}{\dfrac{16}{\sqrt{400}}}\right)$

(C) $P\left(z < \dfrac{\dfrac{270}{365}-\dfrac{267}{365}}{\sqrt{\dfrac{\left(\dfrac{267}{365}\right)\left(1-\dfrac{267}{365}\right)}{400}}}\right)$

(D) $P\left(z < \dfrac{270-267}{16}\right)$

(E) $P\left(z < \dfrac{267-270}{16}\right)$

GO ON TO THE NEXT PAGE

11. A track athlete specializing in the long jump has a median jump of 6.0 meters with an interquartile range of 1.0 meter. What are the median and interquartile ranges of the athlete's jumps in yards, given that there are about 1.1 yards per meter?

(A) Median 6.6 yd, interquartile range 1.0 yd
(B) Median 6.6 yd, interquartile range 1.1 yd
(C) Median 5.5 yd, interquartile range 0.9 yd
(D) Median 5.5 yd, interquartile range 1.0 yd
(E) It is impossible to tell because the quartiles are not given.

12. A small high school has 60 students in the senior class. Of these seniors, 20 compete in the school rodeo, 15 play in the school band, and 5 do both. Which of the following is NOT true for this school's high school seniors?

(A) There are 20 seniors who are involved in neither activity.
(B) Being in the school rodeo and playing in the school band are statistically independent events.
(C) The probability that a randomly selected senior who competes in the school rodeo is also in the school band is 0.25.
(D) The probability that a randomly selected senior competes in the rodeo but is not in the band is 0.25.
(E) The probability that a randomly selected senior is in at least one of these activities is 0.5.

13. A blood test is used to determine if a patient has a certain disease. If the patient tests positive, a painful and expensive course of treatments is administered. If the patient tests negative, no treatment is given. If we take the null hypothesis to be that the patient is healthy, and the alternative hypothesis to be that the patient has the disease, which of the following statements is (are) true?

I. A Type I error consists of a patient who does not have the disease receiving a painful and expensive course of treatments.

II. A Type II error consists of a patient with the disease receiving no treatment.

III. The power of the test is the probability that a healthy patient is correctly diagnosed as being healthy.

(A) I only
(B) II only
(C) III only
(D) I and II
(E) I, II, and III

GO ON TO THE NEXT PAGE

14. From 1982 to 2001, the counts of male and female athletes competing in high school sports in the United States underwent changes. Over these years, the least-squares regression line for number of female athletes versus the number of male athletes, measured in millions of athletes, is found to be

    females = −4.19 + 1.79 males.

    This regression line has the residual plot shown below.

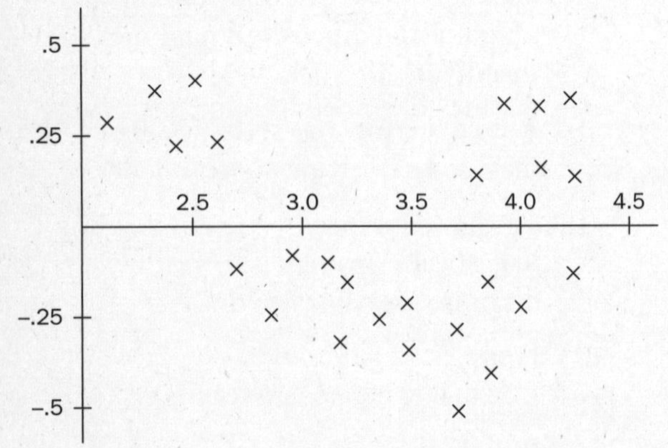

Which of the following conclusions about the regression relationship is best supported by this plot?

(A) The least-squares regression line is a good model for the data because the points are scattered uniformly.
(B) The least-squares regression line is a good model for the data because there are about the same number of data points above 0.00 as below.
(C) The least-squares regression line is a poor model for the data because there is a distinct pattern to the line's tendency to overpredict or underpredict the response variable over the range of values taken by the explanatory variable.
(D) The least-squares regression line tends to underpredict the number of female athletes in years in which there were fewer than 3.5 million male athletes.
(E) The correlation of male athletes vs. female athletes is negative if we consider only the years in which there are more than 3.5 million male athletes.

15. Which of the following is NOT a reason for using a random selection scheme when selecting a sample for an observational study?

(A) Random selection allows the results of analyzing data to be subject to the laws of probability.
(B) Random selection protects against bias.
(C) Random selection makes it likely that the sample resembles the population.
(D) A sample chosen by chance ensures that neither the sampler's preferences nor the respondent's self-selection will skew the data collected from the sample.
(E) Random selection gives all individuals an equal chance of being selected.

GO ON TO THE NEXT PAGE

16. Carlos plans to use a test for the difference between two population proportions p_1 and p_2. The random samples he collects, drawn from very large populations, are both greater than 50 and have sizes n_1 and n_2. His null hypothesis is that $p_1 - p_2 = 0$, and the alternative is $p_1 - p_2 < 0$. If a formula page is not available, which of the following is the formula he should use for the standard deviation of the difference of sample proportions?

(A) $\sqrt{p(1-p)}\sqrt{\frac{1}{n_1}+\frac{1}{n_2}}$, where p is the pooled sample proportion

(B) $\sqrt{p(1-p)}\sqrt{n_1+n_2}$, where p is the pooled sample proportion

(C) $\sqrt{\frac{p(1-p)}{n}}$, where p is the pooled sample proportion and $n = n_1 + n_2$

(D) $\sqrt{np(1-p)}$, where p is the pooled sample proportion and $n = n_1 + n_2$

(E) $\sqrt{\frac{p_1(1-p_1)}{n_1}+\frac{p_2(1-p_2)}{n_2}}$

17. A survey was taken to determine if there is a relationship between the amount of television watched each week and grade level at school. A random sample of 200 students yielded the following results:

	Television Watched	
Grade Level	≤10 Hours/Week	>10 Hours/Week
7–8	11	49
9–10	40	20
11–12	69	11

Which of the following would be the most appropriate inferential statistical test in this situation?

(A) One sample z test
(B) Two sample t test
(C) One sample t test
(D) Chi-square test of independence
(E) Chi-square test of goodness of fit

GO ON TO THE NEXT PAGE

18. To compare the ability of airports to maintain flight schedules, the proportion of flights arriving on time are recorded for each hour of the day from 6:00 AM to midnight at major U.S. airports. Data for the major airports in Atlanta and Boston are compared below to assess whether on-time performances at the two airports are related at corresponding times of day. A regression analysis for this paired data is shown below.

Dependent variable: BOSTON

Predictor	Coef	SE Coef	T	P
Constant	24.07	11.82	2.04	0.059
ATLANTA	0.5566	0.1680	3.31	0.004

S = 5.39629 R–Sq = 40.7% R–Sq(adj) = 37.0%

Which of the following is correct about the correlation coefficient r?

(A) $r = 0.407$, so 40.7% of the variation in Boston's on-time performance is explained by the least-squares regression of Boston's on-time performance on Atlanta's on-time performance.

(B) $r = 0.407$, so according to the least-squares regression line a change of 1 standard deviation in Atlanta's on-time performance corresponds to an increase of 0.407 standard deviations in Boston's on-time performance.

(C) $r = 0.638$, so 63.8% of the variation in Boston's on-time performance is explained by the least-squares regression of Boston's on-time performance on Atlanta's on-time performance.

(D) $r = 0.638$, so according to the least-squares regression line a change of 1 standard deviation in Atlanta's on-time performance corresponds to an increase of 0.638 standard deviations in Boston's on-time performance.

(E) $r = -0.638$, so according to the least-squares regression line a change of 1 standard deviation in Atlanta's on-time performance corresponds to a decrease of 0.638 standard deviations in Boston's on-time performance.

GO ON TO THE NEXT PAGE

19. A student takes a 20-question multiple-choice test and answers every question by random guessing. If each question has five choices, what is the probability of answering at least one question correctly?

(A) $\frac{19}{20}$

(B) $\binom{20}{1}(0.2)(0.8)^{19}$

(C) $1-\binom{20}{1}(0.2)(0.8)^{19}$

(D) $(0.8)^{20}$

(E) $1-(0.8)^{20}$

20. The manager of a factory speculates that employee productivity may have changed from old levels. The manager measures the productivity of a random sample of 31 workers over a period of time and conducts a one-sample *t*-test on the data to see if productivity has changed from the old level. The null hypothesis H_0 is that employee productivity equals the old level, and the alternative hypothesis H_a is that productivity has changed. If the manager computes a test statistic of $t = 1.98$, which of the following is an appropriate conclusion at a 5% significance level?

(A) Fail to reject H_0 since the *p*-value is between 0.025 and 0.05.

(B) Fail to reject H_0 since the *p*-value is between 0.05 and 0.10.

(C) Reject H_0 since the *p*-value is between 0.025 and 0.05.

(D) Reject H_0 since the *p*-value is between 0.05 and 0.10.

(E) Reject H_0 since the *p*-value is between 0.04 and 0.05.

IF YOU FINISH BEFORE TIME IS CALLED, YOU MAY CHECK YOUR WORK ON THIS SECTION ONLY. DO NOT TURN TO ANY OTHER SECTION IN THE TEST. **STOP**

SECTION II

25 Minutes

Directions: Write out answers to the following questions. Clearly show the methods used in obtaining your answer. You will be scored on the soundness of your methods and reasoning and on finding the correct answers.

A pet food company has developed a lower-calorie cat food that they believe will help obese cats lose weight. The company works with a sample of 200 obese cats whose owners have agreed to cooperate with the study. The experiment is completely randomized with 100 cats in each of two groups. One group of cats is fed the lower-calorie cat food and nothing else. The other group is fed the company's regular cat food and nothing else. Each cat's weight is measured both before beginning the experiment and after two months of the prescribed diet. The results are summarized below.

	Sample Size	Mean Weight Loss (lb)	Standard Deviation of Weight Loss (lb)
Cats eating new food	100	5.2	6.3
Cats eating old food	100	1.3	8.2

Do the data support the company's belief that the new lower-calorie food does, in fact, help cats lose weight?

(a) State a hypothesis for a test that will allow you to answer the question above. Be sure to specify the parameter(s) of interest.

(b) Use the data to carry out a *t*-test of the hypotheses you stated in (a).

(c) The results might be more informative for the company if blocking is employed. Describe an appropriate blocking scheme and explain your reasoning.

ANSWERS AND EXPLANATIONS

SECTION I

1. E

The distributions of those never married and those who are married are both clearly skewed right. The distribution of those who are divorced, even including the outlier, is least skewed. The other choices describe the opposite of what the display shows.

2. E

The standard error of $\bar{x}$ is $\frac{s_x}{\sqrt{n}}$, and so is inversely proportional to $\sqrt{n}$. By extension, the margin of error is inversely proportional to $\sqrt{n}$ as well. In order to have a margin of error $\frac{1}{4}$ the size of the original, the sample size will have to be $16n$.

3. C

The null hypothesis is that the current situation meets the desired standards. In other words, it should say that the population of all meter sticks manufactured by the company has a mean length of 1 meter. The alternative is what these students suspect may be true, that the mean length of all meter sticks manufactured by the company is greater than 1 meter.

4. D

There is no assignment of subjects to treatment groups, so this is an observational study and not an experiment. Claiming a cause-effect relationship is not warranted, even if the effects observed are independently replicated.

5. A

The question asks, in a roundabout way, for you to find the 30th and 60th percentiles for a normal distribution with mean 550 and standard deviation 90. The value of z for the 30th percentile is -0.5224, so the 30th percentile for the grammar test is $550 + (-0.5224)(90)$ or about 503. Similarly, the 60th percentile score works out to be about 573.

6. B

For this binomial distribution problem, heads appears 60% of the time and tails appears 40% of the time ($0.4 \times 1.5 = 0.6$). We have $n = 4$ flips, $k = 2$ successes, $p = 0.6$, and $1 - p = 0.4$. The probability is:

$$\begin{aligned} P(k=2) &= \frac{4!}{2!(4-2)!}(0.6)^2(0.4)^2 \\ &= 6(0.24)^2 \\ &= 0.3456 \approx 34.6\%. \end{aligned}$$

7. D

This answer choice is a traditional way to explain a confidence interval in context. The other choices are nonsensical, make false claims about confidence intervals as probabilities, or directly contradict the relationship between a confidence interval and a significance test.

8. C

The mean of the sum of ten independent random variables is the sum of the ten means, so the mean is 3.5 times 10, or 35. The variance of the sum of ten independent random variables is the sum of the ten variances, so the variance is 10 times 1.7^2. Since the variance is 28.9, the standard deviation is about 5.4.

9. C

Since pain level is subjective for each individual, it is important for each test subject to compare the level of pain experienced with each treatment. Only then can the researcher know whether any supposed pain difference that shows up in the data is genuinely felt by the subject. Of the two choices that allow for all test subjects to compare both treatments, the choice that includes randomization of the order of treatment is superior.

10. A

By the central limit theorem, the sampling distribution of $\bar{x}$ is approximately normal with mean $\mu_{\bar{x}} = \mu = 267$ and standard deviation $\sigma_{\bar{x}} = \frac{\sigma_x}{\sqrt{n}} = \frac{16}{\sqrt{400}}$ when n is large. In this problem, $n = 400$, which is certainly large. To find $P(\overline{X} < 270)$,

we only need to find $P\left(z < \frac{270 - 267}{\frac{16}{\sqrt{400}}}\right)$.

11. B

When a variable is transformed by multiplying its values by a constant, the values of measures of center and measures of spread are multiplied by that constant as well.

12. A

All are true except for the claim that there are 20 students who are in neither activity. There are 5 who do both, so only 10 of the band members are in band alone and 15 of the rodeo club members are in the rodeo club alone. That leaves 30 seniors who are in neither.

13. D

Statements I and II correctly describe the effects of Type I errors and Type II errors on the patient. Statement III incorrectly indicates that power is the probability that a healthy patient is correctly diagnosed as healthy. In fact, power is essentially the opposite probability. At a given significance level, the power of this test is the probability that a patient with the disease will be correctly diagnosed.

14. C

The pattern in the residual plot, which is basically curved, makes it clear that the least-squares regression line is a poor model for the data. The correct answer makes an equivalent statement, that there is a pattern in the tendency of the line to overpredict or underpredict.

15. E

All the other choices are reasons why random selection schemes are used when selecting a sample. Despite the claims of choice (E), however, not all random selection schemes give all individuals an equal chance of being selected. Even if they did, this answer choice explains the method for a random selection scheme and is not a reason why random selection should be used.

16. A

Since the null hypothesis assumption is that $p_1 = p_2$, we can pool the samples together as if they were one large sample drawn from a single population. As a result, we use the formula

$$\sqrt{p(1-p)}\sqrt{\frac{1}{n_1} + \frac{1}{n_2}},$$

using p = the pooled sample proportion.

17. D

Answer (D) is correct because a chi-square test for independence is used with a single population to test for an association between two categorical variables; in this case, grade level and the amount of television watched.

Answers (A) and (C) are both incorrect. While each test is used with a single population, they are used to test a hypothesis about a true population parameter, not a hypothesis about association.

Answer (E) is incorrect because a chi-square goodness-of-fit test is used to test whether the distribution of counts for a single categorical variable matches an expected distribution.

18. D

The regression analysis says that $r^2 = 0.407$, so either $r = 0.638$ or $r = -0.638$. The slope of the least-squares regression line is 0.5566, which is positive, so r takes the positive value 0.638. Since the slope of the least-squares regression line is $r \times \frac{s_y}{s_x}$, the line predicts that a change of 1 standard deviation in x results in a change of $r = 0.638$ standard deviations in y.

19. E

The number of correct guesses has a binomial distribution with $n = 20$ and $p = 0.2$. The probability of guessing none correctly is

$$\binom{20}{0}(0.2)^0(0.8)^{20},$$

which simplifies to $(0.8)^{20}$. The probability of guessing at least one correctly is thus $1 - (0.8)^{20}$.

20. B

This is a two-tailed t-test at the 5% significance level with $df = 30$, so the right-tail area is between 2.5% and 5% in the $df = 30$ row. As a result, the p-value is between 0.05 and 0.10, which is not significant enough at the 5% level to reject H_0. Note: Calculator computation shows the p-value to be 0.0569.

SECTION II

(a) μ_{new} = mean weight loss all obese cats would experience after two months on the new food.

μ_{old} = mean weight loss all obese cats would experience after two months on the old food.

H_0: $\mu_{\text{new}} = \mu_{\text{old}}$

H_a: $\mu_{\text{new}} > \mu_{\text{old}}$

OR

H_0: $\mu_{\text{new}} - \mu_{\text{old}} = 0$

H_a: $\mu_{\text{new}} - \mu_{\text{old}} > 0$

(b) **Step 1:** State the name of the test and verify the requirements.

Give the formula using symbols and/or number substitutions.

Two-sample t-test using $t = \dfrac{\bar{x}_{new} - \bar{x}_{old} - 0}{\sqrt{\dfrac{s_{new}^2}{n_{new}} + \dfrac{s_{old}^2}{n_{old}}}}$

Requirements:

1. Random assignments were made to treatment groups so the independence requirement is satisfied.

2. Since $n_{\text{new}} = 100$ and $n_{\text{old}} = 100$ are both large, t procedures can be used even though we do not know whether the population distributions are normal.

Step 2: Mechanics

So $t = 3.77$, $df = 185.7$, p-value = 0.0001. (Note: It is OK to use a conservative df of 99.)

Step 3: Conclusion

Reject H_0. There is convincing evidence that the mean weight loss for obese cats will be greater with the new food. If, in fact, there were no differences to be expected between the two foods, then the probability of seeing our sample results, or something more extreme, is only 0.0001.

(c) Several acceptable blocking schemes are possible. You can block by weight, by degree of obesity, by age, by level of physical activity, by health history, and so on. Any of these variables could confound the results, unless its effects are separated out.

Here is a scheme for blocking by weight: Divide the 200 cats into two blocks of 100 cats each, where one block consists of the 100 heaviest cats and the other the 100 lightest cats. Within each group randomly allocate cats into two 50-cat groups, with one group eating the new food and one the old food.

This is reasonable because the effect of the new food may depend on the size of the cat. Heavier cats may enjoy a greater weight loss since they will be consuming the same number of calories as the lighter cats while having a greater weight for those calories to supply with energy.

DIAGNOSTIC TEST CORRELATION CHART

Use the results of your test to determine which topics you should spend the most time reviewing.

Area of Study	Question Number
Summarizing distributions of univariate data	11
Comparing distributions of univariate data	1
Exploring bivariate data	14, 18
Exploring categorical data	6
Planning and conducting surveys	15
Planning and conducting experiments	9
Generalizability of results	4
Relative frequency	12, 19
Combining independent random variables	8
Normal distribution	5
Sampling distributions	10
Confidence intervals	2, 7, 16
Tests of significance	3, 13, 16, 17
t-distributions and procedures	2

HOW TO USE THE REVIEW SECTION

Each of the following review chapters begin by revisiting the main concepts that apply to the chapter's topic. The chapters will NOT include loads of factual material, but instead will help you tie all the facts together to understand the concepts and discuss briefly how they fit within the thematic emphases designed by the College Board.

In the review chapters, you will be given sample questions that will help you learn and/or review the AP Statistics course material. These questions will also have detailed explanations that include both how these questions address the course concepts and how to use each question to become a better test taker. You will be given the correct answer for each question and be informed of the thought processes you should go through to reach a correct answer. Sometimes you will be given examples of how you might have reached the wrong answer and how to avoid that problem in the future. Again, the emphasis is not only on repetition, but on effective repetition. It's not only about learning the information, but about how to apply that information.

Questions from both sections of the AP Statistics exam, multiple-choice (50% of the exam grade) and "free-response" or essay (50% of the exam grade), are covered in each review chapter. There are two types of questions in the multiple-choice section: Concept Questions and Calculation Questions. After completing all of the questions in the review chapters, you will be prepared to analyze the proper approach to each type of question and how to formulate the proper responses.

Answers to free-response questions must be clearly written and easy to understand, but do not have to be written in complete sentences. Answers that receive full credit by the free-response evaluators are quite comprehensive. When you take the AP Statistics exam, you can expect the first five free-response questions to involve limited topics, and the last free-response question to involve a synthesis of different statistical topics. More importantly, the College Board will expect you to be able to analyze and interpret different concepts and integrate those concepts into a strong, convincing response.

The sample responses in the review chapters are intended to help you learn how to analyze the questions and answer them thoroughly with all your relevant knowledge. The quantity and type of multiple-choice and free-response questions in the review chapters do not represent the exact coverage of material expected on the AP Statistics exam, but they are plausible examples of topics covered.

Read on, explore, discover, and learn!

| Part Three |

AP STATISTICS REVIEW

CHAPTER 3: CONSTRUCTING AND INTERPRETING GRAPHICAL DISPLAYS OF DISTRIBUTIONS OF UNIVARIATE DATA

IF YOU LEARN ONLY THREE THINGS IN THIS CHAPTER . . .

1. There are two main types of data: categorical and quantitative.
2. Categorical data can be depicted graphically using various types of graphs and charts.
3. The most common shapes of data distribution are uniform, symmetric, bell-shaped, skewed right, and skewed left.

WHAT IS ALL THIS DATA ABOUT?

Every day we are bombarded with data telling us why we should purchase certain products, where we should invest our money, why a law should be passed, how much our insurance premiums are going to be, and how our favorite sports team is doing. Our ability to look at this information with a critical eye and separate the truth from the stretched truth is critical. We need to be informed to make sound judgments and decisions.

Statistics is the science of data. Data or information is organized by variables such as height, income, gender, age, breed of dog, color, or political affiliation. The objects described by a set of data are called the individuals or subjects, and the variables are the characteristics of the individual or subject.

Data are classified into two large groups, **categorical** or **quantitative**.

DATA

Categorical data is data that is classified into one of several nonoverlapping groups or categories. Categorical data is also referred to as **qualitative data.**

Quantitative data is data that represents a numerical measure or value.

Examples of categorical data are gender, eye color, political party affiliation, or type of flower. Categorical data can be numerical in nature, such as area code, serial number, or zip code. However, performing numerical operations on this type of data makes no sense.

Examples of quantitative data are height, weight, income, pH level, and test score. With quantitative data, performing arithmetic operations makes sense and has meaning. This includes sums, differences, and averages.

Quantitative data can be further categorized into **discrete** or **continuous**. Discrete quantitative data is data where every possible value can be listed, such as scores on a statistics examination. Continuous quantitative data consists of data such as the exact weight of chickens in grams or the height of adult men in inches. In each of these cases, since we can use the fractional part of grams or inches, there is really an infinite number of possible values the variables can take on.

Data can also be categorized by the number of variables used to describe the subject. **Univariate,** or **one-variable data,** describes a single characteristic of the subject or individual, such as height.

Bivariate, or **two-variable data,** describes two characteristics of the subject or individual, such as height and weight.

Multivariate, or **many-variable data,** describes multiple characteristics of the subject or individual, such as height, weight, percent of body fat, blood pressure, temperature, resting pulse rate, and cholesterol level.

DISPLAYING CATEGORICAL DATA GRAPHICALLY

How do we interpret the data? What does the data tell us? No matter what statistical analysis may be in store for the data you have, you must ensure that you have a good understanding and a good feel for the data first. The first step is always to *graph it.* The expression "a picture is worth a thousand words" is very appropriate when talking about data. Seeing a graphic representation really puts things in perspective, as you can see in Figure 3.1.

Figure 3.1
Bar Graph: Student Enrollment at Eastern High

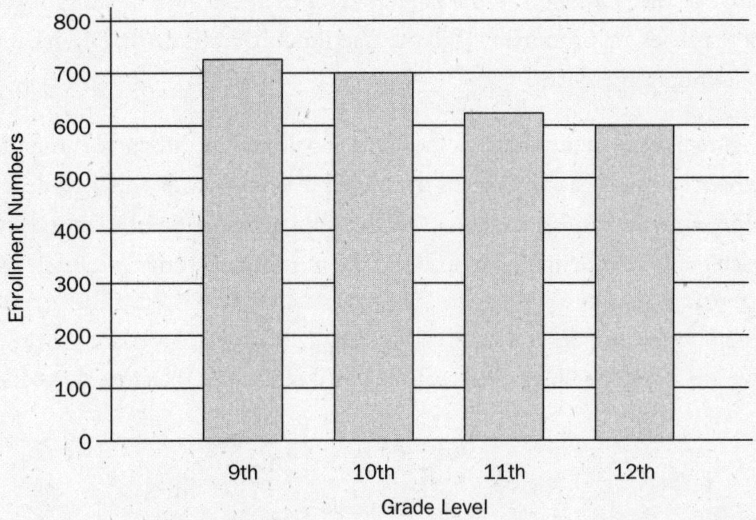

Bar Charts: Bar charts, or bar graphs, show the counts or proportion for each identified category. Each category's information is depicted in a separate bar. Figure 3.1 displays the student enrollment at Eastern High by grade level. Figure 3.2 displays a breakdown of the students in a statistics class by gender.

Figure 3.2
Bar Graph: AP Statistics Class by Gender

> **AP EXPERT TIP**
>
> If a graph is presented for you to analyze, take a moment to identify the labels on the horizontal and vertical axes before you answer the question.

Bar charts are an easy way to compare categories. Note that the individual bars are separated by a space. The heights of the bars represent either counts or proportions. As with all graphical displays, make sure the chart is clearly labeled and leaves no doubt about what the graph represents.

Pie Charts: Pie charts, or circle graphs, show the whole group or population as a circle or pie. The slices of the pie represent each category or subgroup. The sizes of the slices are proportional to the fraction of the whole that each category or subgroup represents.

Figure 3.3 uses a pie chart to show the enrollment numbers for students at a high school by grade. Each piece of the pie represents an individual grade level. Figure 3.4 represents the same data, except it shows the proportion of the total student body each grade represents, as opposed to the number of students in each grade level.

Figure 3.3

Pie Chart: Student Enrollment at Eastern High (raw numbers)

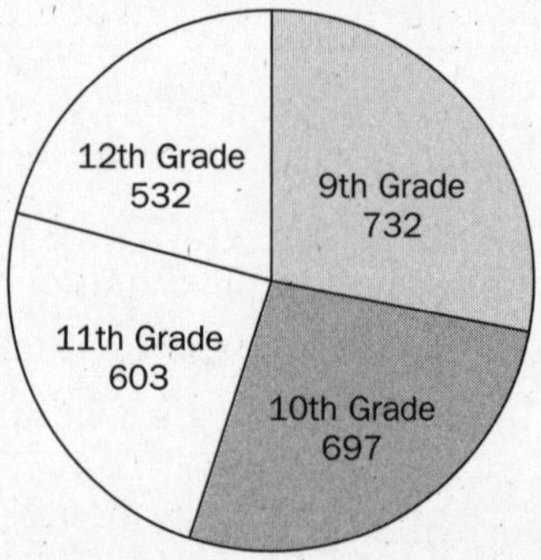

Figure 3.4

Pie Chart: Student Enrollment at Eastern High (percent)

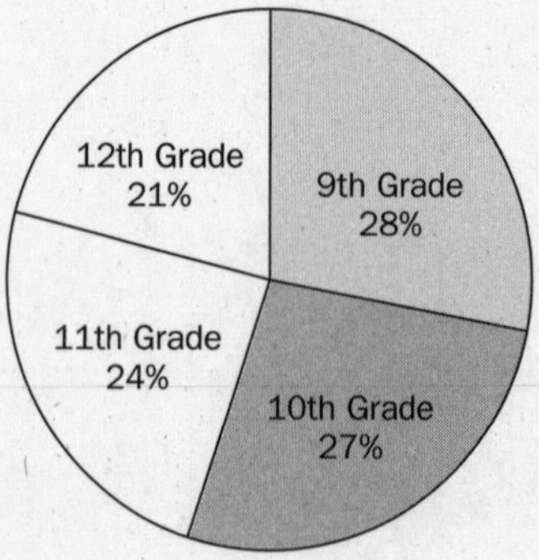

DISPLAYING QUANTITATIVE DATA GRAPHICALLY

It is difficult to tell much from just a listing of data. Seeing the distribution aids our understanding of the data. The distribution of a variable tells us what values the variable takes and the frequency with which these values occur. Several graphic displays provide wonderful visual pictures of the distribution, including dotplots, histograms, stem and leaf plots, and cumulative frequency plots. To understand the data, we must look at the overall pattern of the data depicted in the graph. We must notice the shape, the center (the balance point), the spread, and any data points that lie outside the pattern.

DOTPLOTS

The dotplot is simply a number line representing the possible values of the data, with dots to show the frequency of each data value. Figure 3.5 is a dotplot for the SAT Math scores for a group of 110 high school seniors. Each dot in Figure 3.5 represents one student's score. From the plot, you can see that the range of scores is from about 500 to 800. There are no large gaps in the scores that were achieved, and there were more scores in the 500 to 620 range than elsewhere.

Figure 3.5
Dotplot: SAT Math Scores

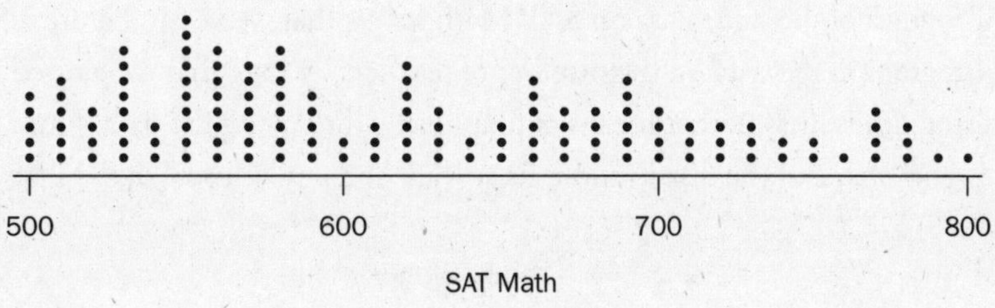

The dotplot in Figure 3.6 represents the number of teachers in grades K through 12 by state. You can see from this graphic that the data is not evenly distributed over the plot. In fact, the data is clustered more to the left of the plot with a few points at the higher end. The three points that are

Figure 3.6
Dotplot: Number of Teachers by State

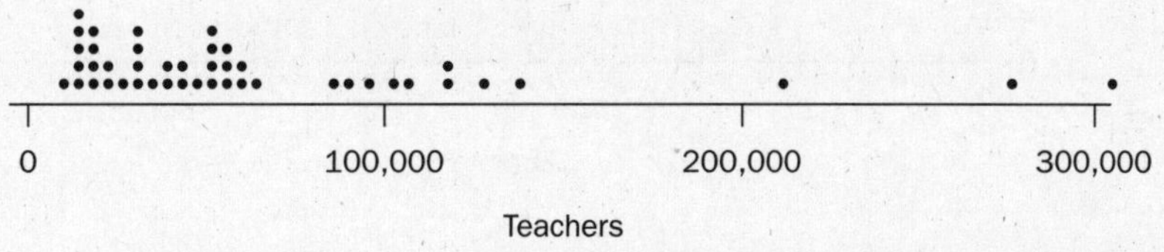

out to the far right, past the 200,000 mark, are what we would call *outliers*. We will examine a more technical definition of outlier in the next chapter, but for now we will use the term to refer to points that are considerably outside the normal pattern of the data.

Constructing a Dotplot: Use the horizontal axis to represent the data values. Construct a horizontal number line that contains all of the values of the data. Depending on the data, some scaling may be required. Place a dot over the appropriate value for each occurrence of that value in the data set.

Dotplots can also be drawn with a vertical scale. Use whatever works best for your data and illustrate your data in the clearest manner. Sometimes the choice is based simply on which fits best in a written report.

HISTOGRAMS

Histograms are another type of graphical representation for data that works well with larger data sets. Like the bar chart, the histogram plots data in bars, where the height of the bar represents the count within the interval depicted on the *x*-axis. (As with most charts, the horizontal and vertical axes can be switched.) A bar represents a range of values, such as the number of students who scored between 85 and 90 on the statistics exam. There are no spaces between the bars on a histogram like between those on bar charts. The axis represents the values that the variable can take on, so each bar will represent an equal range of values.

Figure 3.7 is a histogram of the same data on SAT Math scores that we saw in Figure 3.5. Likewise, Figure 3.8 is a histogram of the data on the number of teachers by state that we plotted in Figure 3.6. Remember: A histogram is just another means of graphically displaying the data. From the histogram, you can see the overall shape of the distribution, its spread, and any unusual data values.

Figure 3.7
Histogram: SAT Math Scores

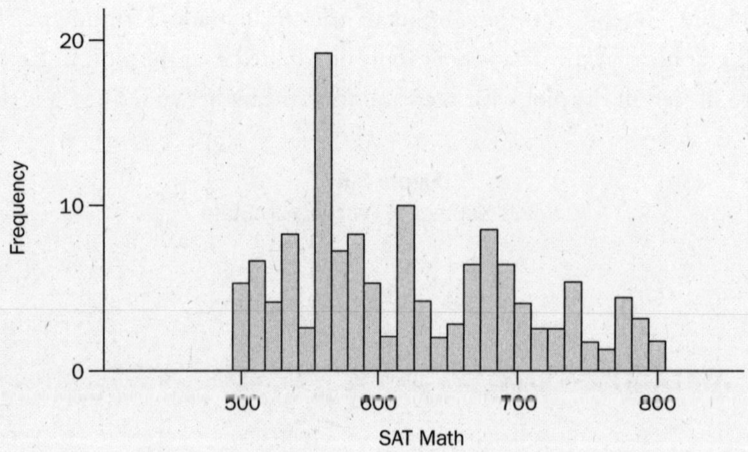

Figure 3.8
Histogram: Number of Teachers by State

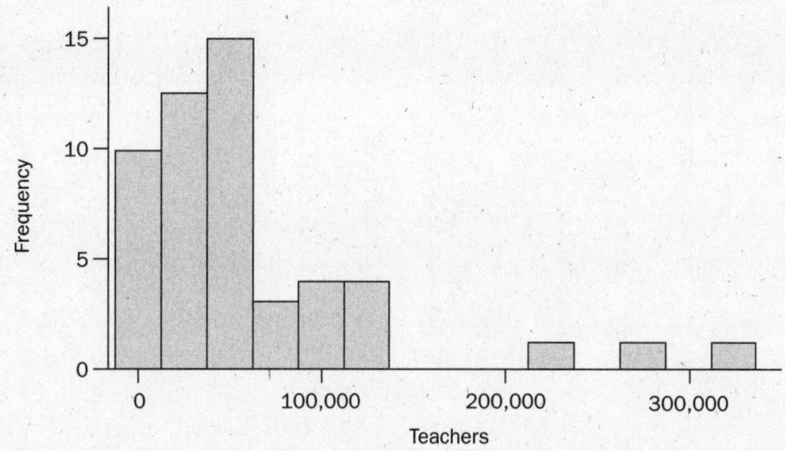

A histogram that has too many bars frequently provides little additional insight into the data. Additionally, a large number of bars sometimes results in a histogram that's too busy. We can collapse our data into fewer bars. In Figure 3.7, there were 26 bars for the SAT Math scores. In Figure 3.9, the number of bars has been reduced to 9. Although you lose some data by collapsing, you still are able to clearly see the overall shape and spread of the data. You can still see that the scores have a heavier concentration in the range from 500 to 620.

Figure 3.9
(a) Histogram: SAT Math Scores (collapsed data)

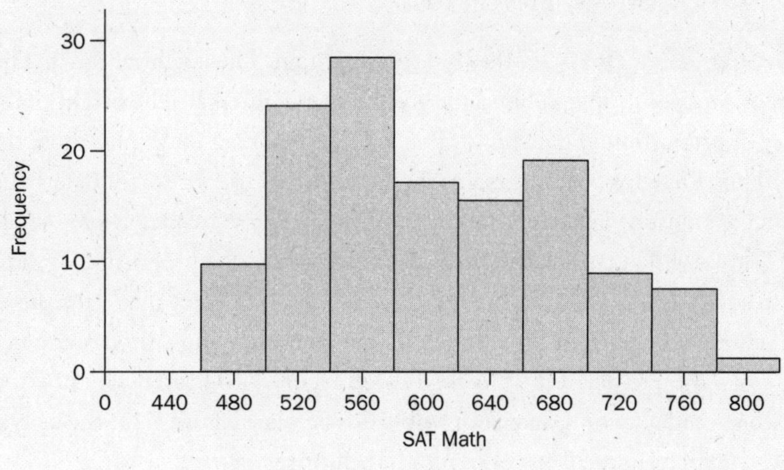

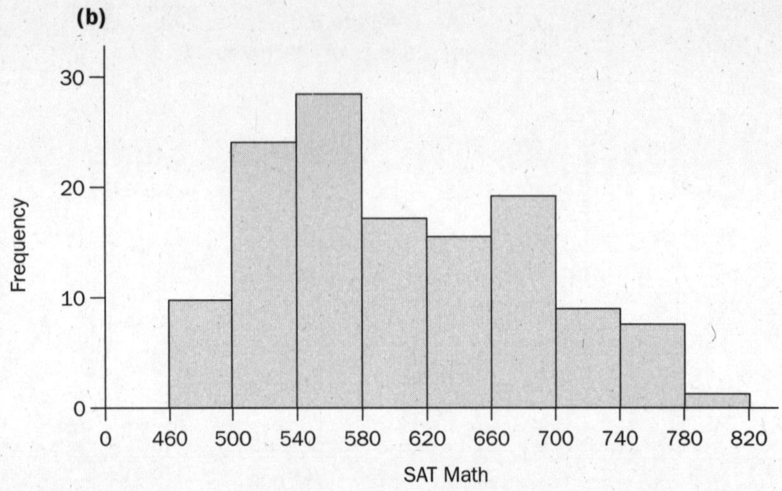

Constructing a Histogram: Use the horizontal axis to represent the data values. Construct a horizontal number line that contains all of the values of the data. Depending on the data, some scaling may be required. The vertical axis represents the frequency with which data values occurred. Construct a vertical axis large enough so that the greatest frequency can be included. Draw bars of equal width that are either centered above the data values on the horizontal axis (as with Figure 3.9.a), or drawn to numbered cutoff points (as with Figure 3.9.b). The height of each bar represents the frequency of that data value.

Often we are interested in the percent or proportion of data points that have a certain value. The term **relative frequency histogram** describes a histogram that displays the percent or proportion of data points that fall into each range or interval.

The **cumulative frequency plot** is another type of histogram. The height of the first bar represents the proportion of data points that fall into the given range or interval. The height of the second bar is the frequency or proportion of data points that fall into that data range plus those that fall into the first range. Thus, any given bar represents the frequency or proportion of data values that fall into that interval or anything less than that value. The cumulative frequency plot would provide a count of the values, while a cumulative percent would provide the proportion of values. These terms are not steadfast. Many texts use the term "cumulative frequency plot" to represent a plot of proportions. Figures 3.10 and 3.11 provide cumulative frequency and cumulative percent plots for our statistics exam scores, respectively. The final bar in Figure 3.10 has a height of 75, since there were 75 total scores, and all scores were at or below this bar. For Figure 3.11, the height of the last bar is 100, since 100 percent of all scores were at or below this value.

Figure 3.10
Cumulative Frequency Plot: Statistics Exam

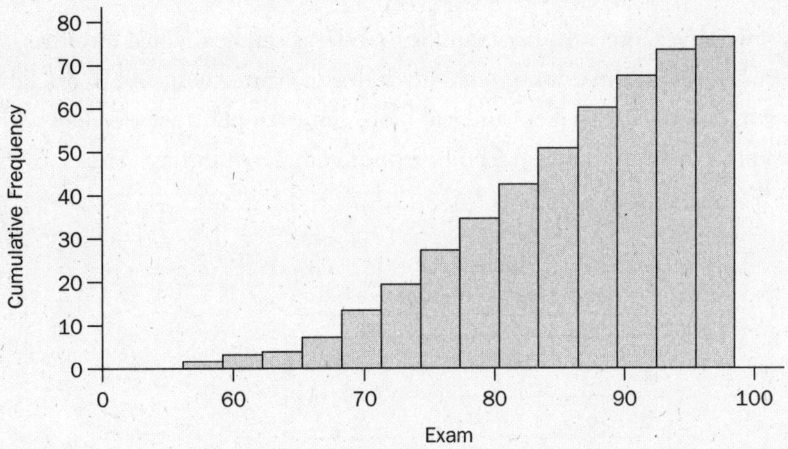

Figure 3.11
Cumulative Frequency Plot: Statistics Exam (percent)

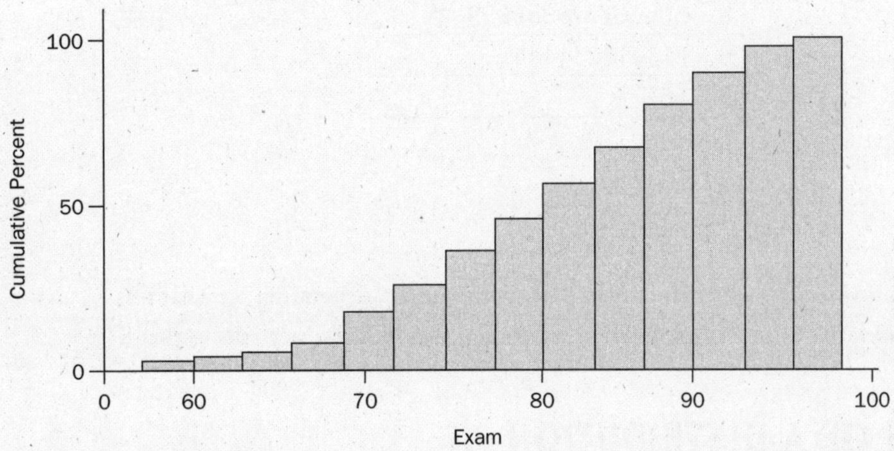

> **AP EXPERT TIP**
>
> For all graphs on the AP exam, clearly examine the vertical axis (is it frequency, relative frequency, or cumulative frequency?) and the horizontal axis before answering any questions.

Stem and Leaf Plot

A third type of graphic display is the **stem and leaf plot**. This plot is similar to a histogram, but it has the advantage that each individual data value is preserved and visible. Thus, it is more appropriate for use with smaller data sets. The stem and leaf plot consists of an orderly arrangement of numbers. Let's consider scores on a statistics examination. The possible scores on the exam would range from 0 to 100. The first digit, or *stem*, would be the "tens" place. Thus, the stems could be 0 to 10. The second digit, or *leaf*, would represent the "units" place. There is always a space or a vertical line between the stem and the leaf. In the example in Figure 3.12, stems are repeated. This is called a *split stem*, and it is used to accommodate a large number of values. The line **8 556667778888999** in the plot represents the scores 85, 85, 86, 86, 86, 87, 87, 87, 88, 88, 88, 88, 89, 89, and 89.

Stems can also be split into larger intervals. For example, ten stems can be divided into five: 00–19 for the first stem, 20–39 for the second stem, 40–59 for the third stem, and so on. Finally, values can be truncated in order to create stem and leaf plots. For example, a set of values with two decimal points can have the second decimal point truncated before plotting. Thus 2.62 is truncated to 2.6 and 5.78 is truncated to 5.7.

Figure 3.12
Stem and Leaf Plot of Exam

5	79
6	0
6	588
7	0000111223344
7	5566666799
8	00000111222334444
8	556667778888999
9	001112234444
9	567

Key
8 | 3 = 83
N = 75

A key is generally provided at the bottom of the plot. For example, depending on what was measured, 8 | 3 could represent 83, 8.3, 0.83, or 8,300. For this plot, the key tells us that 8 | 3 = 83.

THE SHAPE OF A DISTRIBUTION

There are an infinite number of possibilities for the overall shape of a distribution. However, there are several common shapes that are worth noting. (We will discuss how to describe center and spread more thoroughly in upcoming chapters.)

Uniform: A distribution is uniform if the values along the vertical axis all fall roughly on a horizontal line.

Symmetric: A distribution that is symmetric has two halves that are mirror images of each other. A vertical line can be drawn somewhere along the distribution, and when the graph is folded along this line, the two sides will match up.

Bell-shaped: A bell-shaped distribution is a special subset of the symmetric distribution, and is literally shaped like a bell.

Skewed: Distributions are skewed if the majority of their values fall either to the left or the right, and the data is then spread out with a small number of values in the opposite direction. In skewed-right data, the majority of the values are on the left, and the "tail" stretches toward the right. The opposite is true for skewed-left data, with the majority of the values on the right and the "tail" stretching toward the left.

Figure 3.13
Common Patterns of Distributions

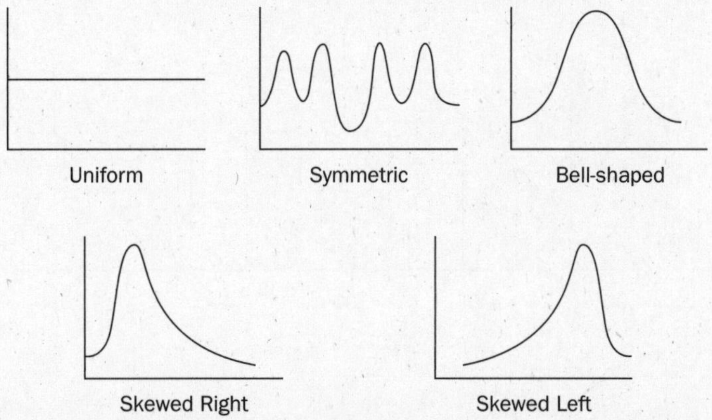

READING GRAPHICAL DISPLAYS

Shape: Overall pattern of the data; unimodal, bimodal, multimodal, uniform, clusters, gaps, symmetrical, skewness

Spread: Range of values, from smallest to largest

Center: Point that divides the data roughly in half

Outliers: Any data points lying outside the general pattern of data

When describing the distribution of data, the key elements are the shape, overall pattern, spread, center, and any unusual features. Figures 3.14 and 3.15 provide a histogram and dotplot for the weight of a sample of bears in the Adirondack State Park. Notice that in both graphs you can see that the overall shape of the data has two mounds. This is called bimodal. From the dotplot, it is apparent that there is an outlier, a very heavy bear. This is not as easy to see in the histogram.

AP EXPERT TIP

If you are asked to describe a distribution (or compare two distributions) on the AP exam, you will not receive full credit if you don't address each of these key elements.

Although you may want to describe a shape as either symmetric or skewed, these terms are not always appropriate. For this particular distribution, the key element of the shape is the two peaks. In skewed-right data, the majority of the values are on the left, and the "tail" stretches towards the right. The opposite is true for skewed-left data, with the majority of the values on the right and the "tail" stretching towards the left.

Figure 3.14

Histogram: Weights of Bears in Adirondack State Park

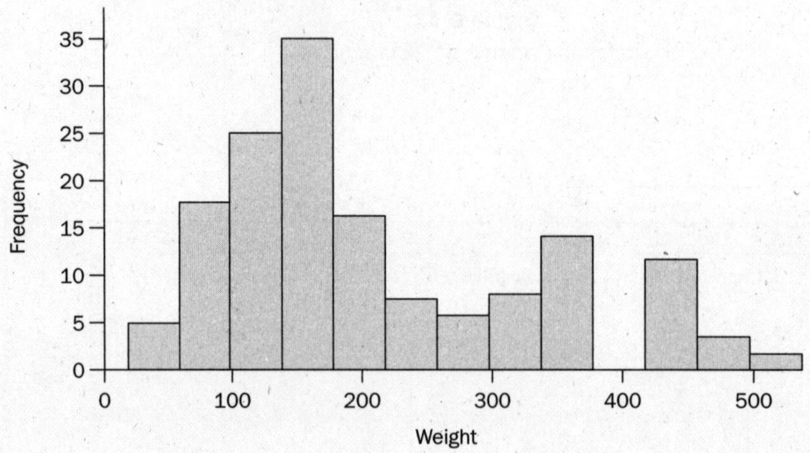

Figure 3.15

Dotplot: Weights of Bears in Adirondack State Park

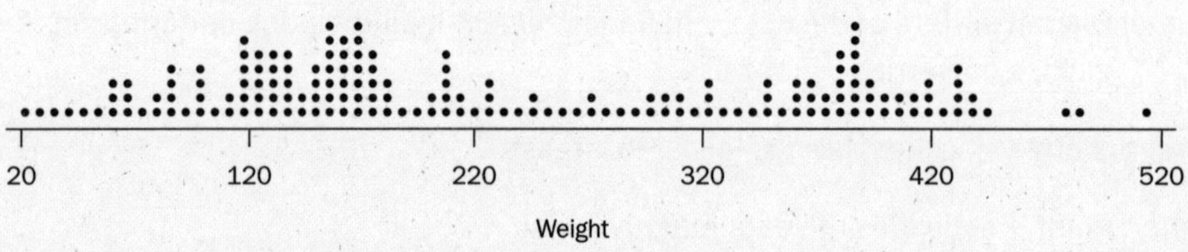

REVIEW QUESTIONS

1. Which of the following are true statements?

   I. Pie charts are useful for both categorical and quantitative data.
   II. Histograms are useful for small and large data sets.
   III. Histograms show the overall shape, center, and spread of the distribution of data.

   (A) I only
   (B) II only
   (C) III only
   (D) I and III only
   (E) II and III only

2. Which of the following is inappropriate for displaying quantitative data?

   (A) Stem and leaf plot
   (B) Dotplot
   (C) Bar graph
   (D) Cumulative frequency plot
   (E) Histogram

3. A graphical display of data that shows the cumulative counts across each of the possible data values or ranges of data values is a

   (A) bar chart.
   (B) pie chart.
   (C) cumulative frequency plot.
   (D) relative frequency histogram.
   (E) cumulative relative frequency histogram.

4. The height of Mrs. Clark's tomato plant is what type of data?

   (A) Categorical
   (B) Quantitative and continuous
   (C) Quantitative and discrete
   (D) Categorical and quantitative
   (E) Categorical and continuous

Free-Response Question

The following table provides the average teacher salary and the number of teachers by state for the 2002–2003 school year. Construct an appropriate graphical representation of the average teacher salary. Provide sufficient rationale as to your choice of graph, and discuss what the graph tells you about the distribution.

State	Average Salary ($)	Number of Teachers
California	55,693	313,330
Michigan	54,020	96,506
Connecticut	53,962	42,754
New Jersey	53,872	105,580
District of Columbia	53,194	5,235
New York	53,017	213,072
Rhode Island	52,879	10,754
Massachusetts	51,942	68,003
Illinois	51,496	133,377
Pennsylvania	51,425	120,015
Maryland	50,410	57,050
Delaware	49,821	7,402
Alaska	49,694	8,120
Oregon	47,463	29,915
Ohio	45,515	119,954
Georgia	45,414	101,977
Indiana	44,966	61,623
Washington	44,961	52,983
Minnesota	44,745	57,471
Virginia	42,778	90,952
Hawaii	42,768	10,943
Colorado	42,679	43,519
North Carolina	42,411	86,325
Vermont	42,038	8,167
New Hampshire	41,909	13,676
Nevada	41,795	20,781
Wisconsin	41,617	59,442
South Carolina	40,362	46,508
Florida	40,281	125,798
Texas	39,972	276,719
Arizona	39,955	45,656

Idaho	39,784	14,505
Alabama	39,524	46,316
Tennessee	39,186	58,984
Maine	38,518	16,895
West Virginia	38,497	20,190
Kentucky	38,486	40,937
Utah	38,268	21,900
Kansas	38,030	32,207
Iowa	38,000	38,000
Nebraska	37,896	21,180
Wyoming	37,789	6,897
Missouri	37,641	62,952
Arkansas	37,536	30,134
Lousiana	37,116	50,727
New Mexico	37,054	20,085
Montana	35,754	10,427
Mississippi	35,135	34,184
North Dakota	33,869	8,300
Oklahoma	33,277	41,180
South Dakota	32,414	8,783

Source: U.S. Bureau of Labor Statistics, May 2002

ANSWERS AND EXPLANATIONS

1. E

Pie charts are only useful for categorical data. Histograms can be used for large or small data sets, and they give a good graphical depiction of the distribution of the data. From the histogram, the overall shape, center, and spread can be seen.

2. C

Answer (C) is correct. Bar graphs are used to display counts or proportions of categorical data, not quantitative data.

Answers (A), (B), (D), and (E) are incorrect. Each of these can be used to display quantitative data.

3. C

The cumulative frequency plot has the height of the first bar representing the count of data points that fall into the given value. The height of the second bar is the count of data points that fall into that data value plus those that fall into the first bar. Thus, any given bar represents the count of data values that fall into that value or anything less than that value. Relative frequencies graph proportions.

4. B

The height of the tomato plant is quantitative. It is a measurement that could take on an infinite number of values, and is thus continuous. Discrete data can only take on a finite set of values.

FREE-RESPONSE ANSWER

Since the average teacher salary is quantitative, a dotplot, histogram, or stem and leaf plot would be appropriate. Because of the nature of the data, it would be appropriate to round to the nearest hundredths prior to graphing. Following are three possible graphical displays for this data.

Histogram

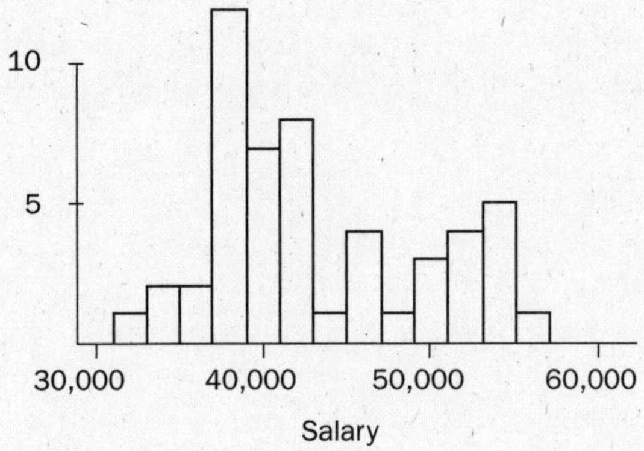

Stem and Leaf

3	233
3	55
3	777777
3	888888999
4	0000111
4	22222
4	45555
4	7
4	99
5	0111

Key
3 | 2 = 32,000
N = 51

Dotplot

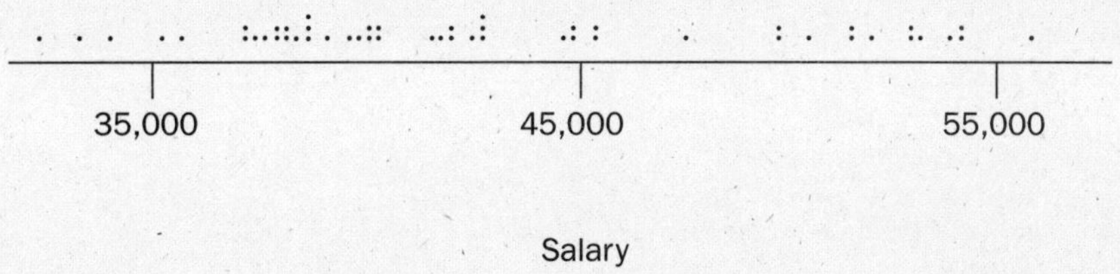

Notice that the histogram and the stem and leaf plot give approximately the same picture. However, due to the scale of the axis in the dotplot, the average salaries appear to be more spread out with several clusters between \$35,000 and \$45,000. From the histogram or stem and leaf plot, you can see that the graph has two modes: one around \$40,000, and another, though smaller, around \$51,000. There are no visible outliers in the salaries.

CHAPTER 4: SUMMARIZING DISTRIBUTIONS OF UNIVARIATE DATA

IF YOU LEARN ONLY FOUR THINGS IN THIS CHAPTER . . .

1. Distributions of data can be compared by measures of center, mean, and median.
2. Measures of spread may be the range, interquartile range, variance, and standard deviation.
3. Common measures of position are quartiles and percentiles.
4. A boxplot focuses on five key measures: minimum value, first quartile, median, third quartile, and maximum value.

MEASURES OF CENTER

How do you compare distributions? Looking at them graphically, we can compare their shape, overall spread, center, and outliers. But rather than simply relying on our guess of the spread and center, some standardized measures have been developed. For the center, these are the mean and the median.

The **mean** of a distribution is the arithmetic average of the distribution. The mean is denoted $\overline{x}$ and is calculated $\overline{x} = \dfrac{\text{sum of the values}}{\text{number of values}} = \dfrac{\sum x_i}{n}$. The mean is a good measure of the center of the distribution if the distribution is approximately symmetric and unimodal.

The **median** of a distribution is literally the middle value, where the data has been ordered. That is, half of the values are above the median and half are below. The median is a good measure of the center of the distribution when outliers are present in the data or when the distribution is skewed.

Example:

Given that Alan scored 87, 94, 90, 82, 93, 85, and 96 on the last seven chapter tests in AP Statistics, we can calculate his mean score as follows.

$$\bar{x} = \frac{\sum x_i}{n} = \frac{87 + 94 + 90 + 82 + 93 + 85 + 96}{7} = \frac{627}{7} = 89.57$$

To find his median score, first order the scores from low to high. {82, 85, 87, 90, 93, 94, 96}. The middle score would be 90, since there are three scores higher than that and three scores lower.

> **AP EXPERT TIP**
>
> Your calculator should be able to find the mean and median of data easily. Read chapter 19 of this book for more information about using a graphing calculator, and make sure you are familiar with all of the statistical functions of your particular calculator.

If there is an odd number of values, the median score would be the value that is in the $\frac{n+1}{2}$ place. In our example, that would be $\frac{7+1}{2} = 4$. Count up from the first data value, and the fourth value is 90. If n is even, the median is the average of the two middle scores. Let's say we only had six test scores: 82, 85, 87, 90, 93, and 94. Since there are six scores, the median would fall between the third- and fourth-place scores, or between 87 and 90. The median is simply the average of these two scores, or 88.5.

Figure 4.1 depicts the relationship of the mean and median in a skewed distribution and in a symmetric distribution. Notice that the values at the left end of the skewed distribution *pull* the mean toward them.

Figure 4.1
Relationship Between Mean and Median

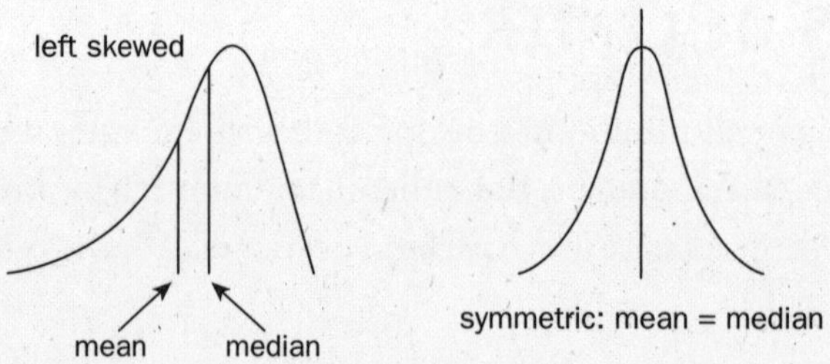

If the distribution is skewed left, the mean < median. If the distribution is skewed right, the mean > median. If the distribution is symmetric, the mean = median.

Here's another way of thinking about the mean and median. The median is the point where, if you drew a vertical line through it parallel to the vertical axis, half of the area under the distribution would be on either side of the line. On the other hand, the mean acts as the balance point, or fulcrum, for the distribution.

MEASURING THE SPREAD

Knowing that the average teacher's salary by state in 2002 was $43,075 doesn't really give a clear picture of the distribution. It doesn't tell that the salaries ranged from a low of $32,414 to a high of $55,693. Recently, there was considerable haggling between the political parties in the United States over proposed tax cuts. Those favoring the cuts touted an average tax cut of $1,083 for individual taxpayers. Those opposing the tax cut were quick to point out that for approximately 60% of the taxpayers, the average tax cut would be $380. Those with incomes over $200,000 would realize an average tax cut of $12,496, and those with incomes over $1,000,000 would have an average tax cut of $90,222. In this instance, knowing the average tax cut is $1,083 really doesn't give a clear picture of what is going on.

Measures of spread or dispersion are used to provide support and clarification to the mean. These measures are range, interquartile range, and standard deviation.

RANGE AND INTERQUARTILE RANGE AS MEASURES OF SPREAD

The **range** is simply the difference between the largest data value and the smallest data value. Returning to the average teacher salary by state, the highest salary was $55,693 and the lowest was $32,414. The difference between these two values, or the range, is $23,279. Knowing whether the data was widely spread out or clustered around the mean provides a better understanding of the distribution and data as a whole. Remember that the range by itself doesn't give much information about the distribution. It only tells the difference between the extreme values.

AP EXPERT TIP

If there are any outliers in the data, the range will be affected.

The **interquartile range** is the difference between the third quartile value and the first quartile value. A quartile, or quarter, is one-fourth of the data. Thus, the first quartile, or Q_1, is the data value that has 25% of the data below it; the third quartile, or Q_3, is the data value that has 75% of the data below it. The interquartile range, or **IQR,** is equal to $Q_3 - Q_1$. The quartiles Q_3 and Q_1 can also be found by finding the median of the lower half of the data and the median of the upper half of the data when the data are listed in order.

The IQR is a measure of the middle 50% of the data. It is used to give a sense of the spread of dispersion of the data and it provides a criterion for identifying **outliers** in the data. Outliers are identified as any data value that is 1.5*(IQR) above Q_3 or 1.5*(IQR) below Q_1.

Example:

Tiffany and Henna are conducting an experiment in biology. They are testing the effect of vitamin supplements on the weight gain of laboratory mice. The weight gain in grams for the eight mice that were given the supplements is as follows: 10, 5, 4, 0, 7, 5, 2, 8. Determine the IQR for weight gain of these mice, and identify any outliers in the data.

Solution: First, place the data in ascending order: 0, 2, 4, 5, 5, 7, 8, 10. There is an even number of values. Since $n = 8$, the median is the average of the 4th and 5th terms. The 4th term is 5, and the 5th term is 5, so the median is $\frac{5+5}{2} = 5$ Now, looking at the lower half of the data (0, 2, 4, 5), Q_1 would be the median of this subset of data, or the average of the two middle terms. $Q_1 = \frac{2+4}{2} = 3$. Likewise, Q_3 would be the median of the upper half of data, or the median of 5, 7, 8, 10. $Q_3 = \frac{7+8}{2} = 7.5$.

$$\text{IQR} = Q_3 - Q_1 = 7.5 - 3 = 4.5$$
$$Q_3 + 1.5 * (\text{IQR}) = 7.5 + 1.5 * (4.5) = 14.25$$
$$Q_1 - 1.5 * (\text{IQR}) = 3 - 1.5 * (4.5) = -3.75$$

Since no data values were at or above 14.25 or below −3.75, there were no outliers in the data.

VARIANCE AND STANDARD DEVIATION AS MEASURES OF SPREAD

The **variance** measures spread by examining the deviation of the data values from the mean. The variance for a data sample is calculated by the formula $s^2 = \frac{\sum_{i=1}^{n}(x_i - \bar{x})^2}{n-1}$.

The **standard deviation** is the square root of the variance, or $s = \sqrt{\frac{\sum_{i=1}^{n}(x_i - \bar{x})^2}{n-1}}$.

It should be noted that there are two basic formulas for variance and standard deviation. One formula is used if the data set represents the entire population, and one is used if the data set represents a sample. Statistical convention is to use the Greek letters to represent the measure of a population, and English letters to represent the measures related to a sample.

	Population	Sample
Variance	$\sigma^2 = \frac{\sum_{i=1}^{n}(x_i - \bar{x})^2}{n}$	$s^2 = \frac{\sum_{i=1}^{n}(x_i - \bar{x})^2}{n-1}$
Standard Deviation	$\sigma = \sqrt{\frac{\sum_{i=1}^{n}(x_i - \bar{x})^2}{n}}$	$s = \sqrt{\frac{\sum_{i=1}^{n}(x_i - \bar{x})^2}{n-1}}$

A small standard deviation indicates that the data is not spread out, but is centered around the mean. Be careful when interpreting the relative size of the standard deviation. Be mindful of the relative size of the data values. If they represent test scores based on a maximum score of 100, the standard deviation would most likely be less than 20. However, if the data were representing the average teacher salaries by state, the standard deviation would be in the thousands of dollars. (For our data on the average teacher salary by state, the standard deviation was $6,371.)

AP EXPERT TIP

While it is important to understand the formulas for variance and standard deviation, your calculator is able to calculate these values from data. Read chapter 19 of this book for more information about using a graphing calculator.

MEASURING THE POSITION

There are many ways for making comparisons within a data set. We all do it every day. Sarah has the highest average in biology. Matt finished in the top 10% of his high school class. The Math SAT scores for the middle 50% of the current freshmen class at Lakeside College are between 550 and 670. The Manchester United soccer team has finished in the top three every year for the past five years in the English Premier League. In all of these cases, we are measuring position of data values relative to the rest of the data. Some of the most common measures are quartiles and percentiles.

Quartiles divide the data into four equal parts, each of which contains 25% of the data. Thus, the first quartile would contain the lower 25% of data and the third quartile would contain all of the data values between 50% and 75% of the data. If Gene scored in the 4th quartile, he scored in the top 25% of all students who took that assessment. If, however, he scored in the third quartile, 25% of all students scored above his quartile and 50% scored below.

The term quartile is also used to identify the points Q_1 and Q_3. That is, the value Q_1, such that 25%, or one-quarter, of the data is below it, and the value Q_3, such that 25%, or one-quarter, of the data is above it.

A **percentile** is one of the 99 values that divide a set of data such that it gives the percentage of the data values that lie below the given value. For example, the 90th percentile is the value of the data set such that 90% of the data values are below the given value. If Leslie scored in the 93rd percentile on a national test, 93% of all of the students who took that test scored below her.

STANDARDIZED SCORES

In order to compare data from different distributions, we need to measure them on a common scale. **Standardized scores,** or ***z*-scores,** provide a means to give the position of a data value relative to its mean and standard deviation. The *z*-score is calculated by the following formula:

Z-SCORES

$$z = \frac{\text{data value} - \text{mean}}{\text{standard deviation}} = \frac{x - \bar{x}}{\sigma}$$

The *z*-score measures how many standard deviations a data value is from the mean. A negative *z*-score indicates that the value is below the mean, and a positive *z*-score indicates that the value is above the mean. The mean itself has a *z*-score of zero.

Example:

The average weight of a male bullmastiff is 120 pounds, with a standard deviation of 9.3 pounds. The average weight of a male mastiff is 182.5 pounds, with a standard deviation of 12.5 pounds. Relative to its breed, is a bullmastiff that weighs 135 pounds bigger than a mastiff that weighs 195 pounds?

Solution: To compare these two dogs, we can calculate the *z*-score for each one, and compare their weights based on how many standard deviations each is from its respective mean.

$$z = \frac{x - \sigma}{\sigma}$$

For the bullmastiff we have: $z_b = \frac{135 - 120}{9.3} = 1.6129$, and for the mastiff we have $z_m = \frac{195 - 182.5}{12.5} = 1$. This tells us that the bullmastiff's weight is 1.6129 standard deviations above the mean, while the mastiff's is only 1 standard deviation above the mean. Thus, the bullmastiff is heavier relative to its breed.

BOXPLOTS

Another very useful graphical display of data is the **boxplot**, or the box and whisker plot. A boxplot graphically represents the distribution of data by focusing on five key measures: minimum value, first quartile (25th percentile), median (50th percentile), third quartile (75th percentile), and maximum value. These five measures are referred to as the **five-number summary**.

DRAWING A BOXPLOT

- Draw and label a number line, scaled appropriately for the data.
- At a reasonable height above and below the number line, draw a rectangular box whose end lines are at Q_1 and Q_3.
- Draw a vertical line in the box at the position of the median.
- Draw two horizontal lines, or whiskers, one from the end of the lower end of the rectangle to the minimum value, and the other from the upper end of the rectangle to the maximum value.

For a recent statistics exam, the five-number summary was:

Min	Q_1	Median	Q_3	Max
57.00	73.00	82.00	88.00	97.00

Using this data, the boxplot would be as shown in Figure 4.2.

Figure 4.2
Boxplot: Statistics Exam Grades

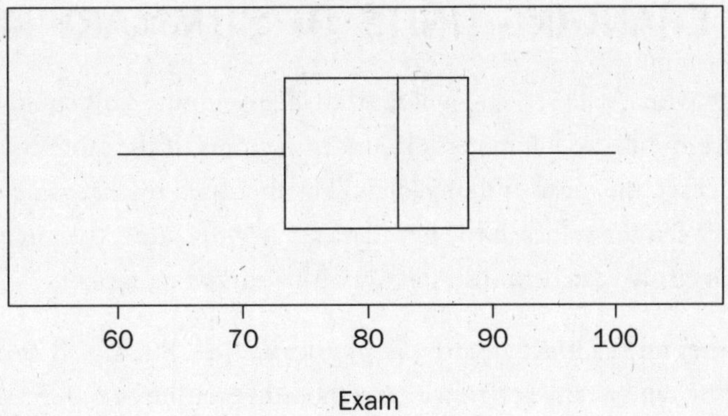

The boxplot displays the center (median) and the spread of the data. From the boxplot above, you can see the range of data (highest value to lowest value). You can also tell something about the skewness of the data. To see this aspect of the boxplot, let's look at the boxplot of our data on the weight of bears in the Adirondack State Park as shown in Figure 4.3.

Figure 4.3
Boxplot: Weights of Bears in Adirondack State Park

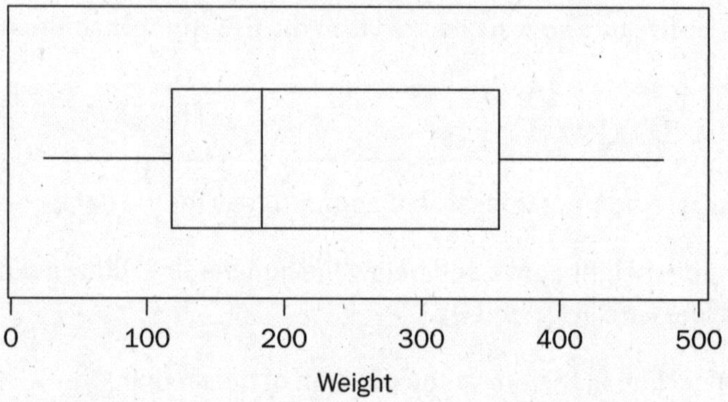

Notice from this plot that the median is very close to Q_1. This indicates a distribution that is skewed right. If the distribution were symmetric, the median would be close to the center of the box.

A **modified boxplot** shows outliers as isolated points. That is, the whiskers will not extend to the minimum or maximum points if those points or others are outliers. As a general rule, points that extend beyond 1.5 times the interquartile range in either direction are not used to define the length of the whiskers, but are labeled separately as dots or asterisks.

Measures such as the median are said to be **resistant**, because they are not influenced by extreme values in the data. The mean and standard deviation are not resistant, since they are both strongly influenced by the extreme values in the data.

EFFECTS OF CHANGING UNITS OF SUMMARY MEASURES

Often, when working with a data set, you will need to change units. This could be something as simple as changing from miles to kilometers or ounces to grams. If the same constant is added to every value in the data set, the mean and median will be increased by that same value. However, since the distance between the values in the new data set stay the same, the range and the standard deviation are unchanged. We can demonstrate this with a simple example.

Let's say your test scores for the fourth quarter in physics are {84, 90, 87, 78, 94}. You have a mean of 86.6, a median of 87, and a range of 16 with a standard deviation of 6.0663. Now, as an end-of-the-year gift, your teacher says he is adding 3 points onto every test grade for the fourth quarter. You now have the grades {87, 93, 90, 81, 97}. Your mean is 89.6, the median is 90.0, the range is 16, and the standard deviation is 6.0663. Note that the mean and median are 3 more than they were with your original test scores.

What if every value in the data set is multiplied by the same constant? In this case, the mean, median, standard deviation, and range will all be multiplied by that constant.

Your history teacher gave you several graded assignments, each worth a maximum of 25 points. Your original scores were {23, 19, 25, 20}. This would give you a mean of 21.75, a median of 21.5, a standard deviation of 2.7538, and a range of 6. Your teacher now decides to grade each of these assignments based on a 100-point scale and to make each a test grade instead of a quiz grade. Each score is multiplied by 4, so you now have {92, 76, 100, 80}. This gives you a mean of 87, a median of 86, a standard deviation of 11.0151, and a range of 24. Note that 87 = 4(21.75), 86 = 4(21.5), 11.0151 = 4(2.7538), and 24 = 4(6). Each measure is four times the original measure.

REVIEW QUESTIONS

1. The mean assessed value of homes in Southern County is \$158,000 with a standard deviation of \$32,000. If the county supervisors decided to increase everyone's assessment by \$5,000, the new mean and standard deviation would be

   (A) \$158,000 and \$32,000.

   (B) \$163,000 and \$37,000.

   (C) \$163,000 and \$32,000.

   (D) \$158,000 and \$37,000.

   (E) Cannot be determined from the information given

2. The mean exam score for the second-period physics class, which had 25 students, was 87.3. The mean exam score for the third-period physics class, which had 19 students, was 92.4. What was the average of all of the students in both classes?

   (A) 89.85

   (B) 89.50

   (C) 90.18

   (D) 91.91

   (E) Cannot be determined from the information given

3. A distribution is skewed right if

   (A) mean = median.

   (B) mean < median.

   (C) mean > median.

   (D) IQR > difference between the mean and the median.

   (E) Cannot be determined from the given information.

4. When a constant is added to every data value in a data set,

   (A) the interquartile range increases by the same constant.

   (B) the interquartile range shifts to the right.

   (C) the interquartile range is multiplied by the constant.

   (D) the interquartile range stays the same.

   (E) Cannot be determined from the given information

FREE-RESPONSE QUESTION

The following stem and leaf plot displays the ages of the first 42 presidents of the United States at the time of their inaugurations.

4	2 3
4	6 7 8 9 9
5	0 1 1 1 1 1 2 2 4 4 4 4 4
5	5 5 5 5 6 6 6 7 7 7 7 8
6	0 1 1 1 2 4 4
6	5 8 9

Key 6 | 5 = 65

(a) Make a boxplot for this data.

(b) Calculate the five-number summary for this data, and determine if there are any outliers.

(c) Describe the shape of the distribution, noting any unusual aspects.

ANSWERS AND EXPLANATIONS

1. C

If the same constant is added to every value in the data set, the mean and median will be increased by that same value. However, since the distance between the values in the new data set stays the same, the range and the standard deviation are unchanged.

2. B

This would be a weighted average, since the classes were not the same size. The mean is calculated $\overline{x} = \frac{\sum x_i}{n}$. However, to account for the difference in class size, we would have

$$\overline{x} = \frac{n_1(\overline{x}_1) + n_2(\overline{x}_2)}{(n_1 + n_2)}$$

$$= \frac{25(87.3) + 19(92.4)}{(25 + 19)} = 89.50.$$

3. C

If mean < median, then the distribution is skewed left. If mean > median, then the distribution is skewed right. And if mean = median, the distribution is symmetric. The mean is pulled to the right by several large data values.

4. D

Since all of the values are increased by the same constant, Q_1 and Q_3 are both increased the same amount. Thus, the difference between them stays the same.

FREE-RESPONSE ANSWER

(a)

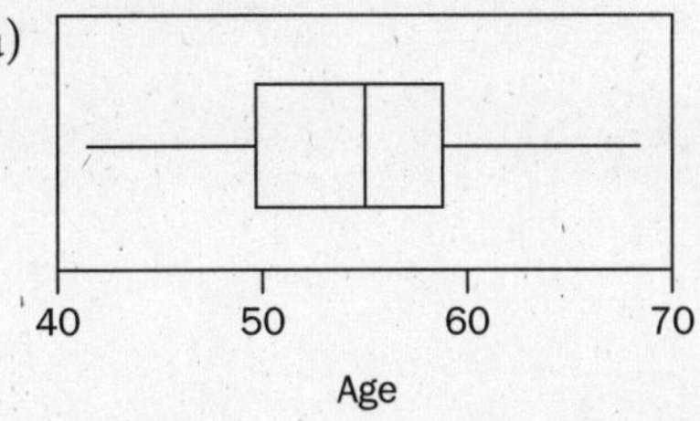

(b) Minimum value = 42; Q_1 = 51; Median = 55; Q_3 = 58; Maximum value = 69

$1.5(IQR) = 1.5(Q_3 - Q_1) = 1.5(58 - 51) = 1.5(7.0) = 10.5$

Outliers would include any president whose age at inauguration was above $Q_3 + 1.5(IQR) = 58 + 10.5 = 68.5$, or any below $Q_1 - 1.5(IQR) = 51 - 10.5 = 40.5$. Since one president at inauguration was aged above 68.5 or below 40.5, there is one outlier.

(c) The data is slightly skewed to the right, but not significantly. Half of the presidents were inaugurated after they were 55, and half before they were 55. There is one outlier in the presidents' ages at inauguration. The middle 50% of ages at the time of inauguration fell between 51 and 58. In addition, 25% of the presidents were inaugurated prior to their 51st birthday, and 25% after their 58th birthday.

CHAPTER 5: COMPARING DISTRIBUTIONS OF UNIVARIATE DATA

IF YOU LEARN ONLY TWO THINGS IN THIS CHAPTER . . .

1. Stem and leaf plots, dotplots, and boxplots all show the center, spread, clusters and gaps, outliers, and shape of the distribution in different and comparable ways.
2. Choosing an appropriate plot depends on which aspects of the distribution are most important to represent.

COMPARING DISTRIBUTIONS WITH STEM AND LEAF PLOTS

Graphical displays of data such as the stem and leaf plot, the dotplot, and the boxplot can be used to compare two sets of data. They are a good "first look" at the similarities and differences between the two data sets. Specifically, they are used to compare the center, spread, clusters and gaps, outliers, shape, and any unusual values in the distributions.

"Back-to-back" stem and leaf plot places two plots on the same stem, with one plot on the left of the stem and one plot on the right. Figure 5.1 provides a plot of the scores on an English exam, splitting the class to compare the performance of males versus females. From the plot, you can see the following:

Center: The center for the females is greater—in the upper 80s—while the center for the males is in the upper 70s.

Spread: Males have a wider range of scores.

Clusters and Gaps: Neither distribution exhibits any clusters or gaps in their data.

Outliers: Neither distribution appears to have outliers.

Shape: From the plot, you can see that the distribution for the females is fairly symmetric and unimodal, but the males' distribution is skewed toward the lower values. The males had more low scores, and fewer high scores. The most frequent scores for both males and females were in the 80s.

Figure 5.1
Stem and Leaf Plot: English Exam, Males vs. Females

Females		Males
	4	7
	5	5 8 8
8 2	6	0 3 6
9 9 8 7 7	7	0 2 2 6 6 7 7
9 9 9 8 8 6 6 6 5 5 3 2	8	1 1 3 3 3 6 8 8 9
8 6 6 6 4 3 1	9	2 3 4 4
0 0 0	10	0

COMPARING DISTRIBUTIONS WITH DOTPLOTS

Distributions can also be compared using dotplots. The individual dotplots should be placed one above the other, and the same scale should be used to compare the distributions.

Figure 5.2 provides dotplots for the weights of bears. One plot provides the distribution of the weights of the male bears, and the other shows the weights of the female bears. From the plots, you can see the following:

Figure 5.2
Dotplot: Weights of Adirondack Bears by Sex

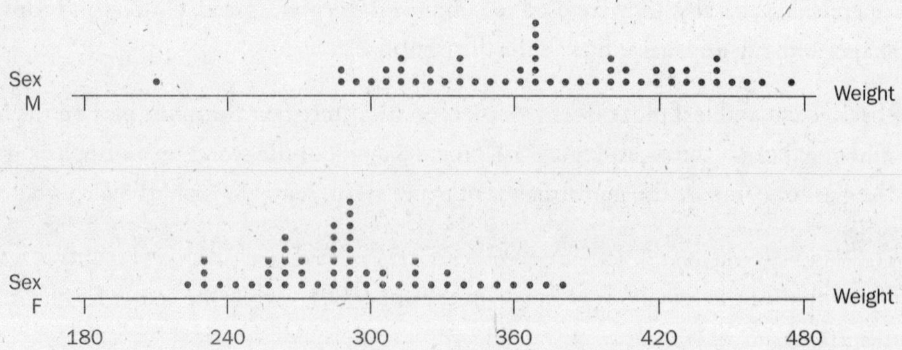

Center: The center of the weights of the males is greater—just above 360 pounds. For the females, the center is just below 300 pounds.

Spread: The spread of weights for the males is considerably greater than that for the females. Male weights range from approximately 210 pounds to just over 470 pounds. For the female, the range is from just about 220 pounds to approximately 385 pounds.

Clusters and Gaps: The dotplots show several clusters of weights in the distribution for the female bears around 260 and 280 pounds. There is only one small cluster in the male weights, at approximately 370 pounds. Neither distribution has significant gaps in their data with the exception of the apparent outlier, a very small male bear. A case could also be made for the largest male bear. There is a slight gap between it and the rest of the male bears.

Outliers: Only the distribution for the weights of male bears shows any apparent outliers. There is one very small male, weighing in around 210 pounds; the largest male might also be considered an outlier.

Shape: Based on the dotplots, the distribution of weights of the females (is skewed very slightly to the right). The distribution of weights of the males is relatively uniform, with the exception of the outlier and the cluster around 370 pounds.

COMPARING DISTRIBUTIONS WITH BOXPLOTS

Using the same data on the bears that we used with the dotplots, we will now look at parallel—also known as "side-by-side"—boxplots as a means of comparing distributions. Figure 5.3 and Figure 5.4 provide the parallel boxplots for the weight of the bears by sex. The difference between the two plots is the orientation of the data. Figure 5.4 simply transposes the x and y axes from Figure 5.3.

AP EXPERT TIP

Keep in mind that not all of the data is represented and, therefore, some information may be missing from these displays.

Since they are based on the five-number summary, parallel boxplots provide an excellent overall comparison of two distributions. From these plots, you can see the following:

Center: The median weight for the male bears is considerably higher than that of the females. The median weight for the male bears is close to 400 pounds, while for the female it is approximately 290 pounds.

Spread: The width of the box, as well as the length of the whiskers, clearly shows that the weights of the male bears are more spread out than those of the female bears. The male bears' weights range from near 210 pounds to 470 pounds. For the female bears, the range is from around 220 pounds to just under 400 pounds.

Clusters and Gaps: Clusters and gaps are not readily apparent from boxplots.

Outliers: The length of the whiskers is an indication of potential outliers. The boxplot for the male bears has an asterisk indicating an outlier for the lowest weight of the males.

Shape: From the boxplots, the distribution of the weights of the male bears appears relatively symmetric. On the other hand, the boxplot for the females indicates the distribution is slightly skewed toward the higher weights, or to the right. The boxplots also show that slightly over 75% of the male bears weigh more than 75% of the females. The median value of the males is approximately level with the largest females, indicating that 50% of the males are heavier than the largest female.

Figure 5.3
Boxplot: Weights of Adirondack Bears by Sex

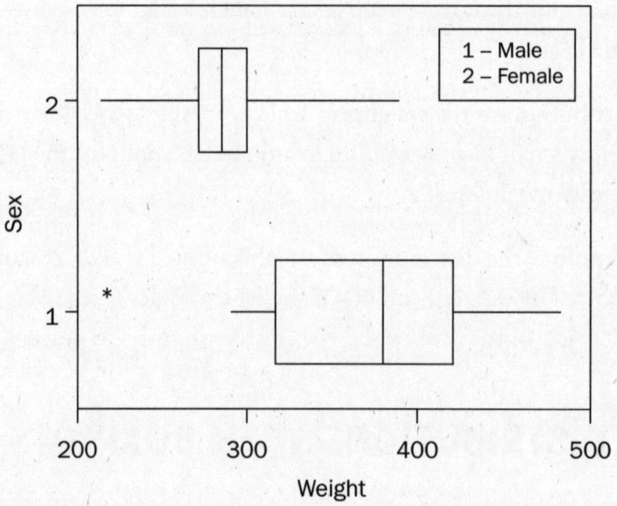

Figure 5.4
Boxplot: Weights of Adirondack Bears by Sex (axes transposed)

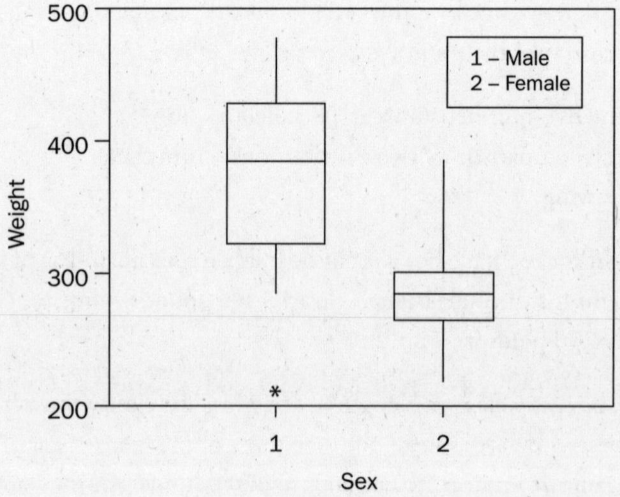

WHICH PLOT?

As with many things, there are no hard and fast rules that say which plot to use. Sometimes it is simply a matter of preference. One plot might show various aspects of the distributions better; one might make a certain relationship more easily identifiable. Following, however, are some general points to consider.

Dotplots and stem and leaf plots display the general shape of the distribution, including clusters and gaps.

The *stem and leaf plot* displays the actual values of all of the data points.

Boxplots display the five-number summary and clearly depict the spread, center, and outliers for the distribution. Remember, though, that boxplots can hide some features of a distribution.

The best strategy is to look at all of the options, and then decide which most clearly provides the information you want to communicate.

REVIEW QUESTIONS

1. Which of the following is true based on the boxplots comparing the average number of hours that male and female students watch TV?

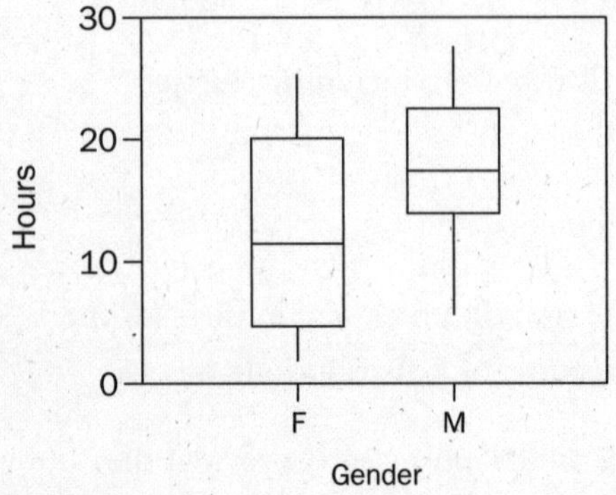

(A) Both distributions are approximately symmetric.
(B) The distribution for females is symmetric, and the distribution of males is skewed to the left.
(C) The medians for both distributions are approximately the same.
(D) The IQR for both distributions is approximately the same.
(E) The distribution for males is skewed to the right and the distribution for females is roughly symmetrical.

2. Consider the following back-to-back stem and leaf plots comparing weight gain in kilograms for male and female horses.

Male		Female
	1	7
7	2	2 3 4 8 8
3 2	3	1 4 5 6 7 8 9 9
8 7 5 4	4	3 4 6 6 8
9 6 5 3 3 0	5	6
6 5 4 3 3 3 1	6	

Which of the following are true statements?

I. The distributions have the same number of observations.

II. The ranges for the two distributions are the same.

III. The means for the two distributions are the same.

IV. The medians of the two distributions are the same.

V. The upper quartiles of the two distributions are the same.

(A) I and II
(B) I and IV
(C) II and V
(D) III and V
(E) I, II, and III

3. Given a set of data comparing the life expectancy in hours for batteries manufactured by two companies, which of the following could not be used to compare these data graphically?

   (A) Back-to-back stem and leaf plot
   (B) Dotplot
   (C) Pie chart
   (D) Parallel boxplots
   (E) All of these can be used to compare the data graphically.

4. Based on the parallel boxplots in the display, which of the following statements about these data sets CANNOT be justified?

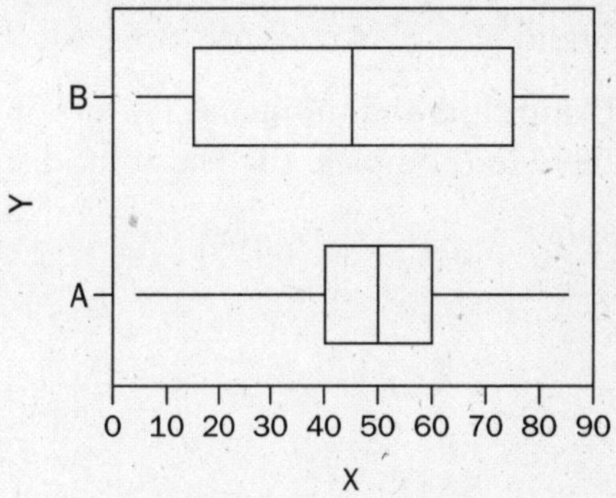

   (A) Set B has more data than set A.
   (B) The range of the two sets of data is the same.
   (C) The interquartile range of set B is more than twice the interquartile range of set A.
   (D) The median of set A is greater than the median of set B.
   (E) The data in set A has less variability than the data in set B.

FREE-RESPONSE QUESTION

The following data represents the hours of continuous use for two brands of batteries.

Brand A: 65, 67, 69, 71, 63, 62, 70, 72, 66

Brand B: 65, 67, 67, 68, 70, 64, 64, 65, 65

Using plots and summary statistics, investigate and report on the comparison of these two batteries. Include a complete analysis of the distributions.

ANSWERS AND EXPLANATIONS

1. E

The lower side of the box represents Q_1 and the upper side Q_3. The median line for males is closer to Q_1, indicating that 50% of the values are closer to Q_1 or below it. The length of the box between the median and Q_3 shows that the values above the median are more spread out. A small number of large values pulls the mean toward the upper end, skewing the distribution.

2. A

Both distributions have 20 observations. The ranges, $66 - 27 = 39$ and $56 - 17 = 39$, are equal. The distribution for the males clearly has a higher mean and median, and with its skewness it also has a larger variance.

3. C

Pie charts are only appropriate for categorical data. The data in this example is quantitative. The stem and leaf plot, dotplot, and boxplot are all appropriate for quantitative data.

4. A

Answer (A) is correct. It is impossible to tell how many data values are in a set using a box and whisker plot. It only provides the five-number summary of the data, not the number of data values.

Answer (B) is a justifiable statement because the range of both sets A and B is approximately 80.

Answer (C) is a justifiable statement because the interquartile range of set B is approximately 60 while that of set A is approximately 20.

Answer (D) is a justifiable statement because the median of set A is slightly larger (to the right in the boxplots) than that of set B.

Answer (E) is justifiable because the interquartile range of set A is about 20. The interquartile range of set B is about 60. Both sets are fairly symmetric. The values in set A are, therefore, on average, closer to the center. Hence, set A has less variability than set B.

FREE-RESPONSE ANSWER

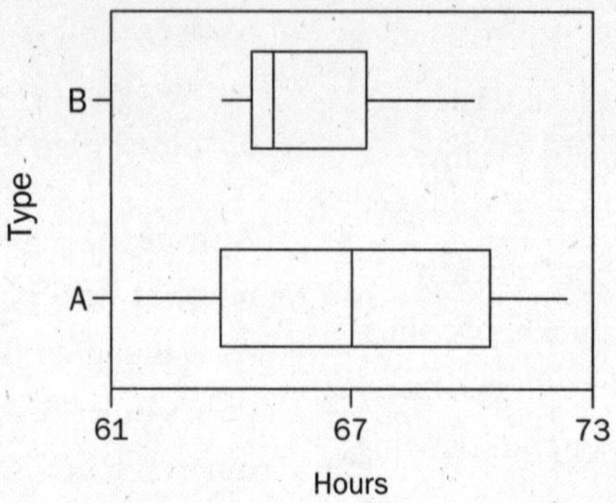

The boxplots clearly show that there is more variation in the distribution of Brand A batteries than in Brand B. The range for Brand A is greater than the range for Brand B, as seen in the measure from the ends of the whiskers. In addition, the middle 50% of values for Brand A is approximately twice as long as the same area for Brand B. The boxplots also show that the median number of hours for Brand A is greater than that for Brand B.

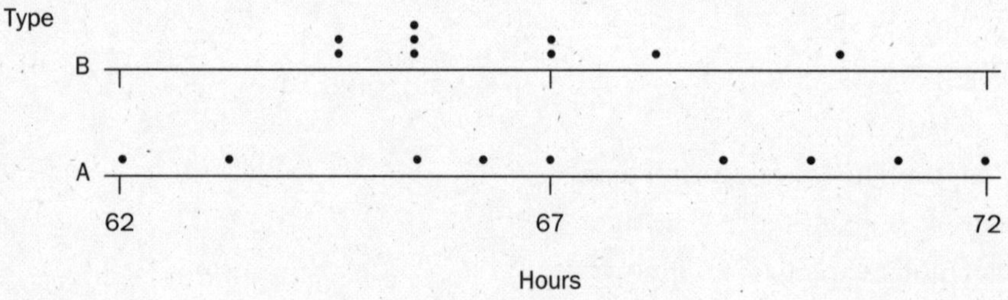

The dotplot shows that the hours for Brand A range from 62 to 72, with no clusters and no duplicates. Each of the nine observations is different. Brand B, on the other hand, is less spread out and shows clusters.

Brand B		Brand A
4 4	6	2 3
8 7 7 5 5 5 5	6	5 6 7 9
0	7	0 1 2

From the stem and leaf plot, the high concentration of values between 65 and 68 for Brand B is easily seen. The stem and leaf plot gives a more symmetric appearance to both distributions. This is partially a result of the small number of observations.

	Min	Q_1	Median	Q_3	Max	Mean
A	62	64	67	70.5	72	67.22
B	64	64.5	65	67.5	70	66.11

The summary statistics bear out that the mean number of hours for Brand A is greater than that for Brand B—67.22 to 66.11, respectively. The summary statistics also validate that the median value of Brand A is 67 hours, while the median value for Brand B is 65. The range for Brand A is $72 - 62 = 10$ and the IQR is $70.5 - 64 = 6.5$. For Brand B, the range is $70 - 64 = 6$ and the IQR is $67.5 - 64.5 = 3$.

Notice from this problem that if you only look at one type of plot and never consider any other, you can get a somewhat limited view of the distribution. While it is not always practical on a timed exam, looking at multiple plots can give you a much clearer picture of the data.

CHAPTER 6: EXPLORING BIVARIATE DATA

IF YOU LEARN ONLY FIVE THINGS IN THIS CHAPTER . . .

1. The knowledge of one variable may help to predict the value of a second variable.
2. A scatterplot graphically represents the relationship between two dependent variables.
3. The linear correlation coefficient is used to measure the strength of a linear relationship.
4. The difference between the *y*-value for an ordered pair and the predicted value for the *x*-value of that ordered pair is called the residual.
5. Not all relationships between two variables are linear; they may also be quadratic or logarithmic.

OVERVIEW: WHY BIVARIATE DATA?

Often when working with real data, a problem will require the analysis of more than one variable. In these cases, we may be interested in knowing if the variables are related and, if so, what type of relationship exists. Although the ultimate goal for the researcher may be to establish some type of cause-and-effect relationship, statistics can also indicate the presence or absence of any mathematical relationship between variables. The knowledge of one variable may help predict the value of a second variable.

Suppose we want to know if a person's height can be used to predict weight, or if a student's academic success can be predicted by his or her score on a standardized test. These situations require the analysis of **bivariate data**, or data that contains values from two ("bi-") different response variables ("-variate") from the same individual. Although these two variables can be 1) both quantitative, 2) both qualitative, or 3) one of each, we will limit the discussion in this section to the first type. See chapter 8 for more information on the second type. Bivariate data of the third type is beyond the scope of this exam.

When both variables are quantitative, we express the two values for each individual as an ordered pair (x, y), where x represents the independent (or explanatory) variable and y represents the dependent (or response) variable. The first variable is considered independent, since it is assumed that the researcher has control over these values. The second variable (y) is referred to as the response variable, since this is the value that will be predicted to some degree by the first variable.

As an example, let's assume that a researcher is attempting to monitor the effects of various levels of a certain drug on blood pressure. The researcher would control the amount of drug administered to the subjects, making dosage (amount of drug) the independent variable. The subject's blood pressure would then be the response variable (or dependent variable).

Often, it is not obvious which variable is independent and which is dependent. For example, when considering the height and weight of individuals, either of the two variables could be considered independent. In these cases, the selection of one variable to be independent would depend on the question being asked. If we were interested in studying the effects of child obesity on height, then weight would be considered the independent variable, and height the response variable. However, if we wish to know if an adult's height is correlated to his or her weight, we would probably make height the independent variable. Keep in mind that the results of the analyses will differ based on which of the two variables is considered independent.

ANALYZING PATTERNS IN SCATTERPLOTS

Once the independent and dependent variables have been determined and the collected data is written as a series of ordered pairs, it is best to graph the data on a scatterplot. A scatterplot is just a graph on the Cartesian coordinate system showing all the points that are represented by all ordered pairs in the data. Keep in mind that the x-values are plotted along the horizontal axis, and the y-values are plotted along the vertical axis. Therefore, the horizontal axis represents the independent variable and the vertical axis represents the response variable.

Once all the ordered pairs are plotted, a quick visual inspection of the scatterplot will provide the first glimpse into the type of relationship that may exist between the variables. If the points tend to cluster around an imaginary line, this indicates that some type of relationship exists—the more the points cluster, the stronger the relationship between the variables.

Take a look at Figure 6.1. The relationship or lack of relationship between two values can be seen in these graphs. The graph on the left indicates a very strong linear relationship between two variables. The graph on the right indicates no relationship between the variables. The middle graph shows that a linear relationship of some type may exist, but if so, it is rather weak.

Figure 6.1

Relationships Between Variables in Scatterplots

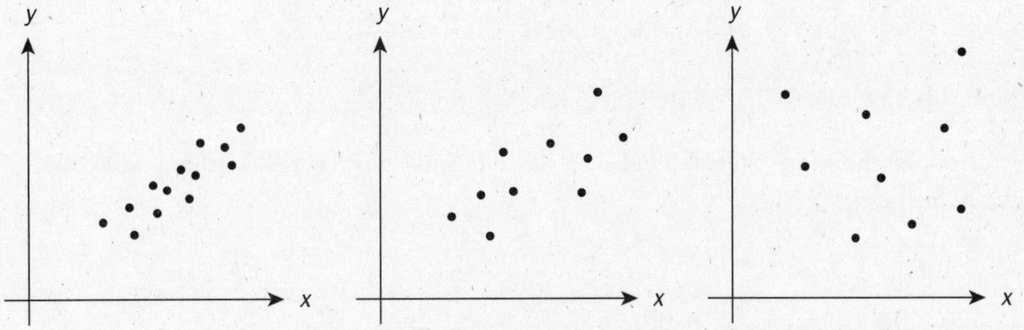

If the cluster of points seems to rise as you move from left to right, then the relationship is a positive one; an increase in the value of the independent variable causes the value of the response variable to increase as well. In other words, both variables increase together or are directly related. However, if the cluster of points seems to fall as you move from left to right, then the relationship is of the negative type. In this case, the variables are indirectly related; as one variable increases, the other decreases as depicted in Figure 6.2.

Figure 6.2

Positive Linear vs. Negative Linear Relationship

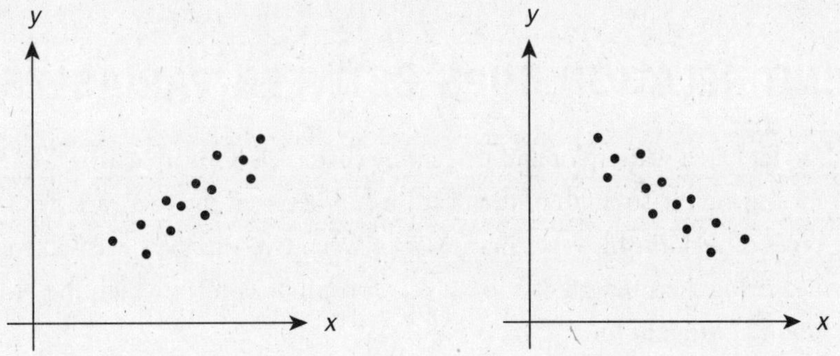

Example:

As part of a military fitness test, soldiers are required to complete as many push-ups and sit-ups as possible in a fixed amount of time. Ten randomly selected soldiers are chosen and the number of push-ups and sit-ups are recorded for each. The data is given on the following page.

	Soldiers									
	1	2	3	4	5	6	7	8	9	10
Push-ups (x)	25	25	17	33	30	50	35	55	40	40
Sit-ups (y)	28	29	25	45	36	23	32	52	48	42

The data, written as ordered pairs with number of push-ups first, is:

(25, 28), (25, 29), (17, 25), (33, 45), (30, 36),

(50, 23), (35, 32), (55, 52), (40, 48), (40, 42)

The scatterplot shown in Figure 6.3 indicates that a possible positive relationship may exist between these variables.

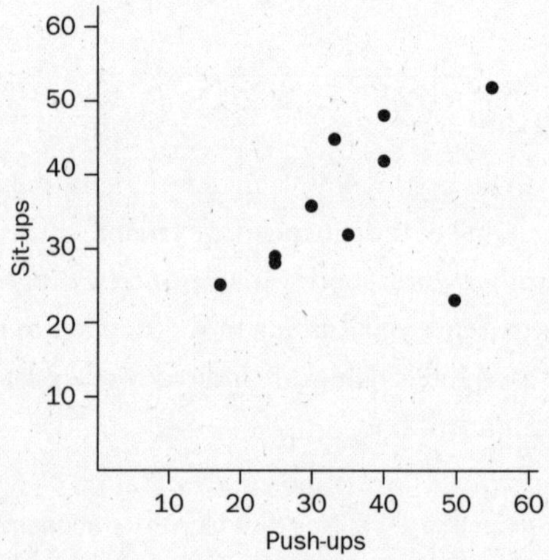

MEASURING RELATIONSHIPS BETWEEN VARIABLES

It is not enough to look at a scatterplot and determine that a relationship exists. We must have some means of measuring the strength of the relationship between the two variables. Although there are many types of relationships that may exist between two variables, we will concern ourselves with measuring the strength of a linear relationship, or one in which the ordered pairs tend to cluster around a straight line.

The **linear correlation coefficient (*r*)** is used to measure the strength of a linear relationship between two variables. This measurement has limits at −1 and +1. The endpoints indicate a perfect correlation: −1 indicates a perfect negative correlation; +1 indicates a perfect positive correlation.

A correlation of zero ($r = 0$) would indicate no correlation. The farther the measurement is from zero, the stronger the linear relationship between the two variables.

Figure 6.4

Strength of a Linear Relationship Between Two Variables

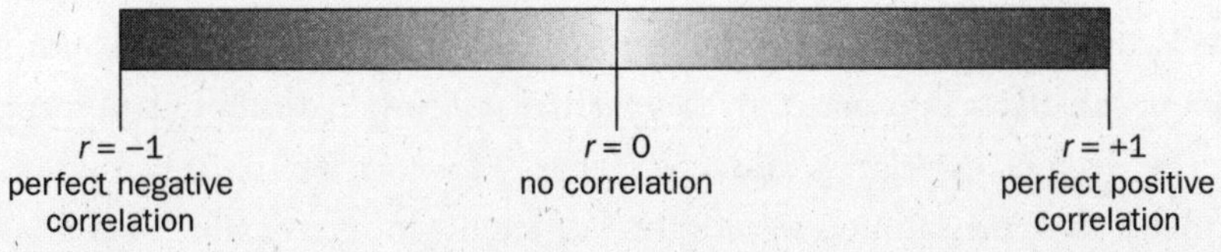

In fact, when the linear correlation coefficient is squared (r^2), we can obtain a value that indicates the percentage of variation in the response variable that can be attributed to the independent variable. For example, if a correlation coefficient is $r = 0.82$, then = $(0.82)^2$ or 0.6724. This indicates that 67% of the variation in the response variable can be attributed to the linear relationship with the independent variable.

Linear correlation coefficients are easily calculated from a set of (x, y) data input into a computer program or statistical calculator. The formula for calculating the correlation coefficient is:

$$r = \frac{n\sum xy - (\sum x)(\sum y)}{\sqrt{n\sum x^2 - (\sum x)^2}\sqrt{n\sum y^2 - (\sum y)^2}}$$

Example (continued):

Returning to our previous example, let us calculate the correlation coefficient for these data. To do so, we create a table as shown below.

Soldier	x	x^2	y	y^2	xy
1	25	625	28	784	700
2	25	625	29	841	725
3	17	289	25	625	425
4	33	1,089	45	2,025	1,485
5	30	900	36	1,296	1,080
6	50	2,500	23	529	1,150
7	35	1,225	32	1,024	1,120
8	55	3,025	52	2,704	2,860
9	40	1,600	48	2,304	1,920
10	40	1,600	42	1,764	1,680
Total	350	13,478	360	13,896	13,145

Now we can substitute into the formula to find the correlation coefficient.

$$r = \frac{10(13{,}145) - (350)(360)}{\sqrt{10(13{,}478) - 350^2}\sqrt{10(13{,}896) - 360^2}} = \frac{5{,}450}{10{,}714.4} = 0.51.$$

As you can see, the formula for r is cumbersome, and you should be familiar with the calculator functions that calculate the correlation coefficient from given data (see chapter 19 for more information about calculator functions). In this particular example, the correlation coefficient is almost halfway between no correlation and a perfect positive correlation. At this point, you may wish to determine if this correlation is significantly different from zero, which would indicate the strength of the relationship. Refer to chapter 17 on hypothesis testing for further information on this procedure.

Finally, on the AP exam you will not receive credit for describing a correlation coefficient as "good" or "bad." You are expected to describe the strength and direction (positive or negative) of the relationship in the context of the problem. In the above example, the correlation coefficient of 0.51 could be interpreted as a "weak positive relationship between the number of push-ups and the number of sit-ups a soldier could complete." Note that $r^2 = 0.26$, so only 26% of the variation can be attributed to the linear relationship.

LEAST-SQUARES REGRESSION LINE

Assuming the test for significance indicates that r is indeed significantly different from zero, we would then be interested in obtaining a formula that would allow us to predict the response variable from the independent variable. This is done through the statistical procedure known as the **least-squares regression**. The procedure fits a line through the data values in such a way that the sum of the squared deviations of all points from the line is minimized.

The outcome of the procedure provides the parameters (slope and y-intercept) for a line on the coordinate plan. The formulas to derive these parameters are rather complex and beyond the scope of this test. However, these parameters can be calculated quickly by a computer or graphing calculator. For testing purposes, the parameters will be given to you.

You must be able to identify the equation of the line (or prediction formula) using these values. Recall from algebra that for the equation of a line $y = ax + b$, the slope is a and the y-intercept is the y-value of the point $(0, b)$.

To form the prediction formula for a set of data after completing a least-squares regression analysis, you merely plug the slope and y-intercept into the formula above. To use the formula to predict a response for a given value of x, plug the value of x into the prediction formula and solve for y.

Example (continued):

Returning to our example for the number of push-ups and sit-ups a soldier can perform in a set amount of time, we run a regression analysis using a computer program and obtain the following results.

	df	*SS*	*MS*	*F*	*Significance F*
Regression	1	241.8770	241.8770	2.7877	0.1335
Residual	8	694.1230	86.7654		
Total	9	936			

	Coefficients	*Standard Error*	*t Stat*	*p-value*
Intercept*	20.4666	9.7586	2.0973	0.0692
Push-ups	0.4438	0.2658	1.6696	0.1335

*Other computer outputs may use the term "constant" in lieu of "intercept."

Most of these results are not applicable to you at this point. The first part of the analysis provides information necessary to test the significance of the linear relationship between the two variables. This is covered further in chapter 17. Our concern is only with the second portion of the output. Here you will see two coefficients, or parameters, that were calculated during the least-squares regression process. The first, *intercept*, provides the value of the y-intercept. The second, *push-ups*, gives the slope of the line. Putting this together, we find the prediction formula for these data to be

$$\text{sit-ups} = 20.4666 + 0.4438 \times \text{push-ups} \approx 20.5 + 0.44 \times \text{push-ups}.$$

We can now predict the number of push-ups we can expect a soldier to perform based on the number of sit-ups she can do. Let's say a soldier completes 20 push-ups. We would expect her to complete 29 ($20.5 + 0.44 \times 20 = 29.3$) sit-ups. Of course, there will be some variation in the number of actual sit-ups completed, but this provides an informed estimate of the number we would expect this soldier to complete.

RESIDUAL PLOTS AND OUTLIERS

If we look at the regression line and scatterplot simultaneously, we see that few—if any—of the ordered pairs from our data fall on the regression line. The difference between the y-value for an ordered pair and the predicted value for the x-value of that ordered pair is called the **residual**. We can calculate the residuals for all our ordered pairs by finding the predicted value for each pair and subtracting the actual y-value of the ordered pair.

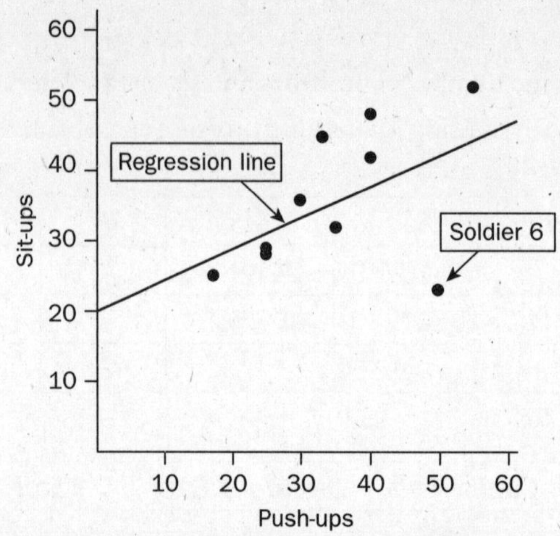

Sometimes it is easier to plot the residuals, which has been done in Figure 6.6. We can see from the list of residuals and the plot of the residuals that one individual, soldier 6, had a residual that is far larger than the others. This indicates that there is an outlier in the data. Since an outlier could have a large effect on the regression line, it is sometimes customary to redo the regression procedure while omitting the outlier. This will provide a better fitting line for the rest of the data. Note that we do not remove the outlier for the correlation calculation unless there is sufficient cause to do so.

Figure 6.6
Push-ups Residual Plot

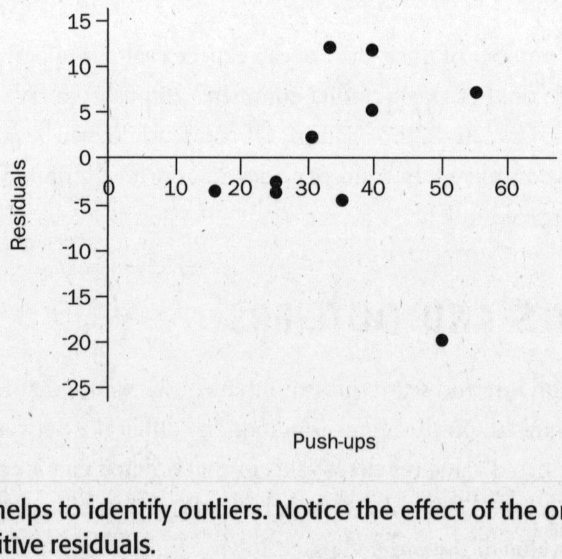

Plotting the residuals helps to identify outliers. Notice the effect of the one extremely low residual on all the positive residuals.

Example:

We will now remove the outlying value from the data in the given example and complete a new least-squares regression analysis with the remaining nine ordered pairs. The resulting output is shown in the following table with a residual plot in Figure 6.7. From the residual plot, we see how much better the line fits the data since the points are very close to the zero line. The new regression equation is sit-ups = $11.5 + 0.78 \times$ push-ups.

Figure 6.7
Push-ups Residual Plot (outlier omitted)

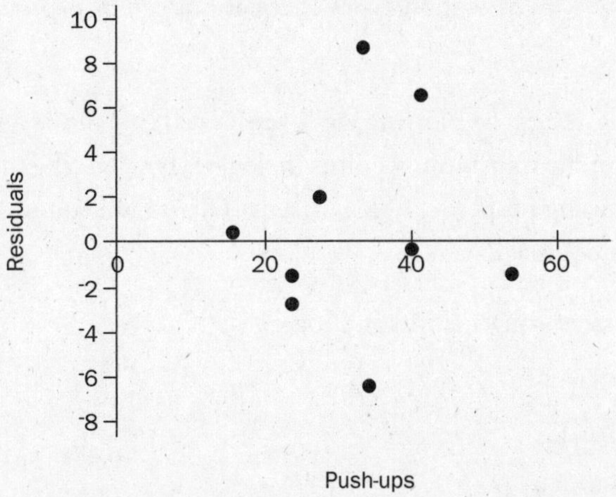

Now the predicted number of sit-ups for a soldier who performs 20 push-ups is 27.1 ($11.5 + 0.78 \times 20 = 27.1$). Again, this predicted value is much more accurate because the prediction equation is not unduly influenced by an outlier.

	df	*SS*	*MS*	*F*	*Significance F*
Regression	1	593.186	593.186	26.783	0.001
Residual	7	155.036	22.148		
Total	8	748.222			

	Coefficients	*Standard Error*	*t Stat*	*p-value*
Intercept	11.484	5.256	2.185	0.065
Push-ups	0.779	0.150	5.175	0.001

TRANSFORMING DATA TO ACHIEVE LINEARITY

The procedures discussed in this chapter pertain to data that exhibits a somewhat linear relationship between the two variables. In the real world, linearity may not be the relationship. Many types of data will present a quadratic or logarithmic relationship. In these cases, transformations can be made to each x- and/or y-value. The resulting ordered pairs might then exhibit a more linear relationship.

Nonlinear data may appear as if it can be modeled with either an exponential ($\hat{y} = a \times b^x$) or a power function ($\hat{y} = a \times x^b$). Each of these types of equations can be expressed in linear form with the aid of logarithms.

For **exponential** functions, begin by plotting $\log \hat{y}$ versus x. If this shows a linear pattern, perform least-squares regression on the transformed points (x, $\log \hat{y}$). Finally, take the linear regression equation and solve for $\hat{y}$ using properties of logarithms. This allows you to see the exponential function that models the original data.

Consider the following linear equation in x and $\log \hat{y}$:

$$\begin{aligned}
\log \hat{y} &= 0.8692 + 0.0115x \\
10^{\log \hat{y}} &= 10^{0.8692+0.0115x} \\
\hat{y} &= 10^{0.8692}10^{0.0115x} \\
\hat{y} &= 7.399 \cdot (1.0268^x).
\end{aligned}$$

Note, when (x, $\log \hat{y}$) is linear, then (x, $\hat{y}$) is exponential.

For **power** functions, begin by plotting $\log \hat{y}$ versus $\log x$. If this shows a linear pattern, perform least-squares regression on the transformed points ($\log x$, $\log \hat{y}$). As before, take the linear equation and solve for $\hat{y}$. This allows us to see the power function that models the original data.

Consider this linear equation in $\log x$ and $\log \hat{y}$:

$$\begin{aligned}
\log \hat{y} &= 0.7480 + 0.66056 \log x \\
10^{\log \hat{y}} &= 10^{0.7480+0.66056 \log x} \\
\hat{y} &= 10^{0.7480}10^{0.66056 \log x} \\
\hat{y} &= 5.5976\left(10^{\log x^{0.66056}}\right) \\
\hat{y} &= 5.5976x^{0.66056}.
\end{aligned}$$

Note when ($\log x$, $\log \hat{y}$) is linear, (x, $\hat{y}$) is a power function.

REVIEW QUESTIONS

1. A scatterplot is obtained by
   (A) scattering the values of x over the values of y.
   (B) scattering the values of y over the values of x.
   (C) plotting the paired values of x and y.
   (D) plotting the dependent variable on the x-axis and the independent variable on the y-axis.
   (E) plotting x and y values along a single axis.

2. A perfect positive correlation means
   (A) the points in a scatter diagram lie on an upward sloping line.
   (B) the points in a scatter diagram lie on a downward sloping line.
   (C) r is equal to -1.
   (D) r is equal to zero.
   (E) there is a direct cause-and-effect relationship between the variables.

3. In a linear regression model, the slope represents
   (A) the value if y where the y-axis intersects the x-axis.
   (B) the value of y where the regression line intersects the y-axis.
   (C) the point where the regression line intersects the x-axis.
   (D) the change in the response variable due to a one-unit change in the independent variable.
   (E) the change in the independent variable due to a one-unit change in the response variable.

4. In the regression equation $y = 12 + 6x$, 12 represents the
   (A) independent variable.
   (B) slope of the line.
   (C) dependent variable.
   (D) y-intercept.
   (E) residual.

FREE-RESPONSE QUESTION

A local gym randomly selected ten members to see if the number of hours spent at the gym was related to the length of time the person had been a member. Number of months as a member and hours spent at the gym in a week were recorded for each of the members. The computer output obtained from fitting a least-squares regression line to the data is shown below. A plot of the residuals is provided as well.

	df	*SS*	*MS*	*F*	*Significance F*
Regression	1	48.800	48.800	33.656	0.000
Residual	8	11.600	1.450		
Total	9	60.400			

	Coefficients	*Standard Error*	*t Stat*	*p-value*
Intercept	9.855072464	0.826	11.925	0.000
Push-ups	−0.664855072	0.115	−5.801	0.000

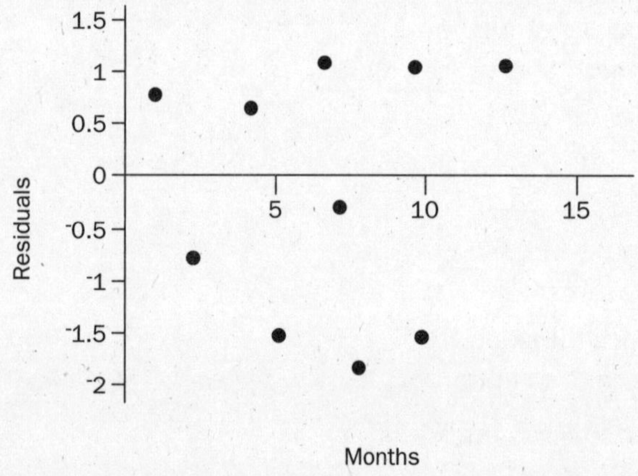

(a) What is the equation of the least-squares regression line given by this analysis?

(b) If this equation is used to predict the hours of usage for someone who has been a member for six months, which of the following would you expect?

- The prediction will be too large.
- The prediction will be too small.
- The prediction will be neither too large nor too small.
- A prediction cannot be made based on the information given on the computer output.

Explain your reasoning.

ANSWERS AND EXPLANATIONS

1. C

Keep in mind that the graph is called a scatterplot, probably due to the fact that the points are scattered about the graph area. We do not "scatter" any values in producing the diagram. We plot the ordered pairs (x, y) for each individual to obtain the scatterplot.

2. A

Positive correlations show an upward trend, while negative correlations show a downward trend. A correlation coefficient of -1 indicates a perfect negative correlation; $+1$ indicates a perfect positive correlation. A correlation coefficient of zero indicates no relationship at all. A perfect positive correlation cannot allow one to determine that a variable is the direct cause of change in another variable.

3. D

The slope of a line is the rate of change of the y-value for a one-unit change in the x-value. The point where the y-axis intersects the x-axis is the origin of the graph. The regression line intersects the y-axis at what is known as the y-intercept. This is usually indicated by the variable name on the computer output. The x-intercept is the point where the regression line intercepts the x-axis. This value is usually not pertinent in regression analysis.

4. D

The regression line is given in the form of $y =$ *y-intercept* + *slope* × *(independent variable value)*. This means that the 12, which is not being multiplied by x, is the y-intercept of the line.

FREE-RESPONSE ANSWER

(a) The least-squares regression line is given by the equation $y = 9.86 - 0.66x$ where y is the response value for the independent value x.

(b) The predicted hours of gym use for a 6-month member would be about 6 hours. Do not expect this value to be too large or too small. There will be variation in the actual number of hours, but the best prediction is 6 hours. In other words, any prediction greater or less than 6 hours would lead to a greater total regression.

CHAPTER 7: EXPLORING CATEGORICAL DATA

IF YOU LEARN ONLY FOUR THINGS IN THIS CHAPTER . . .

1. Categorical data is exhibited using frequency tables and bar charts.
2. Joint frequencies occur where a row category meets a column category; if the frequencies appear in the margins of the table, they are called marginal frequencies.
3. Conditional relative frequencies are expressed in percentages.
4. When graphing two-way contingency tables, one variable is represented by the bars, the other is used to label the *x*-axis.

FREQUENCY TABLES AND BAR CHARTS

Up to this point, we have concerned ourselves with describing and exploring quantitative data. Frequently, statistical studies will generate categorical data, which involves a variable that classifies the subject into one of a number of categories. Since the data will generate descriptions as opposed to numbers, means and standard deviations do not apply. In these cases, we are limited to exploring the data using frequency tables and bar charts.

Creating a frequency table for categorical data is rather simple. Simply count the number of individuals in each class. Then list each of the various classes in one column with the corresponding frequency in the second column. For comparing two frequency distributions, relative frequency (frequency ÷ total) is used.

(a)

Type of Waste	Tons
Paper	11.4
Plastic	5.4
Metal	4.2
Yard Waste	3.3
Food	1.2
Glass	0.6
Other	3.9
Total	30

(b)

Type of Waste	Relative Frequency
Paper	38%
Plastic	18%
Metal	14%
Yard Waste	11%
Food	4%
Glass	2%
Other	13%
Total	100%

Frequency distribution (a) and relative frequency distribution (b) for type of waste found in 30 tons of refuse collected from a random sample of landfills across America. (Source: *USA Today*, 1991)

For a graphical depiction of these tables, a bar chart is used. Each category is represented by a bar. The height of the bar is the frequency of the category. It is customary to place the bars in descending height, except for an "other" category, which is always put last, as shown in Figure 7.1.

Figure 7.1
Bar Graph of Table Data

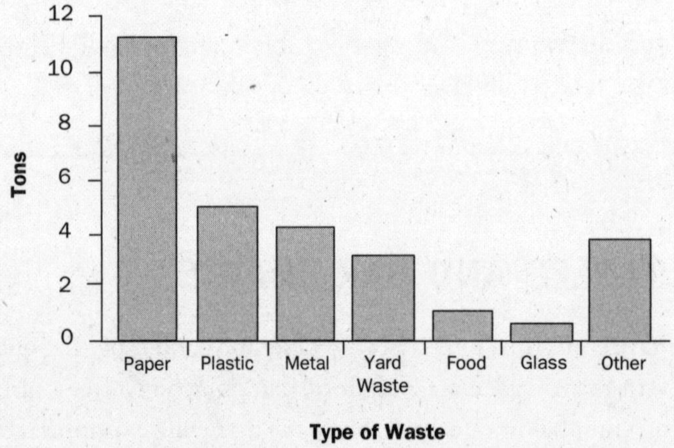

MARGINAL AND JOINT FREQUENCIES FOR TWO-WAY TABLES

Oftentimes, a researcher is interested in studying two categorical variables simultaneously. In these cases, the bivariate categorical data is expressed in a cross-tabulation or two-way contingency table. Imagine that we randomly select 30 students and ask the major of each one. We record the gender and major for each student. Reviewing our results, we see that there are three different majors given: Liberal Arts (LA), Business Administration (BA), and Technology (T). Gender, of course, can be one of two categories. We then proceed to summarize the data in a two times three contingency table; two rows for gender (M or F), and three columns for major (LA, BA, or T). The results of our sample study are organized in the following table.

	Major			
	LA	BA	T	Total
Male	4	6	8	18
Female	5	4	3	12
	9	10	11	30

Each frequency that occurs where a row category meets a column category is referred to as a **joint frequency**. Our table has six joint frequencies. Since the row totals and column totals are given in the margins of the table, they are called **marginal frequencies**. The marginal frequencies represent the frequencies of the categories for the corresponding variable. The total of the marginal frequencies (row or column) is the grand total, and should equal the size of the sample (n).

Contingency tables often show percentages or relative frequencies. These percentages can be based on the entire sample, as in the table below, or on the subsample (row or column) classifications.

	Major			
	LA	BA	T	Total
Male	13%	20%	27%	60%
Female	17%	13%	10%	40%
	30%	33%	37%	100%

With the table expressed as percentages of the grand total, we can easily see that the sample consisted of 60% males and 40% females. As far as majors, 30% were Liberal Arts, 33% were Business Administration, and 37% were studying Technology.

CONDITIONAL RELATIVE FREQUENCIES

If we are interested in comparing the distribution of majors for males versus females, we might give the table as percentages of the marginal row frequencies. In this case, divide each joint frequency by the corresponding marginal row frequency and multiply by 100. Observing the resulting table, you can readily see that the percentage of males and females in Business Administration is equal. This fact is not as noticeable in the other two tables. It will be obvious that a table is given as relative frequencies of the marginal row frequencies when you notice that each row adds up to 100%.

	Major			
	LA	BA	T	Total
Male	22%	33%	44%	100%
Female	42%	33%	25%	100%
	30%	33%	37%	100%

Likewise, the entries in the contingency table can be expressed as a percentage of the marginal column frequencies (major). Below are the results of multiplying each joint frequency by the corresponding marginal column frequency, and then multiplying by 100. Again, you will quickly recognize that the numbers are expressed as percentages of the column totals since each column totals 100.

	Major			
	LA	BA	T	Total
Male	44%	60%	73%	60%
Female	56%	40%	27%	40%
	100%	100%	100%	100%

Be careful, however, when combining data from several groups. There may be a lurking variable that leads to results that seem contradictory. Consider the information in the following tables.

AP EXPERT TIP

These contradictory results are due to a lurking variable, which is the distribution of A or B vs. C or D students in the classes.

Mrs. Smith

Algebra Grades	Number Passing	Total Students	Percent Passing
A or B	70	80	87.5
C or D	5	20	25
Total	75	100	75

Mr. Jones

Algebra Grades	Number Passing	Total Students	Percent Passing
A or B	10	10	100
C or D	40	90	44.4
Total	50	100	50

Mrs. Smith and Mr. Jones are two algebra teachers. If you look at the total percent of passing students, it appears Mrs. Smith is passing a higher percentage of students. However, if you look at each grade category, Mr. Jones has a higher passing rate in each. This is an example of what we call Simpson's Paradox.

GRAPHING TWO-WAY TABLES

Although the frequency tables give detailed information about the data, a graphical representation can provide a quick assessment of the data. Graphs are also better for comparing two or more groups.

To graph a two-way contingency table, you must choose one of the variables to be represented by the different bars. The other variable will be used to label the *x*-axis. The vertical (*y*-axis) will represent the frequencies. There are two different ways to graph a two-way contingency table, since either variable could be used for the bars. The choice will usually be determined by what you wish to compare.

Suppose you wish to compare the distribution of majors between the males and females. You would then use the variable for major as a label on the *x*-axis. Two different shaded bars would represent the gender,—one for males, one for females—as illustrated in Figure 7.2. From the graph, it is easy to detect that the distributions are reversed; majors with more males have fewer females. However, it is not as easy to tell just how different the distributions are, since there is not an equal number of males and females.

Figure 7.2
Distribution of Majors Between Males and Females

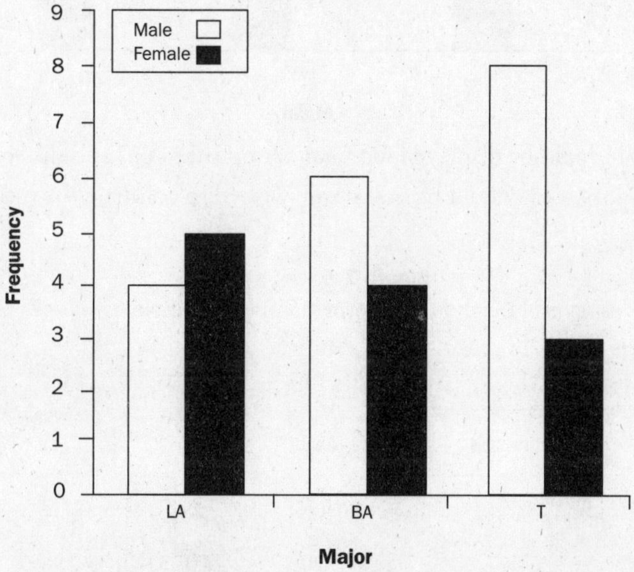

For better comparison, you may wish to graph the conditional relative frequencies calculated from the marginal row frequencies.

Figure 7.3
Conditional Relative Frequencies for Distribution of Majors

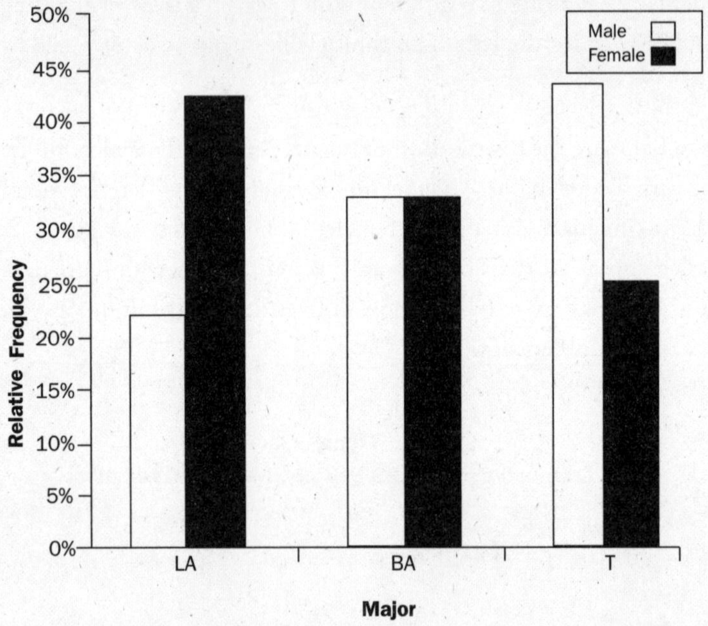

Of course, the same table could be displayed with gender on the x-axis and different bar shading for each of the three majors. This would be best if one wished to compare the majors by gender.

Figure 7.4
Graphical Display of Two-Way Contingency Table

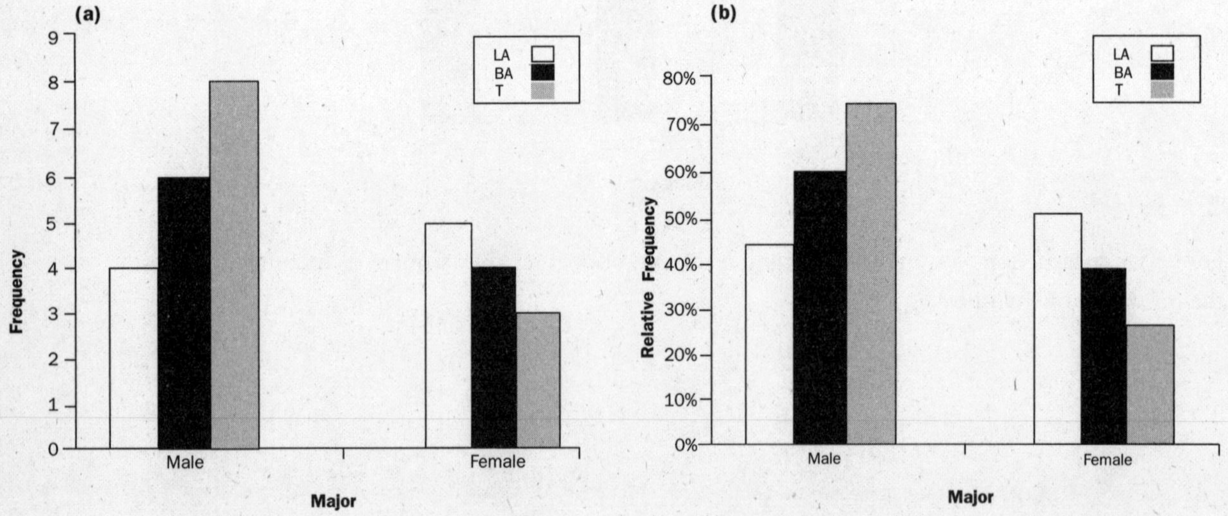

Graphical display of a two-way contingency table with division by gender. Frequencies are shown in graph (a). Relative conditional frequencies based on the marginal column frequencies are displayed in graph (b).

REVIEW QUESTIONS

1. Refer to the table below. The relative frequency for a bachelor's degree is

Highest Degree	*Frequency*
Bachelor's	9
Master's	7
Doctorate	4

(A) 55%.
(B) 45%.
(C) 20%.
(D) 80%.
(E) 9%.

2. In a bar graph, the frequency of a category is given by the

(A) height of the corresponding bar.
(B) width of the corresponding bar.
(C) height multiplied by the width of the corresponding bar.
(D) height divided by the width of the corresponding bar.
(E) None of these

Use the following table to answer questions 3 and 4.

One thousand adults were asked whether Republicans or Democrats have better domestic economic policies. The following table gives the two-way classification of their opinions.

	Republican	**Democrat**	**No Opinion**	**Total**
Male	220	340	40	600
Female	170	200	30	400
Total	390	540	70	1,000

3. What is the joint relative conditional frequency for male Republicans given if the marginal row totals are fixed?

(A) 22%
(B) 37%
(C) 56%
(D) 77%
(E) 58%

4. Which of the following is a joint frequency?

(A) 340
(B) 400
(C) 540
(D) 1,000
(E) None of these

FREE-RESPONSE QUESTION

A nationwide survey was conducted to investigate the relationship between viewers' preferred television news sources (ABC, CBS, NBC, or PBS) and their political party affiliation. The results are summarized below.

	ABC	CBS	NBC	PBS
Democrat	200	200	250	150
Republican	450	350	500	200
Other	150	400	100	50

(a) What are the marginal column totals and what do they represent? What are the row totals and what do they represent?

(b) Rewrite the table showing the conditional relative frequencies based on fixed marginal row frequencies.

ANSWERS AND EXPLANATIONS

1. B

To find the relative frequency, you must first find the total, which is 20 (9 + 7 + 4). You then divide the frequency for a bachelor's degree by the total and multiply by 100, giving 45% (9 ÷ 20 × 100 = 45).

2. A

The height of the bar corresponds to the frequency of the category. The width of the bars is usually determined by the amount of space one has for placing a graph. Height multiplied or divided by width yields area, which, unless the width is set to one unit, is meaningless.

3. B

To find the relative conditional frequency, you must divide by the fixed marginal frequency. In this case, the row totals are fixed, so we divide the male Republicans (220) by the total number of males (600) to obtain 0.37. Converting this to a percentage gives us the answer: 37%.

4. A

A joint frequency occurs at the point where columns and rows of data intersect. Remember that frequencies involving totals—found at the edges or "margins" of a data table—are known as marginal frequencies.

Free-Response Answer

(a) The marginal column totals are 800 for ABC, 950 for CBS, 850 for NBC, and 400 for PBS. These numbers represent the total number of people who watch each of the networks for their news information. The row totals are 800, 1,500, and 700, and they represent the total number of people in each political party affiliation that were polled in the survey.

(b)

	ABC	CBS	NBC	PBS	Total
Democrat	25%	25%	31%	19%	100%
Republican	30%	23%	33%	13%	100%
Other	21%	57%	14%	7%	100%
Total	**27%**	**32%**	**28%**	**13%**	**100%**

CHAPTER 8: OVERVIEW OF METHODS OF DATA COLLECTION

IF YOU LEARN ONLY FOUR THINGS IN THIS CHAPTER . . .

1. There are two primary types of studies: observational and experimental.
2. Observational studies measure attributes with no intervention by the person(s) conducting the study.
3. Experiments impose a deliberate treatment on a set of individuals.
4. An attribute of a sample is called a statistic; an attribute of a population is called a parameter.

PLANNING A STUDY: OVERVIEW

When you collect data, you must pay careful attention to the design of your study. You can't count on a haphazard collection of data, nor can you rely on "expert opinion" to lead to results that provide an accurate report of the situation you are investigating. In real-world applications of statistics, you confront a research problem in a series of steps. You plan and execute a study. You then represent the data you collect from your study using summary statistics, visual displays, and/or equations. Finally, you use the rules of probability to make inferences about the larger population based on the sample you measured in your study.

The proper design and execution of a study is of fundamental importance in the real world, but it does not give rise to many direct questions on the AP Statistics exam. You will find that only 4 to 6 of the 40 multiple-choice questions and 1 of the 6 free-response questions are specifically about planning a study. Most of these questions will treat the subject in more depth than the discussion in this chapter, but that doesn't mean you can skip to the next one. The foundations established here are important. The mechanics of how to design and execute a study show up in other questions, too, especially in free-response questions about significance tests and confidence intervals. The hypotheses you state, the assumptions you check, and the conclusions you write will often require you to understand what kind of study is being conducted.

For all types of studies, the primary goal of the investigator is to eliminate bias. **Bias** is any systematic tendency to favor certain outcomes at the expense of others. This may or may not be intentional on the part of the investigator. Unintentional bias is generally the bigger problem because it can be so difficult to detect. The techniques of effective design are grounded in an effort to eliminate bias.

All studies can be placed in one of two broad categories: **observational studies** or **experiments**. Experiments can show direct cause and-effect relationships, while observational studies cannot. On the AP exam, it is of paramount importance for you to be able to tell them apart, because many of the questions in this category are actually testing whether you can differentiate between the two types.

The easiest way to decide which type of study you are dealing with is to ask one simple question: Are you deliberately imposing a treatment or not? If the answer is yes, then you are conducting an experiment. If the subjects themselves are selecting their treatment, or if there is no treatment at all, then you have an observational study.

OBSERVATIONAL STUDIES

When you conduct an observational study or survey, you are measuring an attribute without intervening. This could involve asking voters for their political opinions, or determining whether smokers have higher rates of lung disease than nonsmokers. In these examples, you are not imposing a treatment or influencing behavior. Even though your *results* may be used to make decisions or influence others, the actual steps you take in collecting your data do not promote a political position or stop people from smoking. You are simply measuring what is happening on its own.

Census vs. Sample Survey

The term **population** refers to the set of all individuals under consideration. When your observational study examines the entire population, it is called a **census**. It can be hard or even impossible to reach every individual in a population, so it is often preferable to examine a sample of individuals from the population of interest. When you only look at a sample of your population, your observational study is called a **sample survey**.

A census provides a flat factual account of a population. Making use of a census gives an analyst the most reliable information possible. A census is unfeasible to conduct in many circumstances, however. If you are testing cars for their ability to protect passengers in crashes, you cannot crash every single car off the production line.

EXPERIMENTS

When you conduct an experiment, you deliberately impose a treatment on a set of individuals. These individuals, called subjects if they are people, or experimental units if they are not, should resemble the population of interest. (Note: It is common to use *subjects* as the collective term for subjects and/or experimental units.) The purpose of an experiment is to determine whether the treatment imposed causes the effects measured in the subjects. These measured effects are the response variables in an experiment. The explanatory variables are often referred to as factors, each of which can take multiple values, called levels. A particular treatment, then, is described by a combination of factors and levels. The set of all possible treatments can be represented using a treatment matrix.

Example:

Suppose you are investigating the effects of various amounts of sunlight and moisture on the growth of tulip plants. The plants would be your experimental units, and you could measure the height of the tulips as your response variable. You could subject samples to direct sunlight, partial shade, and full shade. You could also water them either every day or every other day. The treatment matrix below shows the combinations of factors and levels.

Figure 8.1

Treatment Matrix for Light and Moisture

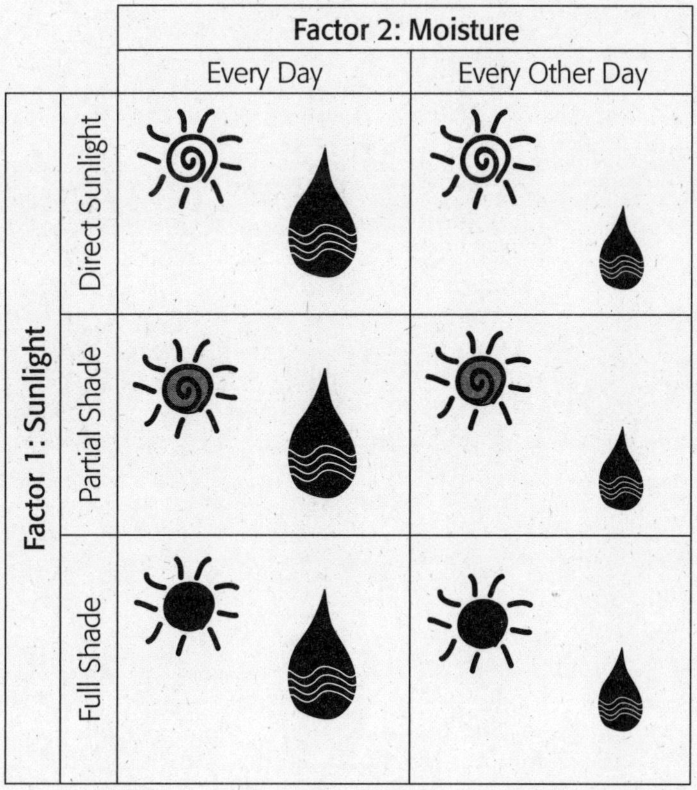

The factors are light and moisture, and so these are the explanatory variables. There are three levels of light and two levels of moisture. Altogether there are six treatments.

STATISTICS AND PARAMETERS

An attribute of a sample is called a statistic. The mean, standard deviation, median, quartiles, and so on, of a sample are all statistics. An attribute of a population is called a parameter. The mean, standard deviation, median, quartiles, and the like, of a population are all parameters. Keeping this terminology straight is easier if you notice the alliteration: sample statistic and population parameter.

Different samples of the same population do not provide the same value for a statistic. This property is known as sampling variability. One of the major goals of statistical analysis is to predict this variability, and sampling variability can be predicted only if a sample survey is properly conducted. Statistical inference is the name given to the process of drawing conclusions about parameters based on statistics.

REVIEW QUESTIONS

1. Ben conducts a study in which 100 subjects, randomly chosen from the population of all students at a school, guess when 60 seconds have elapsed. He records the actual number of seconds that have elapsed when the subject thinks it has been 60 seconds. The subjects make their guesses while listening to music. They have a choice of two treatments: fast music or slow music. Fifty-five of Ben's subjects choose fast music and 45 choose slow music. What kind of study is this?

   (A) Experiment, because the subjects are responding to treatments

   (B) Experiment, because there is a response variable

   (C) Experiment, because the subjects are randomly chosen from the population

   (D) Observational study, because the participants select their own treatments

   (E) Observational study, because the treatment groups are different sizes

2. Which of the following statements about observational studies is true?

   I. A census is always preferable to a sample survey since it includes the entire population.

   II. A neutral designer of a survey who has no predisposition towards any particular conclusion can still produce biased data.

   III. Statistical inference is not necessary when a census is conducted properly.

   (A) I only

   (B) II only

   (C) I and II

   (D) I and III

   (E) II and III

3. A consumer investigative agency wants to determine how gas mileage varies for a particular type of car under different driving conditions and using different grades of fuel. For the experiment, 120 cars of this type are available. Experimental trials will be conducted using two driving conditions: stop-and-go city driving, and high-speed highway driving. The trials will also use three fuel grades: regular, plus, and premium. All possible combinations of these driving conditions and fuel grades will be used in this experiment, and the cars will be randomly assigned to treatment groups. Which of the following properly describes the factors, levels, and/or treatments in this experiment?

   (A) There are 2 factors, one with 2 levels and one with 3 levels.

   (B) There are 2 factors and 5 treatments.

   (C) There are 2 factors and 6 levels.

   (D) There are 5 factors and 6 treatments.

   (E) There are 5 factors and 120 treatments.

FREE-RESPONSE QUESTION

An experiment concluded that low-dose aspirin is an effective treatment to prevent heart attacks in men who have already experienced and survived one heart attack. In the experiment, 1,000 volunteers were randomly assigned to one of two treatment groups. One group took a low-dose aspirin every day, and the other group took a pill that looked exactly like the aspirin pill but contained no medicine. The subjects were given the treatment over a three-year period. The experimenters then determined how many in each group suffered a second heart attack over this period. The experimental evidence for the effectiveness of the aspirin treatment was so convincing that all the participants were put on the low-dose aspirin regimen at the end of the experiment.

(a) What measures in this experimental design ensure that the collection of data is not haphazard?

(b) Describe the explanatory and response variables in the experiment.

(c) If there were no second group and all of the subjects had been given the low-dose aspirin, would this still be an experiment?

Hint: Part (a) of this question asks you to take note of the aspects of this experiment that are well-designed. No particular jargon or vocabulary is indicated, but you should feel free to use it. With this type of question, the AP graders are looking for concepts more than vocabulary. If they want vocabulary, they'll ask for it. Part (b) does. It specifically checks if you recognize what is meant by explanatory and response variables, and if you can apply this understanding in context. Part (c) presents you with a new situation and asks you to make an assessment based on the fundamentals of what makes an experiment and what makes an observational study.

This is indeed a description of a well-designed and well-executed experiment. It would be nice to know more details, such as whether the subjects are truly representative of the population of all men who have already experienced and survived one heart attack. You don't know for sure based on the information provided, but this situation is typical in medical experiments. The subjects are volunteers, and they might not be representative of everyone who could use this treatment. The description of the experiment ends with a statement that all the subjects were put on the low-dose aspirin treatment. This is bonus information that has nothing to do with parts (a), (b), or (c). It describes events after the conclusion of the experiment, and your answers need to stay focused on the experiment itself.

ANSWERS AND EXPLANATIONS

1. D

The question starts out with a discussion of treatments, so you might be tempted to think of this as an experiment and as a result only consider answers (A), (B), and (C). Perhaps (A) would tempt you with its mention of treatments, or (B) with its acknowledgment of a response variable, or (C) with its suggestion that randomization is key. None of these is correct, though, because the treatments are not assigned by the experimenter. This is the single most important detail to consider when deciding if a study is an experiment or an observational study: Is the experimenter imposing the treatments? Choice (D) gets it right. It's always worth your while to consider all choices, so take a look at (E) as well. It turns out to be wrong. The relative size of treatment groups is irrelevant to the question of whether or not a study is an experiment. It is common for experiments to have treatment groups of unequal size.

2. E

On a triple true-false question like this one, it's a good strategy to examine one statement at a time and narrow down your remaining answer choices with each conclusion. An examination of statement I shows it is false. Sometimes a study will destroy the experimental units it investigates and sometimes a census is impossible, so it's not always better to conduct one. Once you have eliminated statement I, the answer must be either choice (B) or choice (E). Both of these choices say that statement II is true. It's worth taking a look at it to make sure you like your reasoning so far. It *is* true; bias is quite often unintentional. Now consider statement III. It's true. A properly conducted census actually computes the parameter; it doesn't estimate it. No inference is necessary. Statistical inference is the process of making a prediction about a parameter using sample results.

3. A

This question is easy if you know the vocabulary. It's a straight-up example of factors and levels. What if your recollection is hazy? You still may be able to work out the right answer. Notice that there are 2 kinds of driving conditions and 3 grades of fuel. That's a list of two things, driving conditions and fuel grade, one with a sublist of 2 and one with a sublist of 3. Choice (A) is looking mighty tempting. Let's consider the other choices, though. The question states that every possible combination of the driving conditions and fuel grades will be used, so that's a list of 6 possibilities. What are these 6 things called? Levels, treatments, or something else? You should be able to deduce that it's treatments, since a treatment describes what is being done to an experimental unit. This knocks out choices (B), (C), and (E), and leaves (D) as the only possible alternative to (A). You will have to decide whether there are 5 factors or 2. You might have to resort to guessing, but guessing from two choices is better than guessing from five.

Free-Response Answer

(a) There are three primary measures in the design of this experiment that ensure the data collection is systematic rather than haphazard: there are two different treatments being compared; the subjects are randomly assigned to treatment groups; and there are many subjects in each group. The fact that the two treatments seemed identical to the participants, and that the subjects were randomly assigned to treatment groups, made sure that any difference in the subjects' health came from the aspirin.

(b) The explanatory variable is the treatment given: low-dose aspirin for one group and a pill without any medicine for the other group. The response variable is the number of men in each group who had a heart attack in the three years of the study.

(c) This would still be an experiment even with only one treatment group. Without a control group, it is not a very good experiment, but it is still an experiment. The ultimate question, whether a treatment is being imposed on the subjects, receives the answer yes. There is no indication that the subjects would have chosen this treatment on their own without the intervention of the experimenters. You can also look at it from the opposite perspective. Is this an observational study? Observational studies seek to measure a situation without intervening and changing it in any way. If this were an observational study, the only men who would have taken the aspirin treatment would be the men who found out about it on their own.

CHAPTER 9: PLANNING AND CONDUCTING SURVEYS

IF YOU LEARN ONLY FOUR THINGS IN THIS CHAPTER . . .

1. There are two crucial questions to ask when evaluating an observational study: 1) Does the sample truly resemble the population? 2) Are you actually measuring the variable you think you're measuring?
2. For a successful study, the sample must represent the population.
3. Among the many types of bias that may affect the success of the study are the following undercoverage, overcoverage, nonresponse, convenience sampling, bad sampling frame, and judgment bias.
4. It is *not* always true that a sample in which each individual has the same chance of being selected is a simple random sample.

COLLECTING DATA: OBSERVATIONAL STUDIES AND SURVEYS

An observational study measures a variable in a sample in order to make an inference about that variable in the population. An observational study that attempts to measure opinions or attitudes using a question or series of questions is called a survey. Only 2 or 3 of the 40 multiple-choice AP questions will be about observational studies or surveys. Test-takers typically do very well on these questions. You are less likely to see a free-response question on this topic than on the topic of experimental design. Historically, there has been a free-response question on observational studies or surveys once every three or four years.

Characteristics of Well-Designed and Well-Conducted Observational Studies and Surveys

When you're making an assessment of an observational study, it boils down to two crucial questions:

1. Does your sample truly resemble the population?
2. Are you actually measuring the variable you think you're measuring?

The underlying concept in both questions is bias. As mentioned in chapter 8, bias is any systematic tendency to favor certain outcomes at the expense of others. There are specific terms for the various types of bias that cause problems in sample selection, and there are additional terms for the types of bias that cause incorrect measurement of your variable of interest. This second category is a particular issue with surveys. Since they try to determine a respondent's opinions or attitudes from that person's responses to a question or series of questions, accurate measurement can be a problem.

POPULATIONS, SAMPLES, AND RANDOM SELECTION

The data you collect about a sample will not tell you anything about your population of interest unless your sample resembles the population. There are no statistical tricks that can rescue data drawn from a badly selected sample. An obvious method to select a representative sample is to select individuals at random from a list of the members of the population. Surprisingly often, it is impractical or impossible to do this. Suppose your population of interest is all the residents of Chicago. There is no list of all the people who live in Chicago.

A common alternative is to select your sample from a **sampling frame**. This is a list of individuals from which the sample is actually drawn. For example, if you select a sample of people from the phone book, then you are using the phone book as your sampling frame. However, the phone book will not include people with unlisted numbers, people who have recently moved into the area, or people who do not have a land-line telephone. As this example illustrates, using a sampling frame can introduce bias. A well-chosen sampling frame resembles the population.

When selecting a sample, it is crucial to use randomization. This is critical for eliminating bias, since randomizing makes sure that individual preferences, either yours or those of your potential participants, are not a part of sample selection. Second, random selection is crucial because it ensures that your sample results are subject to the laws of probability. This will allow you to make probability-based inferences about the population.

A table of random numbers is commonly used to select random samples. On the AP exam, you may be asked about the structure of a table of random numbers. The table consists of one very long string of digits. Every digit 0 through 9 occurs with roughly equal frequency over the entirety of the table. So does every two-digit number 00 through 99, every three-digit number 000–999, and so on. In addition, the digits occur in such a way that the entries are independent of each other. That is, knowing what numbers occur in one part of the table will not help you predict how the numbers occur in other parts of the table.

Although you may be asked about the structure of a table of random numbers, it's more likely that you'll be asked to use one to select a sample. It's a two-step process: label and table. First, assign a number to every individual in your population or sampling frame. Every numerical label must have the same number of digits, so you would use 01, 02, 03, ..., 19, 20 for a population with 20 individuals and not 1, 2, 3, ..., 19, 20. Make sure your written work shows your numerical assignments clearly. That's all there is to the label step of the process. In the table step, you'll select numbers that start in a designated row of the table of random numbers. If you are picking two-digit numbers, select two-digit segments of the table and ignore spaces. (Note: The only reason the digits are in groups of five is to make the table easier to read.) If a number you select does not match one of your labels, ignore that number and move to the next block of digits. If the number you select does match one of your labels and you haven't already selected that number, that individual is now officially in your sample. Continue until you have the sample size you set out to select.

AP EXPERT TIP

On the AP exam, you can show work directly on the random digits table that is provided with the problem. Your final result must be clearly stated.

SOURCES OF BIAS

You will need to know the technical terms for the types of bias common to observational studies. Some types of bias are particular to surveys and do not occur in other types of observational studies. The various types of bias fit into two categories that correspond to the two crucial criteria about observational studies already mentioned. That is, the sample should resemble the population, and the variable you are actually measuring should be the same as the variable you intend to measure.

SOURCES OF BIAS IN SAMPLE SELECTION

Bias in a sampling method is any tendency to produce samples that do not resemble the population. Two commonly confused bias sources of this type are **undercoverage** and **nonresponse**. Undercoverage is the exclusion of a part of your population from the sampling process. This occurs when your sampling frame does not include all of the population. Nonresponse occurs when individuals that were selected in your sample do not participate. This is most commonly associated with opinion surveys in which participants refuse to answer a question.

AP EXPERT TIP

An example of undercoverage that has already been discussed is when a phone book is used to choose individuals from a population.

Two other types of bias occur due to sampling methods that are not random. **Convenience sampling** is the term for selecting a sample that is easy to collect rather than one that is representative of the population. **Voluntary response bias** occurs when participants self-select whether or not to participate. Phone-in polling and Internet voting are typical examples. In surveys, voluntary response samples tend to attract participants with strong opinions rather than participants who represent the population as a whole.

In addition, there are some other bias sources that are mentioned in some textbooks but not others. Many AP Statistics students will not have learned about them, so they are less likely to occur on your AP exam. They could show up on multiple-choice questions as wrong answers, though.

Overcoverage is the inclusion of individuals in the sampling process who are not actually in the population. **Bad sampling frame** is a general term that includes undercoverage, overcoverage, duplication of entries, and so on in a sampling frame. Bad frames lead to bad samples. **Size bias** is any tendency to favor either larger or smaller individuals in your sample. Finally, **judgment bias** is the name given to samples selected using "expert judgment" rather than random sampling techniques.

Sources of Bias from Incorrect Measurement

Often an observational study or survey does not measure what it claims to measure. This can simply be a matter of making bad measurements of some physical quantity, especially if the quantity is hard to measure (e.g., the distance between stars). More often, though, this occurs in surveys that attempt to collect information about opinions or behavior. It is hard to write good survey questions. The personal interaction between interviewer and participant can come into play. People may be reluctant to admit their opinions or behaviors if they are embarrassed about them. These are all potential pitfalls.

One bias source particular to surveys is the use of poorly worded questions. This can literally involve questions worded so badly that the respondent gives an answer contrary to his or her actual opinion. It can also result from a question that pushes the respondent to give a particular response. A final instance of this kind of bias has to do with the order of questions in a survey. A line of questioning can push the respondent to give a particular response even though each question taken individually appears to be appropriate. In all of these varieties, the survey is not getting an accurate read on how the respondent really feels.

Another broad term for a set of problems in the conduct of a survey is **response bias**. This refers to any bias that results from the conduct of the interviewer in a survey, the circumstances in which the interview is conducted, or the interaction between the interviewer and the participant. Response bias is an issue if a person smoking a cigarette conducts a face-to-face survey with people about their attitudes towards smokers, or if an interviewer asks someone about marital infidelity in the presence of that person's spouse.

One final source of bias encountered in surveys occurs when the respondent deliberately gives an incorrect response. Sometimes respondents lie about their behavior or attitudes. Participants in medical studies, especially dietary studies, may lie about following doctor's orders or about following a particular eating regimen. Families that keep journals of their viewing habits for TV ratings companies often shade the truth about the shows they watch. A participant could even deliberately try to influence the results of a survey.

SIMPLE RANDOM SAMPLING

Finding a **simple random sample (SRS)** is a fundamental, commonplace task in inferential statistics. The name seems to promise that it's simple, and for the most part it is. If you want an SRS of n individuals from your population, just randomly select one individual at a time without replacement until you have n of them. This means that every individual in the population has the same chance of being selected. That's how most encounters with simple random samples work, and that's probably all you'll need to know about them on the AP exam.

However, you may be asked a little bit more. A standard question asks, "Is it always true that a sample in which each individual has the same chance of being selected is an SRS?" The answer, surprisingly, is NO. Consider the following example. A population of 100 students consists of 50 boys and 50 girls. A sample of size 50 will be selected by flipping a coin. Heads, the 50 boys are selected. Tails, the 50 girls are selected. Every student has the same chance of being selected, but you wouldn't call it a simple random sample. You couldn't get a sample that includes some of the boys and some of the girls. This example illustrates why we have the following definition of a simple random sample: for a simple random sample of n individuals from a population, every different sample of size n must have the same chance of being selected.

AP EXPERT TIP

A good way to think of a simple random sample is putting everyone's name in a hat and selecting n names to make up your sample. Every possible sample of size n is equally likely.

STRATIFIED RANDOM SAMPLING AND OTHER SCHEMES

Simple random sampling is the primary sampling scheme you will encounter on the AP exam, but it is not the only one. It is only a special case of a more general framework for sampling known as **probability sampling**. This is the name given to any sampling scheme in which each member of the population has a given, fixed probability of being selected in a sample. There are other commonly used probability sampling schemes in addition to simple random sampling.

Stratified Random Sampling

If you divide your population into nonoverlapping strata and then select an SRS from each stratum, then you are using **stratified random sampling**. For example, if your school population included 500 freshmen, 400 sophomores, 300 juniors, and 200 seniors, you might select a stratified random sample of 50 freshmen, 40 sophomores, 30 juniors, and 20 seniors.

You use stratified random sampling when you can divide your population into strata so that the individuals within each stratum are alike. This has two key advantages. First, your sample will more closely resemble the population since you will not get odd samples that underrepresent particular strata, such as a sample of high school students that by chance happens to include very few sophomores. Second, you can make inferences about each stratum of the population using subsamples. The key disadvantage to stratified random sampling is that your judgment in creating the strata may be faulty; you could be creating a biased sample.

Other Sampling Schemes

The AP exam will probably only question you about simple random sampling and stratified random sampling, but there are some additional schemes that could make an appearance. **Systematic sampling** is the term for any regular, systematic way of selecting individuals from a sampling frame, such as taking every tenth name from a list of the population. **Cluster sampling** involves selecting entire groups at random. Each group, or cluster, should resemble the population and each other. For example, suppose you wanted to sample first grade students in a school system. You could sample random classes of first graders out of all the first grade classes in a school system.

A **two-stage sampling** design combines cluster sampling with simple random sampling. Divide the population into clusters, select a random sample of those clusters, and then take simple random samples from each chosen cluster. That may sound like three stages, but there are only two stages of random sampling. **Multistage sampling** refers to any random sampling scheme with at least two steps that incorporates elements of stratified sampling, cluster sampling, and/or simple random sampling. For example, you could create a multistage sample of U.S. citizens by dividing the country into strata by geographical region, selecting a random sample of neighborhoods (these are the clusters) within each region, and then interviewing a simple random sample within each selected neighborhood.

REVIEW QUESTIONS

1. A U.S. government researcher wants to select a sample of tax returns that will include returns from a variety of different income levels. He divides the set of all the different incomes shown on the forms into 10 nonoverlapping ranges, then he randomly selects 100 tax returns from each. Which of the following best describes the sampling scheme used in this example?

   (A) Stratified random sampling
   (B) Simple random sampling
   (C) Convenience sampling
   (D) Two-stage sampling
   (E) Cluster sampling

2. Which of the following is NOT a property of a large table of random digits?

   (A) The table will contain, somewhere, the sequence of digits 1234.
   (B) Consecutive rows do not start with the same digit.
   (C) Each digit 0 through 9 occurs with equal frequency.
   (D) Each three-digit number 000 through 999 occurs with equal frequency.
   (E) The contents of one section of the table are independent of other sections of the table.

3. The owner of a factory that employs half the citizens in a small town is trying to decide whether to take a public stand on a controversial issue. He realizes that he would benefit from knowing how the townspeople feel. He randomly selects 50 of the townspeople from a list of all the town's population. He personally contacts all 50 and asks them their opinion on the issue. Most give him an answer, but 12 townspeople decline to participate. He decides to summarize his results based on the 38 responses. Which of the following list the most significant sources of bias in this survey?

   (A) Voluntary response bias and undercoverage
   (B) Response bias and undercoverage
   (C) Nonresponse and undercoverage
   (D) Response bias and nonresponse
   (E) Voluntary response bias and nonresponse

Free-Response Question

Administrators at a high school hire an independent research firm to conduct a survey of the high school's alumni to find out how satisfied they are with their education at the school. Once or twice a year, the high school contacts alumni in order to keep records of current contact information. The research firm selects its sample from a list of alumni who have responded to the school's communications and have provided current contact information.

(a) Point out a potential source of bias, and how you believe that source of bias will affect the survey's ability to estimate the level of satisfaction for all alumni.

(b) The research firm wants to make sure that separate results are available for alumni who have graduated in the last 10 years, 11 to 20 years ago, and 21 or more years ago. Describe a sampling method that will meet this requirement.

Hint: If you have a free-response question on an observational study, it will probably look like this. The question begins with a basic description of an observational study with a few critical clues. Then you will see some follow-up questions that ask about the sampling method and/or sources of bias, and a question supplying a new twist. The question with the new twist will ask you to adapt the study as given in order to meet the new requirements.

The critical clues here begin with the fact that the high school administrators are not conducting the survey themselves; they are hiring an independent research firm. The neutrality of the results would be suspect if the administrators were conducting the survey themselves. That probable bias source is now a nonissue as you try to answer the follow-up questions. Second, the problem makes clear that the sampling frame is an issue. The administrators have made an effort to maintain a contact list that is as thorough as possible, but it isn't perfect. There really isn't much other information offered about the sampling, so the sampling frame must be the critical issue. Resist the temptation to speculate about potential bias sources that could occur if mistakes were made in ways not hinted at in the description of the observational study. For example, don't go off on a tangent about the problems that might arise if the sampling weren't random. Stick to the issues presented.

ANSWERS AND EXPLANATIONS

1. A

You will probably be asked to identify a sampling scheme somewhere on your AP exam. This is a clear case of stratified random sampling. The strata are the ten different income ranges, and a random sample is being selected from each range. Note that the strata are made up of individuals that are like each other. You might briefly consider cluster sampling, but cluster sampling selects entire clusters, not samples from each cluster. In addition, clusters are chosen so that the members of a cluster are not like each other. The answer isn't two-stage sampling since that would involve cluster sampling as a first step. The other choices aren't very tempting. Simple random sampling would skip over dividing the incomes into ten different ranges, and convenience sampling wouldn't include anything so systematic, nor would it include randomization.

2. B

Be sure to notice when a question asks for the answer choice that is NOT true. You should circle the word NOT in your test booklet to make sure you notice it, and cross out the choices that definitely ARE true as you read through them. Sometimes with this kind of problem you can find two choices that are directly contradictory, which means one of those choices is right and the other is wrong. That can simplify things tremendously. You may have noticed in this question that choices (B) and (E) cannot both be true. If the contents of one section of the table are independent of other sections, then the digit starting one row cannot exclude that digit from starting another row. Choice (E) is part of the definition of a table of random digits, so (B) must be the false statement and the correct answer.

Suppose you don't recognize this. There are other ways to figure out the right answer. As you look through the choices, you should recognize right away that (C) and (D) are both true. They are part of the definition of a table of random digits. Choice (E) is true as well, since tables of random digits are designed to have different sections be independent of each other. That leaves choices (A) and (B). A large table of random digits will contain all the numbers 0000 through 9999 with equal frequency, so 1234 will show up somewhere. That means (B) is NOT true, and so it is the correct answer.

3. D

The scenario has multiple sources of bias, so you may have trouble sorting out which are the most prominent. The factory owner clearly has a lot of power and influence in this town, so the citizens may be reluctant to speak honestly about their opinions on a controversial issue. That's called response bias; we know it's a problem, but maybe there are other more significant sources. The other bias source that should grab your attention is the fact that a significant portion of the sample did not participate. That's nonresponse. Choice (D) seems to be the best choice, but you should contemplate the other bias sources mentioned in the choices, voluntary response bias and undercoverage. Since the interview subjects did not select themselves, there is no voluntary response bias. Undercoverage would be a problem if the sampling frame omitted some of the townspeople, but the factory owner used a list of all of the small town's residents. That settles it; the answer must be (D).

Free-Response Answer

(a) The principle source of bias to watch out for is undercoverage. Even though the administrators have tried to keep a complete record, there will be alumni left out of the contact list. It is likely that most of the alumni who have a very positive rating about their education will keep their contact information current, so a sample drawn from the contact list will probably overestimate the satisfaction level of alumni.

(b) In order to make sure separate results are available for alumni in the 0 to 10 group, 11 to 20 group, and 21 and over group, use stratified random sampling. The sampling frame (the contact list) needs to be stratified into three strata using the same criteria. Then select a large sample from each stratum so that there is minimal sampling variability in each subgroup.

CHAPTER 10: PLANNING AND CONDUCTING EXPERIMENTS

IF YOU LEARN ONLY THREE THINGS IN THIS CHAPTER . . .

1. The hallmarks of a well-designed experiment are control, randomization, and replication.
2. If an experiment is poorly designed, a researcher may be subject to sources of bias such as confounding variables, lurking variables, and/or the placebo effect.
3. Important experimental designs include a completely randomized design, a randomized block design, and matched pairs.

THE FUNDAMENTALS OF EXPERIMENTAL DESIGN

Randomized comparative experiments are a fundamental part of statistics, and they appear in a substantial number of questions on the AP Statistics exam. Only a few questions, however, ask directly about the design and execution of experiments. On the multiple-choice part of the AP Statistics exam, you will be asked two or three questions about experiments. In addition, you may have to design an experiment or analyze an experimental design as one of your free-response questions. Understanding experimental design will also be relevant to any questions about significance testing.

Well-Designed and Well-Executed Experiments

The hallmarks of a well-designed experiment are **control**, **randomization**, and **replication**. Control means that the experiment should compare multiple treatments. Randomization means that the subjects should be randomly assigned to treatment groups. Replication means that each treatment group should include multiple subjects. More replication leads to more reliable results, because larger samples exhibit less sampling variability. The term "replication" can also refer to the repeating of an entire experiment to confirm the validity of the results.

In some class other than AP Statistics, you may have heard about the importance of using a control group in an experiment. This usually refers to a group that receives no treatment at all. This is one of the ways that an experiment can be considered in control, but statistical control only requires having more than one treatment.

Example:

A researcher wants to find out whether mice, after training, will run through a maze more quickly during the day or during the night. He has 100 mice available for the experiment, which he randomly allocates to two groups of 50. The first group is trained to run through the mazes in the day, and the second at night. The training consists of putting a mouse in the maze and rewarding the mouse with food when it completes the maze. Each mouse goes through many trials, and the average time for each mouse's last three trials is recorded.

This experiment exhibits control since two treatments are being compared. There is randomization because the mice are randomly assigned to treatment groups. The requirement for replication is met since there are many mice in each treatment group.

CONFOUNDING AND SOURCES OF BIAS

Without proper attention to control, randomization, and replication, an experiment is subject to a number of possible sources of bias. Inadequate attention to any of these aspects of an experiment can lead to confounding of the effects of the explanatory variables with other factors, called **confounding variables**. Two variables are confounded if their effects on the response variable cannot be distinguished. You might recall the similar-sounding term "lurking variable." A **lurking variable** is any variable that might have an effect on the explanatory or response variable, through either common response or confounding.

One confounding variable that is commonly encountered in experiments with human subjects is the **placebo effect**. This occurs when a change is observed in the response variable simply because the subjects believe they are receiving a treatment, whether or not the treatment is having any genuine effect. Usually the placebo effect is managed in experiments by including a treatment that resembles, from the subjects' point of view, an actual treatment—even though it has no effect at all.

In a medical study, this could be a pill that has no active ingredient. In such cases, this nontreatment is called a placebo. Do not confuse placebo with placebo effect.

Bias can still be a problem even when control, randomization, and replication do, in fact, receive proper attention. One source is **lack of realism**. That is, the experiment might not be a realistic instance of the real-world phenomenon the experimenters are investigating. For example, an experiment may attempt to measure how different learning environments affect how subjects learn a foreign language. If the experiment does not measure full language acquisition, but instead only measures how well the subjects memorize vocabulary words from a list, then the experiment is not realistic. Language acquisition also includes knowledge of grammar as well as speaking and listening skills.

Another area that demands attention when designing an experiment is **blinding**. An experiment is blinded when the subjects do not know which treatment they are receiving. Blinding is important, because the subjects may have some idea about which treatment they would prefer. If they do not get the treatment they prefer, then their displeasure with the treatment could alter their response to the treatment, and could even lead them to seek out other treatments. The fact that the subjects know what treatment they are receiving becomes a lurking variable that will be confounded with the effect of the treatment.

An experiment is **double blinded** when the experimental staff who are in contact with the subjects do not know which treatment is being used for each subject. If the staff know which treatment is being used, it can affect their interactions with the subjects. As before, the fact that the staff know which treatment is being employed becomes a lurking variable that can be confounded with the effect of the treatment.

EXPERIMENTAL DESIGN SCHEMES

The simplest experimental design is a completely randomized design. The subjects are randomly allocated to treatment groups. Then each treatment group receives a different treatment. Finally, the response variable (or variables) is measured for each group and a statistical analysis is conducted on the results. All instances of completely randomized design can be diagramed as shown in Figure 10.1. If you are asked to explain the design of an experiment on your AP exam, you will find it much easier to simply draw a diagram with perhaps a couple of extra explanatory comments. Make sure to include the fact that group assignments are decided by random allocation. In addition, be sure to include all the information indicated below and make your entries specific to the context of the experiment.

AP EXPERT TIP

Randomization is a key element of experiment design, and you will not receive full credit on an AP problem if random allocation is not indicated.

Figure 10.1
Diagram for a Completely Randomized Experiment

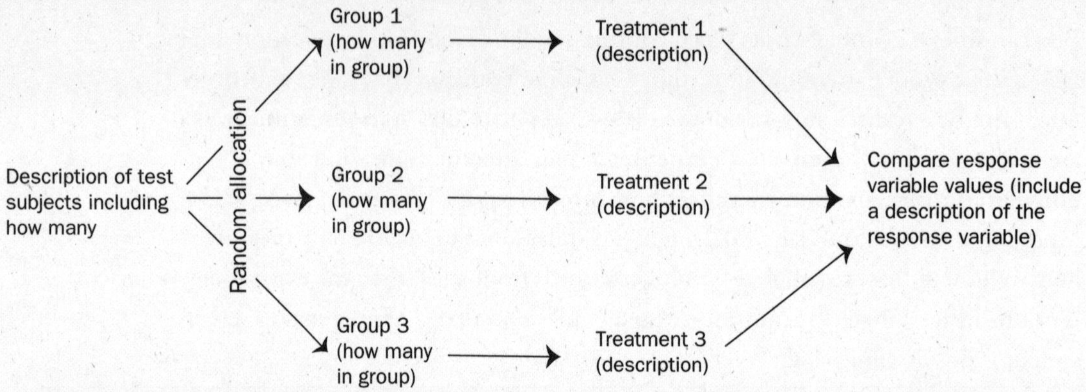

Standard form for the diagram of a completely randomized experiment. This example compares three treatments.

Randomized Block Design

Often the response variable will show variability due to the effect of a confounding variable. The design of an experiment can reduce this variation by splitting the test subjects into blocks. For example, you might be conducting a completely randomized experiment using students at your school. If you expect that the results would be very similar for subjects who are in the same grade but different for subjects in different grades, you could set up the experiment so that the results for each grade will be analyzed separately from each other. This is known as **randomized block design**, and it is typically diagramed as shown in Figure 10.2.

There are some common mistakes that students make when designing randomized block experiments on the AP exam that you will need to avoid. The biggest and most common mistake is forming the blocks incorrectly. It is crucial that the members of each block resemble each other so that each block will show minimal variation. The blocks must be homogeneous. Often, though, students mistakenly set up blocks so that each block contains as much diversity as possible. That completely defeats the purpose of forming blocks in the first place. Some students go so far as to assign subjects to blocks randomly. If you are going to do that, there is no point in having blocks at all. The idea of using block factors in the first place is to remove factors that unintentionally affect the results, such as the operator who ran the experiment, the time of day, and the room temperature at the time the data was collected. Utilizing block factors in a controlled way is the best way to remove or reduce these factors. Only after this is accomplished can randomization be effective.

Figure 10.2

Diagram for a Randomized Block Design

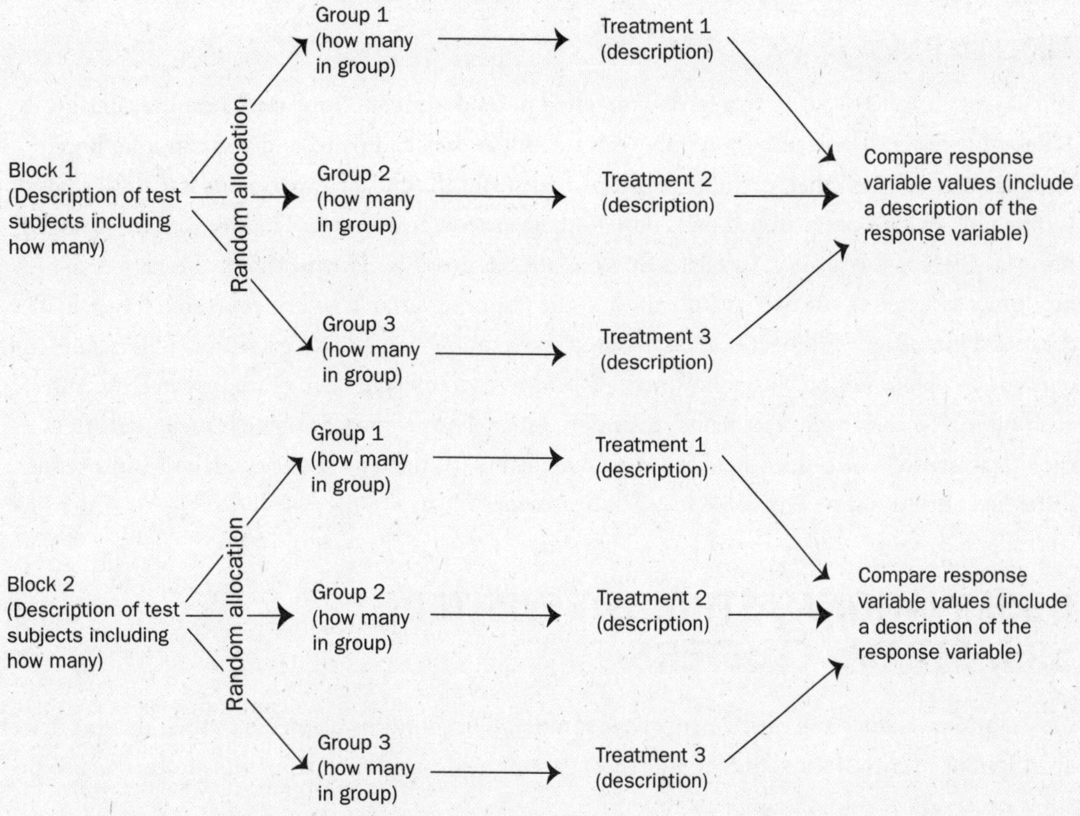

Standard form for the diagram of a randomized block experiment. This example uses two blocks and compares three treatments.

Another common mistake is to label different names for each corresponding group and treatment from each block. Note that in the diagram there are Groups 1, 2, and 3, and there are Treatments 1, 2, and 3, even though it might at first seem better to label them Groups 1 through 6 and Treatments 1 through 6. The topmost group of Block 1, however, is getting the same treatment as the topmost group of Block 2, so they should both be labeled Group 1 and they should both be getting Treatment 1. In the end, you will analyze the results separately, but the subjects themselves need not ever be aware that they are in separate blocks. If you use Treatments 1 through 6 as labels instead of 1 through 3, you are implying that the treatments Block 2 is getting are in fact different from the treatments for Block 1. If that were the case, this wouldn't be a randomized block design. It would be two entirely separate experiments.

Another common mistake is taking the analysis one step too far. It is not necessary to draw arrows comparing the results of the blocks with each other. Note that in the diagram above, the results are only analyzed within each block. No arrows are drawn to combine all the results together in the end. An experimenter may choose to perform this additional analysis, but it's not a proper part of randomized block design.

Matched Pairs

Paired comparisons occur in two forms: **matched pairs design** and **repeated measures design**. Many statistics textbook authors simply refer to both as matched pairs, and you can, too. Both designs are used when there are only two treatments, and the end-of-experiment statistical analysis is the same. In a proper matched pairs design, there are two treatments. The subjects are arranged into pairs, where the subjects in each pair are as alike as possible. The members of each pair are randomly assigned to the two treatments, and the response variable will be measured for each. The data used for analysis is the set of differences in the response variable for each pair. This is similar in concept to a block design with many, many blocks of two subjects each. With repeated measures, each subject receives both treatments in random order. The response variable is measured after each treatment. As with the other version of matched pairs, the data used for analysis is the set of differences in the response variable for each treatment.

OBSERVATIONAL STUDIES REVISITED: SRS, STRATA, CLUSTERS

Observational studies and experiments have some similar sounding elements to their designs that are different in critical ways. Sometimes these get jumbled together in students' minds, so here are a few observations to disentangle them.

The subjects of an observational study are selected at random, but it is often impossible to select the subjects of an experiment at random. Your sample of experimental subjects should resemble the population of interest, but on the AP exam you will not have to worry about how the subjects of an experiment were selected. You do need to make sure, though, that the subjects are randomly assigned to treatment groups.

To many students, the blocks in randomized block design sound a lot like the strata in stratified random sampling. They are similar in one respect: Blocks should be composed of similar units and so should strata. They are different in that blocks are a way of splitting up a sample, while strata are a way of splitting up a population.

Finally, the blocks in randomized block design may sound similar to clusters in cluster sampling. They are quite different. Blocks are made up of similar individuals, while each cluster in cluster sampling should show the same diversity as the whole population. What's more, every member of a block will participate in an experiment. Only certain randomly chosen clusters will be a part of an observational study.

REVIEW QUESTIONS

1. A researcher is trying to find the best soil composition for growing a new variety of tomato. The researcher is considering 4 different soil nutrients, each with 5 possible levels. A large field is split into 625 different plots, one for each possible combination of nutrient levels. The placement of each combination of nutrients is random. The researcher plants one tomato plant in each plot, and the success of each nutrient combination will be measured by the size of the tomatoes produced by each plant. Which of the following is the most serious problem with this design?

    (A) Lack of replication
    (B) Lack of control
    (C) Lack of randomization
    (D) The experiment isn't double blinded
    (E) Placebo effect

2. Which of the following is the least important way in which the designer of an experiment can guard against confounding?

    (A) Matching
    (B) Randomization
    (C) Replication
    (D) Control
    (E) Blocking

3. David knows that dancers are trained to spin many times without losing their ability to move in a straight line after spinning. He wonders whether this ability is dependent on the number of spins. He wants to design an experiment that will compare the ability of experienced female dancers to walk a fixed distance in a straight line after 5 spins with their ability after 10 spins. Which of the following is the most appropriate design for this experiment?

    (A) Completely randomized design
    (B) Stratified design
    (C) Randomized block design
    (D) Cluster design
    (E) Matched pairs design

Free-Response Question

Product testers for a carpet manufacturer want to discover which of two brands of stain-proofing chemicals, A or B, is more effective at combating routine wear and tear when used on the company's carpets. An experiment is to be designed to decide which brand to use. The testers believe, however, that two different carpet types made by the company, Berber and cut pile, might not respond the same to the two types of stain proofing. There are 200 Berber and 200 cut pile samples, all the same color, available for testing. After treatment with stain proofing, the product testers will subject all samples to a simulation of one year's wear and tear. The effectiveness of the stain proofing will be measured by digital imaging techniques that measure the color change in the samples.

(a) Design a randomized block design for this experiment that will allow the product testers to decide the best design to use for each type of carpet.

(b) Replication is a part of this experiment. Describe the replication taking place here, and explain why it is important.

Hint: You will probably have to answer a free-response question about experimental design, and this question is representative of what you are likely to see. There is a description of the experimental scenario, a few helpful details, and then an instruction to describe the details of the experiment given a particular design scheme. Then you have a follow-up question asking you to explain a major aspect of experimental design in context. This question asks about replication, but you should also be prepared for a question about randomization or control.

Your description can be in paragraph form or in diagram form. If you are familiar with diagrams from your coursework, you will find it easier and faster than paragraph form. The major details the graders will look for, whether you use a diagram or not, are the following: splitting the population of carpet samples into blocks, random allocation to treatment groups for each block, the number of experimental units in each treatment group, a clear description of each treatment, and a description of the response variable.

ANSWERS AND EXPLANATIONS

1. A

You will probably have a question describing a flawed experimental design. You will have to figure out the most significant shortcoming of the design or execution. At first glance, this experiment seems to have control, randomization, and replication. There are 625 plants, 625 treatments are being compared, and plots are located randomly in the field. Double blinding and the placebo effect make little sense here, since they are issues that occur with human subjects. You might wonder whether all the plants are getting the same amount of water and sunlight, but these possibilities aren't among the answer choices.

A second look (or perhaps you spotted it on your first look) reveals that this experiment has no replication. It might seem like there is since we're putting in 625 different plants, but there is only one plant in every treatment group. That's a complete lack of replication.

2. A

An experiment can't determine cause and effect if the effects of the explanatory variables are confounded with other variables. You should be prepared to deal with confounding as an issue in any question involving an experiment, whether it is about design or about testing hypotheses. In this question, you are confronted with the most basic issue about confounding: How do you prevent it? Randomization is crucial since it makes sure that the treatment groups resemble each other in all ways, taking into account confounding variables you know about and confounding variables that you don't. Replication is important too, since the effects of randomization will only benefit you with sufficiently large treatment groups. Running a controlled experiment is essential, since the effects of the treatment will be indistinguishable from the effects of any other lurking variable without comparison of treatments. Finally, blocking is important since it allows you to separate out the effects of a variable that you know could confound results. That leaves matching. Matching can be an effective strategy for increasing the precision of an experiment but you wouldn't use it as a design in order to combat confounding.

3. E

You should know the names of the three major design schemes for experiments, and as a result, you should know there are no such things as choices (B) and (D), stratified design and cluster design. These are sampling schemes, not experimental designs. Choice (C), randomized block design, doesn't make much sense here, since the population consists of experienced female dancers. There are no blocks you might use. That narrows it down to choices (A) and (E). Matched pairs makes more sense than completely randomized design. You want to know if each individual dancer is affected differently with a different number of spins, so it would be helpful to have each subject experience both treatments.

FREE-RESPONSE ANSWER

(a)

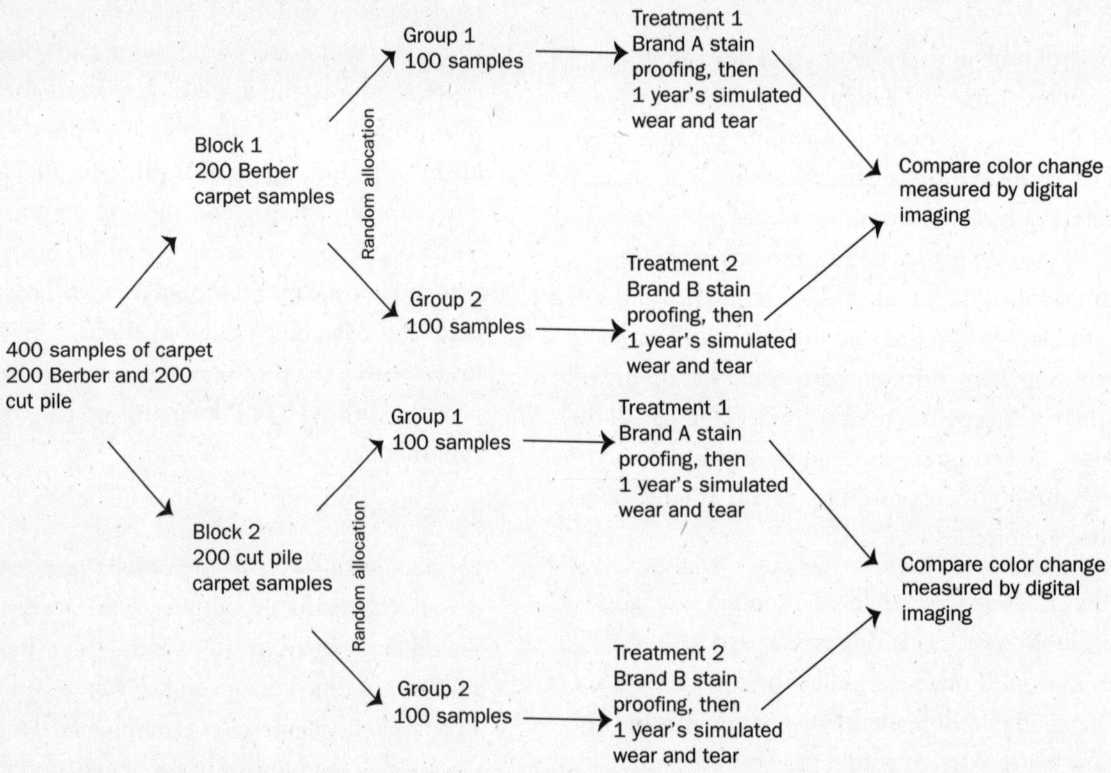

(b) Replication is provided in this experiment by the 100 samples in each treatment group in each block. Replication is important here because it ensures that any observed color change in the carpet samples is due to the stain proofing and not some oddity about the construction of a few of the samples. In other words, it protects against the influence of confounding variables. Replication is also important because it allows for less sampling variability in the results for each treatment group. Statistical significance can only be determined when the variation between treatment groups is greater than the variation within the treatment groups, so the precision of the experiment is improved by replication.

CHAPTER 11: GENERALIZABILITY OF RESULTS FROM EXPERIMENTS, OBSERVATIONAL STUDIES, AND SURVEYS

IF YOU LEARN ONLY THREE THINGS IN THIS CHAPTER . . .

1. A larger sample size leads to better generalizations.
2. A well-executed observational study allows two kinds of generalizations: sample statistics can be used to 1) draw conclusions about the corresponding population parameters, and 2) establish an association between two or more population parameters.
3. The conclusion of an experiment is dictated by the statistical significance of the results.

GENERALIZABILITY

On the AP Statistics exam, there will be four to six multiple-choice questions about observational studies and experiments. All of these questions require a solid understanding of these topics as reviewed in the preceding chapters. Some of them, though, will be about the generalizability of results from observational studies and experiments. These questions will ask you to distinguish between the conclusions that can be drawn from experiments and the conclusions that can be drawn from observational studies and surveys.

The most important thing to know for the AP exam is that a well-designed and well-executed experiment can establish a cause-and-effect relationship. An experiment can prove, for example, that a particular medical treatment is the remedy for a disorder, or that a particular teaching technique is responsible for higher test scores. No observational study, no matter how well-designed, well-executed, or thoroughly analyzed, can do this. For each kind of study, proper attention to design can allow inferences to be made about the population of interest.

SAMPLE SIZE AND POPULATION SIZE

With observational studies, large sample sizes lead to better generalizations. With experiments, this same concept is known as replication. Large treatment groups lead to more reliable conclusions. This is because larger samples experience less sampling variability. This requires, however, that the samples in question be selected at random. Randomizing makes the results subject to the laws of probability. It's worth noting that questions on the AP exam may tempt you into claims that larger samples reduce bias, reduce effects of confounding variables, and decrease variability in the population. They certainly do not. In fact, populations have no sampling variability. Samples vary; populations do not.

The size of the population is largely irrelevant to the process of making generalizations. As long as the population is much larger than the sample, the population size does not matter. A typical figure for "much larger than" is a population that is at least ten times the sample size. You may have heard the soup analogy as an explanation of this. A cook who is preparing soup can tell how it is with just a taste, assuming the soup is well-stirred. It doesn't matter if the cook is preparing a pot or a large vat. In the same manner, a statistician can use a simple random sample of 500 individuals to draw conclusions about a population of 1,000,000 just as reliably as a population of 100,000.

OBSERVATIONAL STUDIES

A well-designed and well-executed observational study allows two kinds of generalizations: sample statistics can be used to draw conclusions about the corresponding population parameters, and sample statistics can be used to establish an association between two or more population parameters. An association between two variables simply means that knowledge of the value of one variable makes it easier to know the value of the other. It does not mean that one causes the other. For example, knowing how many firefighters are at a fire helps you to know how severe the fire is. This does not imply that the presence of more firefighters causes a fire to be more severe.

How far can the requirements of a well-designed observational study be bent? Can you make generalizations about a population even when an observational study is not as well-designed or well-executed as you would wish? Ultimately, you have to be confident about two fundamental questions for any observational study: Does the sample truly resemble the population, and are you actually measuring the variable you think you're measuring? If the answer to both questions is yes, you can probably draw generalizations even though the requirements are not perfectly met.

It is critically important, though, that there is a minimal amount of bias in the observational study. No inference is possible if either the sampling method or the data collection shows significant bias. In particular, convenience samples and voluntary response samples lead to suspect data. These sorts of samples almost never resemble a population.

EXPERIMENTS AND STATISTICAL SIGNIFICANCE

The three principle criteria of a well-designed experiment—control, randomization, and replication—each play a role in your ability to make generalizations from the results of an experiment. Nothing can be concluded from a poorly designed experiment. Poor design can lead to confounding of the effects of the treatment with some other variable, and then establishing cause and effect is impossible. Just as with observational studies, there are no mathematical tricks to rescue data obtained from a poorly conducted experiment.

Experiments can establish a cause-and-effect relationship because the subjects are randomly assigned to treatment groups. Cause and effect can be established even when the subjects are not randomly selected from the population of interest. In such cases, though, inference is limited to the segment of the population resembling the sample. If you conduct an experiment in which the subjects are drawn randomly from the population of interest, then you can make safe inferences about the population. Most experiments with human subjects don't select subjects at random. They use volunteers.

The ultimate conclusion of an experiment is dictated by the **statistical significance** of the results. Experimental results are called statistically significant if the difference in values of the response variable from one treatment group to another is too large to be attributable to chance alone. In order to establish whether the results of an experiment are statistically significant, it is helpful to minimize the variation of results within each treatment group. In other words, the variation between groups must be larger than the variation within each group.

There are two standard ways to minimize the variability within each treatment group. One is to make sure the experiment shows good replication. Larger sample sizes exhibit smaller sample variability. The second way to minimize variability is to use a randomized block design. If you suspect that some subjects in your experiment will respond differently to the treatment than others, you partition them into blocks. The treatment groups within each block should then exhibit far less variability than if the experiment were not blocked. As a result, proper use of blocking reduces variability in sample results and makes it easier to establish whether your results are statistically significant.

REVIEW QUESTIONS

1. High cholesterol levels can lead to heart disease and other health problems. A new medicine is being introduced to lower cholesterol levels, but it is suspected of harming the livers of some patients. A study is conducted on 244 patients with high cholesterol. The subjects are randomly assigned to two treatment groups with 122 subjects in each group. One group is given the new medicine, and the other is given a placebo. After two years, the study showed more liver problems among the group given the new medicine. Which of the following statements about the generalizability of the study is best supported by these results?

   (A) Cholesterol is harmful to a person's health, so it is worth risking liver problems in order to reduce it.

   (B) The experiment demonstrates that the new medicine causes an increase in liver problems.

   (C) This is an observational study, so no conclusion can be drawn about cause and effect.

   (D) It is not clear whether the difference in liver problems is statistically significant, so no conclusion can be drawn.

   (E) Two years is too short of a time for the experiment, so no conclusion can be drawn.

2. Which of these best explains why large samples tend to provide more reliable inference about a population than small samples?

   (A) The response variable will be measured more effectively.

   (B) There is a smaller probability of bias.

   (C) It will be less of a problem if the randomization is imperfect.

   (D) Large samples are more diverse.

   (E) Large samples show less sampling variability.

3. An opinion survey is going to be conducted at two colleges. College A has 5,000 students and College B has 20,000 students. Each survey will be conducted with a simple random sample of 200 students. The results from each survey will be used as an estimate for the opinions of the student body at each college. Which college is likely to have its students' opinions estimated more accurately by the survey?

   (A) College A since it has a larger percentage of its students surveyed

   (B) College A since the larger school is likely to have more diversity of opinion

   (C) Neither college is more likely to have a more accurate estimate.

   (D) College B since the students with out-of-the-mainstream opinions will have less of an influence on the average student opinion at a larger school

   (E) College B since there is less sampling variability with a larger population

Free-Response Question

Claire is investigating whether gender is a factor in the relationship between the memorizing skills of third grade students and the same students' belief in their memorizing skills.

Claire performs a study with a large sample of 8-year-old children, equally split between boys and girls. Assume this is a representative sample of the population of all 8-year-old boys and girls in the United States. She tells them that they will be given a list of 20 nonsense words and that they will have five minutes to memorize as many words as they can. Then they will be given five minutes to write down as many of the words as they can remember in any order they like. Before she shows them the list, she asks each of them to write down the number of words they expect they will get right. She then has them go through the memorization and recall steps. Then she computes, for each child, the actual number of words right minus the predicted number of words right. Then she computes the mean of these values for boys and the mean for girls. Claire finds that this mean difference is greater for girls than it is for boys.

(a) What does this result tell you about the sample?

(b) If Claire's result about the comparison of mean differences is statistically significant, does this prove that gender is the cause of the discrepancy in performance for the population of interest?

Hint: This is a sophisticated structure for a study, and a big part of sorting through the question is getting a handle on it at a fundamental level. In particular, part (a) requires you to understand just what the results are saying about the relative performance of boys and girls in the study. In order to answer part (b), you have to know whether this is an observational study or an experiment. Only experiments can provide cause-and-effect conclusions, and only well-designed and well-executed ones at that.

ANSWERS AND EXPLANATIONS

1. D

There are several lines of analysis you might choose to follow after reading the description of the experiment, so it is important to consider only the available choices rather than giving your imagination free reign. In other words, stick to the text. Choice (A) sounds like a reasonable real-world analysis, but it's not supported by the results of the experiment. You don't know how severe the liver damage might be. It could be worse than the health problems associated with high cholesterol. Choice (B) might at first seem appealing, since it talks about experiments providing a cause-and-effect conclusion, but consider it in comparison to choice (D). You should notice, if you haven't already, that there is no mention as to whether the result is statistically significant. That rules out choice (B), and makes choice (D) seem pretty tempting. Choice (C) can't be right since this is, in fact, an experiment; it would have been the best option if this had been an observational study. Choice (E) is another loser. If it were the right answer, then it would be testing how much you know about medical issues and not how much you know about statistics. Choice (D) is best.

2. E

The relation between sampling variability and sample size is one of the fundamental issues of this curriculum, so you should be alert for its appearance on the AP exam. Smaller sampling variability is exactly the reason why large samples tend to produce more reliable inference about a population. Choices (C) and (D) are both true, but they are less important than sampling variability. They're not central to the situation in the way sampling variability is. Neither answer by itself provides the clear connection between sample size and reliability, as does (E). Don't be dazzled by buzz words such as *response variable* and *bias* in choices (A) and (B), respectively. Those two explanations don't make sense.

3. C

As long as the population is much larger than the sample size, samples of the same size provide population estimates just as reliably for a small population as a large population. Both populations are at least 10 times the sample size of 200, so choice (C) is correct. The other choices may sound plausible, but each has a reason it is not correct. If choice (A) seems tempting, consider taking it to an extreme. It would claim that a sample of 1 student tells you just as much about a population of 25 as a sample of 200 tells you about 5,000 since they measure the same percentage of their respective populations, which is obviously untrue. As for choice (B), it's not clear that the larger school will have more diversity; even if it does, those diverse opinions are also less likely to be included in the sample. Similar issues are brought up in choice (D), but this is wrong, too, since the influence of out-of-the-mainstream opinions will likely be the same in a sample as a population. Choice (E) is wrong because it incorrectly associates sampling variability with population size when it is actually a function of sample size.

Free-Response Answer

(a) For the sample, a girl will tend to perform better at memorizing the nonsense words relative to her expectations than a boy. To arrive at this conclusion, notice that a child with a positive value for the actual number of words right minus the predicted number got more words right than he or she expected to get. If one child has a larger difference than another, then that child did better relative to his or her expectations than the other. Since the mean value of these differences is greater for girls, then girls tended to do better relative to their expectations than boys.

(b) The observed gap between the mean difference for girls and the mean difference for boys is statistically significant, but this is not an experiment. It is an observational study. No cause-and-effect conclusion is in order. The cause-and-effect relationship referred to in the question implies that gender is the treatment under consideration, and Claire did not impose this treatment. She simply measured performance on an identical memorizing task for all the children in the study with no randomization. An observational study cannot establish cause and effect because there is no control for potential confounding variables, such as the different ways boys and girls are socialized in the United States.

CHAPTER 12: PROBABILITY

IF YOU LEARN ONLY FIVE THINGS IN THIS CHAPTER . . .

1. In general, $P(A) = \dfrac{\text{the number of times the desired outcome occurs}}{\text{the total number of trials}}$.
2. Events are independent if the outcome of one event does not influence the outcome of any other event.
3. Events are mutually exclusive if they cannot occur together.
4. Addition rule: $P(A \text{ or } B) = P(A) + P(B) - P(A \text{ and } B)$
5. Multiplication rule: If A and B are independent events, $P(A \text{ and } B) = P(A)P(B)$.

WHAT ARE THE CHANCES?

We are a data driven society. We collect data and use it to draw conclusions and make decisions. We have seen previously that in the interest of cost, time, and logistics, sampling can provide us with information about a population of interest. Drawing conclusions about the population based on a sample is called statistical inference. Statistical inference looks at the basic question, "How often would this method give a correct answer if I used it many, many times?" In other words, "What are the chances?"

Chance behavior is unpredictable in the short run, but has a regular and predictable pattern in the long run. Consider tossing a fair coin. We know intuitively that there is a 50/50 chance of getting a head on any single toss. But if we toss the coin 10 times, it is not unlikely that we would see as many as 8 heads or as few as 2 heads. What if we toss the coin 100 times? 1,000 times?

10,000 times? Over the long run, we will continue to hone in on the expected 50/50 split of heads and tails, with large deviations in the percentage of each increasingly unlikely.

Probability is the branch of mathematics that studies patterns of chance. The study of probability was driven at various times in history by the desire to understand gambling and games of chance. Today, however, it is the foundation for decisions in medicine, insurance, manufacturing, business, law, social science research, politics, and much of what we do on a daily basis.

Probability and Randomness

In statistics, random describes the kind of order that emerges only in the long run, over many, many trials.

The idea of probability is based on observation. Probability describes what happens over many, many trials.

Random Phenomenon: A phenomenon is random if individual outcomes are uncertain, but there is nonetheless a regular distribution to the outcomes over a large number of repetitions.

In probability, an **experiment** is any sort of activity whose results cannot be predicted with certainty. Roll a die. You could get a 1, 2, 3, 4, 5, or 6. You have no way of determining what you will get on any one roll.

The **sample space, S,** of an experiment is the set of all possible outcomes.

Flip a coin: { H, T }
Roll a die: { 1, 2, 3, 4, 5, 6 }
Roll two dice and find the sum: { 2, 3, 4, 5, 6, 7, 8, 9, 10, 11, 12 }

An **outcome** is one of the possible results that can occur as a result of an experiment, such as getting a 5 when you roll a die.

A **trial** is a single running or observation of a random phenomenon. Rolling a die only once is considered a trial.

An **event** is any outcome or set of outcomes of a random phenomenon. Thus, it is a subset of the sample space.

Probability: The probability of any outcome of a random phenomenon or experiment is the *proportion* of times the outcome would occur in a long series of repetitions. That is, probability is the long-term relative frequency.

Common Notation: If *A* is defined as the event of rolling a die and getting a 5, then ***P(A)*** is defined as the probability of *A* or the probability of rolling a 5.

Calculating Probability

Enumerating the outcomes in the sample space is critical to determining probabilities.

In general: $P(A) = \dfrac{\text{the number of times the desired outcome occurs}}{\text{the total number of trials}}$.

Let A = rolling a 5, then $P(A) = \dfrac{\text{the number of times you roll a 5}}{\text{the total number of times you roll the die}}$.

Example:

Ryan rolls the die 20 times and gets a 5 on 7 of the rolls. Let A be the event of rolling a 5. Then,

$$P(A) = \frac{\text{the number of times you roll a 5}}{\text{the total number of times you roll the die}} = \frac{7}{20}.$$

This is sometimes called *experimental probability.*

Equally likely outcome: What does it mean if events are equally likely to occur? If a random phenomenon has k possible outcomes, all equally likely, then each individual outcome has probability $\frac{1}{k}$. Think about rolling a fair die. There are six possible outcomes: 1, 2, 3, 4, 5, or 6. Since each one of these is equally likely to occur, the probability of any one of them occurring is $\frac{1}{6}$. This is sometimes called *theoretical probability.*

MODELING PROBABILITY

When we toss a coin or roll a die, we do not know the outcome in advance. Because we believe we have a fair coin or a fair die, we believe that the chance of getting a head when we toss the coin is $\frac{1}{2}$ and the chance of getting a 5 when we roll the die is $\frac{1}{6}$. Determining both of these probabilities requires two elements: a list of possible outcomes, and a probability for each outcome.

Probability Model: A probability model is a mathematical description of a random phenomenon consisting of two parts: a sample space S and a means of assigning probabilities to events.

Thinking about how we calculate probability, we have the following general rules:

(a) Probability is a number between 0 and 1. If something has no chance of ever happening, its probability is 0. If something will definitely happen, it has a probability of 1.

(b) The sum of the probabilities of all possible outcomes in a sample space is 1.

(c) The probability that an event does not occur is 1 minus the probability that it does occur. The probability of not A, denoted A^c, is also called the **complement** of A. Simply, if an event has the probability of 0.3 of happening, then it has a probability of 0.7 of not happening ($1 - 0.3 = 0.7$).

These basic concepts of probability can be expressed mathematically as follows:

$$0 \le P(A) \le 1$$
$$P(S) = 1$$
$$P(A^C) = 1 - P(A)$$

The concepts of independence and mutually exclusive events are critical in calculating probabilities.

Independent: If the outcome of an event or trial does not influence the outcome of any other event or trial, then the events or trials are said to be independent. (Roll a die twice, the second time is independent of the first. The outcome of the first roll does not affect the probability of the outcome on the second roll.)

Mutually Exclusive or Disjoint: Two events are mutually exclusive or disjoint if they cannot occur together. Sam either passed the test or failed the test. He couldn't do both at the same time.

Set Notation

Sometimes we use set notation to describe events. The event $\{A \cup B\}$, read "*A* union *B*," is the set of all outcomes that are either in *A* or in *B*. So, $\{A \cup B\}$ is another way to indicate the event $\{A \text{ or } B\}$.

The event $\{A \cap B\}$, read "*A* intersect *B*," is the set of all outcomes that are both *A* and *B*.

If two events *A* and *B* are disjoint (mutually exclusive), we can write $\{A \cap B\} = \varnothing$ which is read "*A* intersect *B* is empty."

Venn Diagrams, as shown in Figure 12.1, are often used to give a visual representation to the sample space and the possible outcomes.

Figure 12.1
Venn Diagrams: Possible Outcomes for Events *A* and/or *B*

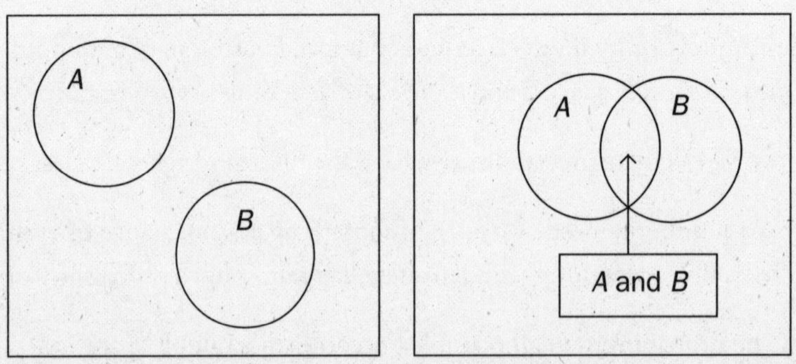

The diagram at the left shows two events *A* and *B* that have no outcomes in common. The diagram on the right depicts two events *A* and *B* that share some outcomes. The box itself represents the entire sample space.

RULES OF PROBABILITY

GENERAL ADDITION RULE

$P(A \text{ or } B) = P(A) + P(B) - P(A \text{ and } B)$ or in set notation

$$P(A \cup B) = P(A) + P(B) - P(A \cap B)$$

Suppose we are randomly selecting a student from Pine County High School. If we want to determine the probability of selecting a senior or a girl, we would need the probability of selecting a senior and the probability of selecting a girl. But what about a senior who is a girl? We would have to subtract this probability out so that senior girls are not counted twice.

Figure 12.2
Probability for Events *A* and *B*

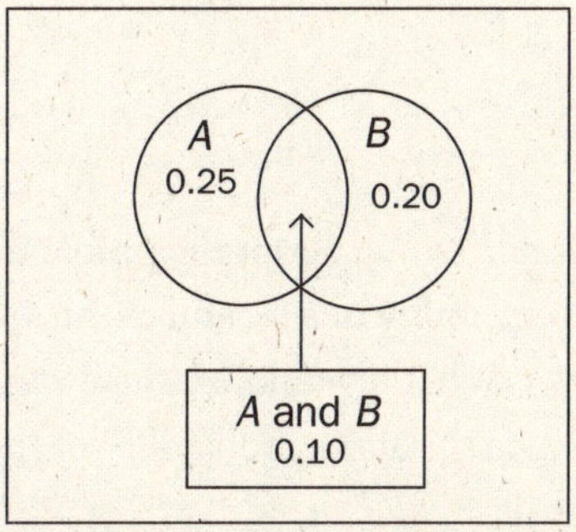

In Figure 12.2, you can see that events *A* and *B*, represented by the circles, have some common outcomes. The diagram makes it easy to calculate probabilities. The $P(A)$ equals the probability of those outcomes that are only *A* (0.25) plus those outcomes that are both *A* and *B* (0.10). So, $P(A) = 0.25 + 0.10 = 0.35$. Likewise for *B*, $P(B) = 0.20 + 0.10 = 0.30$. If we simply add $P(A) + P(B)$ to obtain $P(A \text{ and } B)$, then we would be including the piece that is both *A* and *B* (0.10) twice. Therefore, we subtract it to eliminate the double counting. $P(A \text{ or } B) = P(A) + P(B) - P(A \text{ and } B) = 0.35 + 0.30 - 0.10 = 0.55$.

If two events *A* and *B* are disjoint, $\boldsymbol{P(A \text{ or } B) = P(A) + P(B)}$. This is a special case of the addition rule, since $P(A \text{ and } B) = 0$. Let's consider the probability of selecting a senior or a junior from Pine County High School. Since these two events are mutually exclusive (they don't overlap), there is no doubling when counting, and the probability of being a senior and a junior is zero. Therefore, P(selecting a senior or a junior) = P(selecting a senior) + P(selecting a junior).

MULTIPLICATION RULE

If A and B are independent events, $P(A \text{ and } B) = P(A)\,P(B)$.

If we wanted to know the probability of drawing a heart from a standard deck of cards and rolling a 3 with one roll of the die, we would multiply the probabilities of these two events together. Think about tossing a coin and getting heads. If you toss it once, the probability of getting heads is $\left(\frac{1}{2}\right)$. If you toss it twice, you could get {HH, HT, TH, TT}. The probability of getting two heads is thus 1 out of 4 or $\frac{1}{4}$. This result is also the product of the probability of getting a head on the first toss $\left(\frac{1}{2}\right)$ and a head on the second toss $\left(\frac{1}{2}\right)$. Since the coin has no memory, the outcome of the first toss does not influence the outcome of the second toss and P(getting a head on two consecutive tosses) $= \left(\frac{1}{2}\right)\left(\frac{1}{2}\right) = \left(\frac{1}{4}\right)$.

Conditional Probability

Sometimes we already have some information about the probability of a certain event. Knowing this information may change the probability of a second event. Consider the probability that a person chosen at random has cancer. What if you knew there was a history of cancer in the person's family? Does that change your assessment of the probability of this person having cancer?

Conditional probability describes the situation where the probability of a second event is dependent upon a first event having occurred.

CONDITIONAL PROBABILITY

Conditional probability is defined as $P(B \mid A)$, which is read "the probability of B given A." The probability is calculated as follows. $P(B \mid A) = \dfrac{P(A \cap B)}{P(A)} = \dfrac{P(A \text{ and } B)}{P(A)}$

The following table shows the number of hours the students in the senior class at Academy High worked on their senior research project.

	Less Than 20 Hours	Between 20-40 Hours	40 Hours or More	Total
Males	21	154	182	357
Females	11	138	223	372
Total	32	292	405	729

From the table, we can calculate the probability that a student, chosen at random, worked between 20 and 40 hours. $\frac{292}{729} = 0.4005$

What is the probability of a student working between 20 and 40 hours, given that the student is a female?

$$P(\text{worked between 20–40 hrs.} \mid \text{student is female}) = \frac{P(\text{worked 20–40 hrs. and female})}{P(\text{female})}$$

$$= \frac{\frac{138}{729}}{\frac{372}{729}} = 0.37097.$$

This formula can also be expressed as $P(A \cap B) = P(A)P(B \mid A)$

$$P(A \text{ and } B) = P(A)\,P(B \mid A).$$

Proving Independence

Conditional probability also gives us a means of establishing the independence of two events. Remember that if two events are independent, the outcome of one does not influence the outcome of the other. To prove two events are independent, show that $P(B \mid A) = P(B)$.

This results from the fact that if two events A and B are independent, then $P(A \text{ and } B) = P(A)P(B)$. So, $P(B \mid A) = \frac{P(A \text{ and } B)}{P(A)} = \frac{P(A)P(B)}{P(A)} = P(B)$.

AP EXPERT TIP

The most common way to check for independence is simply to check that $P(A) \times P(B) = P(A \text{ and } B)$, as this will prove that $P(B) = P(B \mid A)$.

POSSIBLE OUTCOMES AND COUNTING TECHNIQUES

If you can do one task in a ways and a second task in b ways, then both tasks can be done in $a \times b$ ways.

Flip a coin and toss a die ⇒ (2)(6) = 12 possible outcomes.

Tree diagrams are often useful for displaying all possible outcomes, as shown in Figure 12.3.

Figure 12.3
Tree Diagram: "Flip a Coin and Toss a Die"

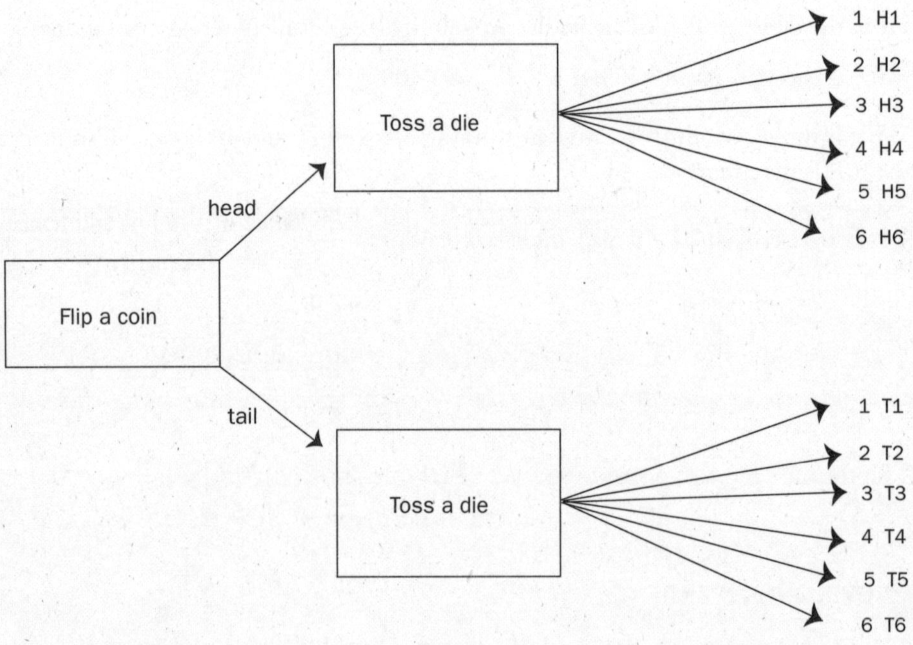

From the diagram, it is easy to see that there are 12 possible outcomes.

How many ways can we take n objects r at a time without regard to order? When order doesn't matter, it is a **combination** and is calculated

$${}_nC_r = \binom{n}{r} = \frac{n!}{r!(n-r)!} \text{ where } n \geq r.$$

When the order does matter (a, b is different from b, a), it is called a **permutation** and is calculated ${}_nP_r = \dfrac{n!}{(n-r)!}$.

AP EXPERT TIP

While you should know the formulas for combinations and permutations, most calculators have functions that will do these calculations for you. See chapter 19 for more information about the statistical functions of the calculator.

Example:

Faked numbers in tax returns, accounting records, expense account claims, and many other settings often display patterns that aren't present in legitimate records. Some patterns, like too many round numbers, are obvious and easily avoided by a clever crook. Others are more subtle. It is a striking fact that the first digits of numbers in legitimate records often follow a distribution known as Benford's Law.

One might be tempted to think that a distribution of first-digit numbers is purely random, so that the probability of occurrence for each digit between 1 and 9 (keeping in mind that a first digit can't be 0) is about 0.011. However, patterns of numbers from a large variety of sources show that the digit 1 is by far the most common first digit, and the probability of larger digits being first decreases as the digit value increases. The following table shows the probabilities for first digits.

	1	2	3	4	5	6	7	8	9
***P*(*x*)**	0.301	0.176	0.125	0.097	0.079	0.067	0.058	0.051	0.046

The use of Benford's Law is common in the accounting and auditing fields. Theoretically, investigators can detect fraud by comparing these probabilities with the first digits in records. When people make up data, they often choose first digits fairly uniformly.

$A = \{\text{first digit is } 1\}$ $B = \{\text{first digit is 6 or greater}\}$

$P(A) = P(1) = 0.301$

$P(B) = P(6) + P(7) + P(8) + P(9) = 0.067 + 0.058 + 0.051 + 0.046 = 0.222$

$P(\text{the first digit is odd}) = P(1) + P(3) + P(5) + P(7) = P(9) = 0.609$

Remember the addition rule only applies to disjoint events.

Often it is easier to compute the probability of the complement. For example, what is the probability that the first digit is anything other than a 1?

$P(A^c) = 1 - P(A) = 1 - 0.301 = 0.699.$

REVIEW QUESTIONS

1. Based on concerns over the eating habits and fitness level of school-aged children, the school board of a large district decided to offer healthy choices in the school cafeterias. They randomly selected students from all grade levels and provided them with proposed menus for the healthier lunches. Students were asked if they would purchase these lunches. The results of the survey are summarized in the table below by grade level.

	K-5	6–8	9–12	Total
Yes	6,231	5,964	3,493	15,688
No	2,016	1,912	3,939	7,867
	8,247	7,876	7,432	23,555

What is the probability that a high school student selected at random would not plan to purchase the proposed healthier lunches?

(A) 0.1672
(B) 0.3339
(C) 0.5300
(D) 0.5007
(E) 0.9447

2. According to a recent national survey of college students, 55% admitted to having cheated at some time during the last year. What is the probability that for two randomly selected college students, one or the other would have cheated during the past year?

(A) 0.5500
(B) 0.7975
(C) 0.3025
(D) 0.2475
(E) 0.2025

3. Given two events, A and B, if $P(A) = 0.37$, $P(B) = 0.41$, and the $P(A \text{ or } B) = 0.75$, then the two events are

(A) independent but not mutually exclusive.
(B) mutually exclusive but not independent.
(C) mutually exclusive and independent.
(D) neither mutually exclusive nor independent.
(E) It cannot be determined from the given information if the two events are independent or mutually exclusive.

4. Security procedures at the U.S. Capitol require that all bags—meaning briefcases, backpacks, shopping bags, any carrying bag, and purses—must be screened. Currently, it is reported that 95% of all bags that contain illegal items trigger the alarm, and 12% of the bags that do not contain illegal items trigger the alarm. If 3 out of every 1,000 bags entering the Capitol contain an illegal item, what is the probability that a bag that triggers the alarm will contain an illegal item?

(A) 0.0233
(B) 0.0029
(C) 0.9500
(D) 0.1140
(E) 0.1225

5. Suppose your teacher's stash of calculators contains 3 defective calculators and 17 good calculators. You select two calculators from the box for you and your friend to use on the AP Statistics exam. What calculations would you use to determine the probability that exactly one of the calculators drawn will be defective?

(A) $\frac{17}{20}+\frac{3}{19}$

(B) $\left(\frac{17}{20}\right)\left(\frac{3}{20}\right)+\left(\frac{3}{20}\right)\left(\frac{2}{20}\right)$

(C) $\left(\frac{17}{20}\right)\left(\frac{3}{19}\right)$

(D) $\left(\frac{17}{20}\right)\left(\frac{3}{19}\right)+\left(\frac{3}{20}\right)\left(\frac{17}{19}\right)\left(\frac{17}{20}\right)\left(\frac{3}{19}\right)$

(E) $\left(\frac{17}{20}\right)\left(\frac{11}{19}\right)\left(\frac{3}{18}\right)\left(\frac{2}{17}\right)$

FREE-RESPONSE QUESTION

The United States Youth Soccer Association estimates that 2% of those youths who play travel or select soccer, programs designed to develop soccer skills beyond a recreational level, try out for their state's Olympic Development Program (ODP). Of those who try out, 93% go on to play in college, while only 81% of those who don't try out end up playing in college. Of those who try out for ODP and go on to play in college, 5% receive some scholarship money to play. This is compared to 2% of those who don't try out for ODP and go on to play in college.

(a) Ryan's coach has encouraged him to try out for ODP. Ryan wants to know the chances of getting some money to play in college if he tries out. Determine the probability that Ryan would get money to play in college if he tried out.

(b) If a college player is selected at random, what is the probability that he tried out for ODP?

(c) Given that a college player chosen at random receives scholarship money to play, what is the probability that the player did not try out for ODP?

ANSWERS AND EXPLANATIONS

1. C

This is a conditional probability. We know that the student selected at random is a high school student. Thus, we can focus just on the population of high school students. Now of the total number of high school students (7,432), we want to look at how many responded "no" to the survey. That number is given as 3,939. So,

$$P(\text{responded no} \mid \text{high school student}) = \frac{3{,}939}{7{,}867} = 0.5300.$$

2. B

The question asks for the probability that at least one student chosen at random would have cheated. That is the probability that the first student chosen or the second student chosen would have cheated.

Let A = the event that the first student selected will have cheated.

Let B = the event that the second student selected will have cheated.

The questions asks for the probability that either A or B cheated.

$$P(A \text{ or } B) = P(A) + P(B) - P(A \text{ and } B).$$

Since the students are selected randomly, the events A and B are independent. So the $P(A \text{ and } B) = P(A)P(B)$ and then, $P(A \text{ or } B) = 0.55 + 0.55 - (0.55)(0.55) = 0.7975$.

3. D

We are given that $P(A \text{ or } B) = 0.75$ and the $P(A) = 0.37$ and $P(B) = 0.41$.

Since $P(A \text{ or } B) = P(A) + P(B) - P(A \text{ and } B)$, we have $0.75 = 0.37 + 0.41 - P(A \text{ and } B)$ or $P(A \text{ and } B) = 0.03$ and the $P(A) + P(B) = 0.78$. Since the $P(A \text{ or } B) \neq P(A) + P(B)$, the two events are not mutually exclusive. Further, if the two events were independent, $P(B \mid A) = P(B)$. The $P(B \mid A) = \frac{P(A \text{ and } B)}{P(A)} = \frac{0.03}{0.37} = 0.081$, while the $P(B) = 0.41$. Since $P(B \mid A) \neq P(B)$, the two events are not independent.

4. A

This is a conditional probability problem. The question is, what is the probability that a bag contains an illegal item, given that it sets off the alarm?

$$P(\text{a bag contains an illegal item} \mid \text{triggers the alarm}) = \frac{P(\text{it triggers the alarm and contains an illegal item})}{P(\text{it triggers the alarm})}.$$

$$P(\text{it triggers the alarm and contains an illegal item}) = (0.003)(0.95).$$

$$P(\text{it triggers the alarm}) = (0.003)(0.95) + (0.997)(0.12).$$

$$P(\text{a bag contains an illegal item} \mid \text{triggers the alarm}) = \frac{(0.003)(0.95)}{(0.003)(0.95) + (0.997)(0.12)} = 0.0233.$$

A tree diagram, as shown in Figure 12.4, works nicely here. You can easily account for all possible outcomes and assign the probabilities along the path.

Figure 12.4

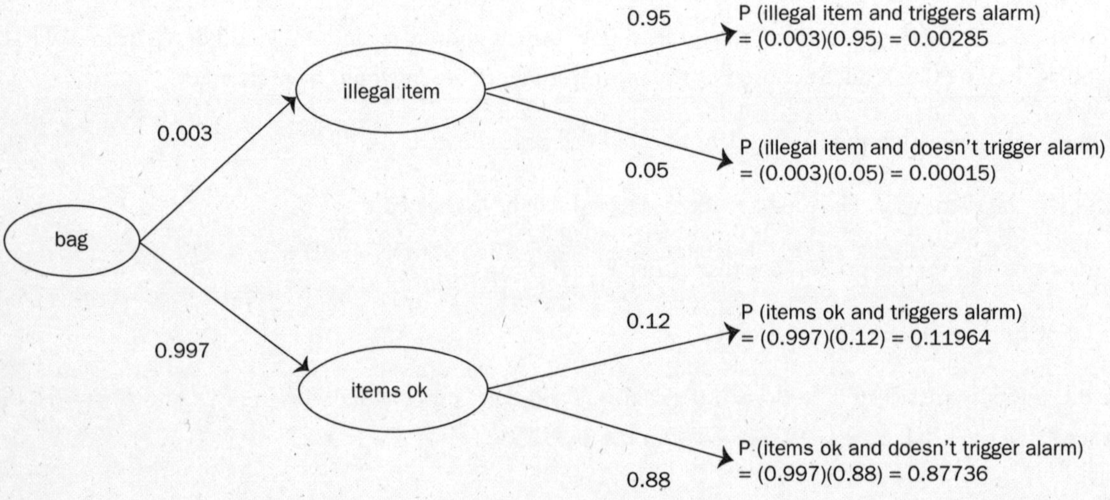

5. D

The question asks for P(first is good and the second is defective or the first is defective and the second is good). The calculators are chosen without replacements. So the sample space for the second calculator is reduced.

Remember $P(A \text{ or } B) = P(A) + P(B)$ and if A and B are independent, $P(A \text{ and } B) = P(A)P(B)$. In this problem, $P(A) = P$(first is good and the second is defective)$=\left(\frac{17}{20}\right)\left(\frac{3}{19}\right)$ and P(first is defective and the second is good) $=\left(\frac{3}{20}\right)\left(\frac{17}{19}\right)$. Thus, P(first is good and the second is defective or the first is defective and the second is good) $=\left(\frac{17}{20}\right)\left(\frac{3}{19}\right)+\left(\frac{3}{20}\right)\left(\frac{17}{19}\right)$.

FREE-RESPONSE ANSWER

Before we begin, let's enumerate the probabilities of each combination of events.

$$P(A) = P\left(\frac{\text{select player tries out for ODP}}{\text{plays travel or select}}\right) = 0.02.$$

$$P(B) = P\left(\frac{\text{select player does not try out for ODP}}{\text{plays travel or select}}\right) = 0.98.$$

$$P(C) = P\left(\frac{\text{select player tries out for ODP and plays in college}}{\text{plays travel or select}}\right) = (0.02)(0.93) = 0.0186.$$

$$P(D) = P\left(\frac{\text{select player doesn't try out for ODP and plays in college}}{\text{plays travel or select}}\right) = (0.98)(0.81) = 0.7938.$$

$$P(E) = P\left(\frac{\text{select player tries out for ODP and plays in college and gets scholarship to play}}{\text{plays travel or select}}\right)$$
$$= (0.02)(0.93)(0.05) = 0.00093.$$

$$P(F) = P\left(\frac{\text{select player doesn't try out for ODP and plays in college and gets scholarship to play}}{\text{plays travel or select}}\right)$$
$$= (0.98)(0.81)(0.02) = 0.0159.$$

(a) The probability that Ryan would get scholarship money if he tried out is the product of the following:

Probability he tries out for ODP.

Probability he plays in college GIVEN he tried out for ODP.

Probability he gets scholarship money GIVEN he tried out for ODP.

Since we know he is trying out for ODP, the first probability is 1 and the second two are independent events. The product becomes $(1)(0.93)(0.05) = 0.00465$, or 0.465%.

(b) Every college player either did or didn't try out for ODP. Looking at the list above, we see the probability a player tries out for ODP and plays in college is 0.0186, and the probability a player doesn't try out for ODP and plays in college is 0.7938. Therefore, the total percentage of select players who play in college is $0.0186 + 0.7938 = 0.8124$.

Therefore, the probability a college player tried out for ODP is

$$\frac{\text{P(tried out for ODP and played in college)}}{\text{P(played in college)}} = \frac{0.0186}{0.8124} = 0.029.$$

(c) This is a conditional probability, P(college player chosen at random did not try out for ODP given that he receives scholarship money). This would equal P(college player didn't try out for ODP and receives a scholarship) divided by P(a college player receives scholarship money).

$$P(\text{college player didn't try out for ODP} \mid \text{he receives scholarship money})$$
$$= \frac{P(\text{college player didn't try out for ODP and gets scholarship money})}{P(\text{player gets scholarship money})}$$
$$= \frac{(0.98)(0.81)(0.02)}{(0.98)(0.81)(0.02) + (0.02)(0.93)(0.05)} = 0.9447.$$

This problem could also be done using a tree diagram.

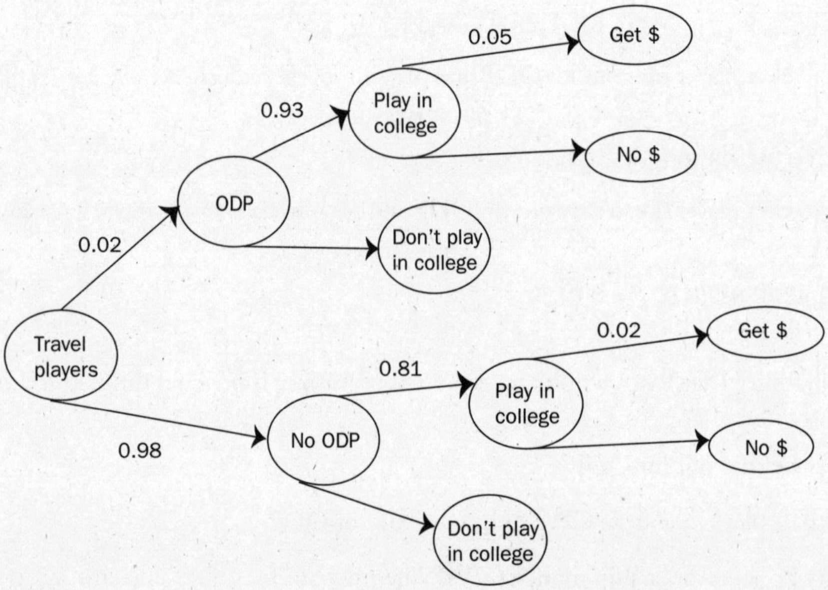

CHAPTER 13: RANDOM VARIABLES

IF YOU LEARN ONLY FIVE THINGS IN THIS CHAPTER . . .

1. A random variable is a numerical measure of the outcomes of a random phenomenon.
2. If X is a random variable and a and b are fixed numbers, then
   $\mu_{a+bx} = a + b\mu_X$ and $\sigma^2_{a+bX} = b^2\sigma^2_X$
   (multiply or add the same fixed number to every outcome).
3. If X and Y are random variables, then $\mu_{X+Y} = \mu_X + \mu_Y$.
4. If X and Y are independent random variables, then $\sigma^2_{X+Y} = \sigma^2_X + \sigma^2_Y$ and $\sigma^2_{X-Y} = \sigma^2_X + \sigma^2_Y$.
5. As the number of trials in a binomial distribution gets larger, the binomial distribution gets closer to a normal distribution $N\left(p, \sqrt{\frac{p(1-p)}{n}}\right)$.

DISCRETE AND CONTINUOUS RANDOM VARIABLES

The driving force behind many decisions in science, business, and other aspects of everyday life is the question, "What are the chances?" To answer that question, we need to know the sample space. What are the possible outcomes? A **random variable** is a numerical measure of the outcomes of a random phenomenon. Random variables are denoted by a capital letter and the values that the random variable can take on are denoted by lowercase letters.

Pick a student at random (this is the random phenomenon). The student's grades, SAT scores, height, or weight are the random variables describing the properties of the student. There are two categories of random variables, discrete and continuous. Discrete random variables have a finite or countable number of outcomes and a continuous random variable has an infinite number of possible values it can take on.

The grades one can receive on the AP exam, {0, 1, 2, 3, 4, 5}, represent a discrete random variable. The weight of the mice in a lab experiment would be a continuous random variable.

A general form for working with random variables and their probabilities is shown in the table below.

X	x_1	x_2	x_3	x_4	x_5	...	x_n
$P(X)$	P_1	P_2	P_3	P_4	P_5	...	P_n

At one time, it was a common practice for college professors to grade on the bell-shaped curve. Using a 4.0 grading scale, the probability of getting any particular grade is provided in the table below.

X	0	1	2	3	4
$P(X)$	0.025	0.135	0.68	0.135	0.025

Here the random variable X represents the grade in the course. The table shows that the $P(3) = 0.135$. That is, the probability of getting a B in the course is 0.135.

A histogram can be used to display the probabilities as well. The height of each bar would represent the probability of the outcome depicted at its base. This is also referred to as graphing the relative frequency or the probability histogram.

Figure 13.1
Probability Histogram

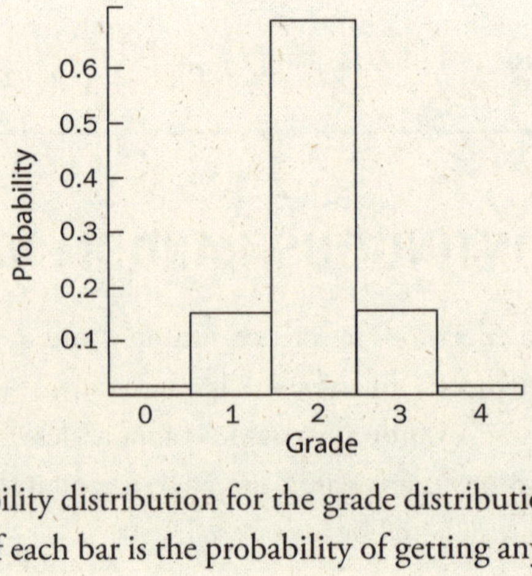

Figure 13.1 shows the probability distribution for the grade distributions based on a normal curve. Note that the height of each bar is the probability of getting any of the particular grades. A **probability distribution** is a listing or graphing of the probabilities associated with a random variable.

Example:

What is the probability distribution of the discrete random variable X that counts the number of heads in five tosses of a fair coin? First, recall that there are only two possible outcomes, heads or tails, and we have five tosses of the coin. This means that we have 2^5 — 32 possible outcomes.

$X = 0$	$X = 1$	$X = 2$	$X = 3$	$X = 4$	$X = 5$
TTTTT	HTTTT	HHTTT	TTHHH	THHHH	HHHHH
	THTTT	HTHTT	THTHH	HTHHH	
	TTHTT	HTTHT	THHTH	HHTHH	
	TTTHT	HTTTH	THHHT	HHHTH	
	TTTTH	THHTT	HTTHH	HHHHT	
		THTHT	HTHTH		
		THTTH	HTHHT		
		TTHHT	HHTTH		
		TTHTH	HHTHT		
		TTTHH	HHHTT		

Summarizing these outcomes, we can calculate the probabilities for the number of heads one would get in five tosses of a fair coin. These are provided in the table below.

X	0	1	2	3	4	5
$P(X)$	$\frac{1}{32} = 0.03125$	$\frac{5}{32} = 0.15625$	$\frac{10}{32} = 0.3125$	$\frac{10}{32} = 0.3125$	$\frac{5}{32} = 0.15625$	$\frac{1}{32} = 0.03125$

With the probability distribution, probabilities of various outcomes are easy to compute.

What is the probability of at least two heads?

$$P(X \geq 2) = P(X = 2) + P(X = 3) + P(X = 4) + P(X = 5).$$

$$P(X \geq 2) = 0.3125 + 0.3125 + 0.15625 + 0.03125 = 0.8125.$$

The probability distribution of a continuous random variable assigns probabilities as areas under a density curve. All continuous probability distributions assign the probability of 0 to any individual outcome. Essentially, it is a vertical line and there is no area under it.

NORMAL DISTRIBUTION AS PROBABILITY DISTRIBUTION

A density curve describes an assignment of probabilities. The normal distributions are thus probability distributions. If the random variable X has the $N(\mu, \sigma)$ distribution where μ is the population mean, then the standardized variable $Z = \frac{X - \mu}{\sigma}$ is a standard normal random variable with distribution $N(0, 1)$. Remember that Z tells you how many standard deviations you are from the mean.

MEAN AND VARIANCE OF RANDOM VARIABLES

In describing data, we have looked at graphs of the distributions and then numerical measures to further describe and assist in analyzing the data. These numerical measures are the mean, median, mode, and standard deviation. We can apply these same concepts to the distribution of random variables.

We cannot calculate the mean of a random variable in the same manner as we do for a set of data such as test scores. Remember that a random variable is a numerical outcome of a random phenomenon, and not all outcomes are necessarily equally likely.

The center of the histogram is the point where the histogram is balanced. Thus, for the normal distribution, it would be the middle, with equal areas to the left and right. However, for the skewed distribution, it would be further right of center.

Figure 13.2 Normal and Skewed Distributions

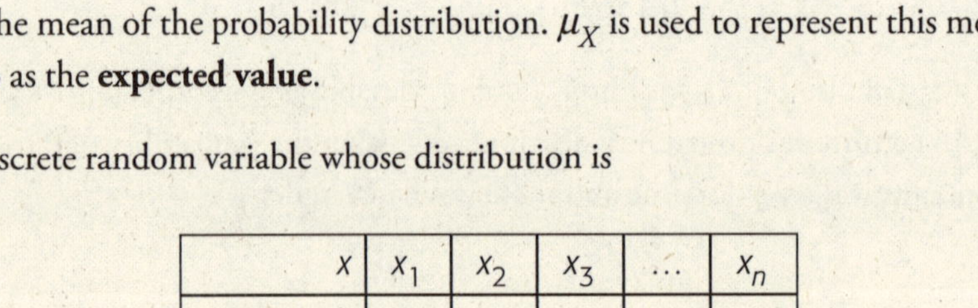

The center is the mean of the probability distribution. μ_X is used to represent this mean, and is also referred to as the **expected value**.

Given X is a discrete random variable whose distribution is

x	x_1	x_2	x_3	...	x_n
$P(X = x)$	p_1	p_2	p_3	...	p_n

then $\mu_X = \sum x_i p_i$.

The mean of a random variable is really a weighted average.

Example:

Benford's Law (see chapter 12) is widely in the accounting and audit world. Many feel that the problem with Enron's books should have been noticed long before the company went under by simply examining the first digit of the numbers in their financial records. If the first digits in a set of data appear "at random," the nine possible digits 1 to 9 all have the same probabilities. The probability distribution of the first digit *X* is shown below.

X	1	2	3	4	5	6	7	8	9
$P(X)$	$\frac{1}{9}$	$\frac{1}{9}$	$\frac{1}{9}$	$\frac{1}{9}$	$\frac{1}{9}$	$\frac{1}{9}$	$\frac{1}{9}$	$\frac{1}{9}$	$\frac{1}{9}$

$$\mu_x = 1\left(\frac{1}{9}\right) + 2\left(\frac{1}{9}\right) + 3\left(\frac{1}{9}\right) + 4\left(\frac{1}{9}\right) + 5\left(\frac{1}{9}\right) + 6\left(\frac{1}{9}\right) + 7\left(\frac{1}{9}\right) + 8\left(\frac{1}{9}\right) + 9\left(\frac{1}{9}\right) = 5.$$

If, on the other hand, the data obey Benford's Law, the distribution of the first digit *X* is

X	1	2	3	4	5	6	7	8	9
$P(X)$	0.301	0.176	0.125	0.097	0.079	0.067	0.058	0.051	0.046

$$\mu_x = 1(0.301) + 2(0.176) + 3(0.125) + 4(0.097) + 5(0.079) + 6(0.067) + 7(0.058) + 8(0.051) + 9(0.046) = 3.441.$$

Notice that the mean under Benford's Law is smaller than if all digits have the same probability. This reflects the greater probability of smaller first digits under Benford's Law. Given *X* is a discrete random variable whose distribution is

x	x_1	x_2	x_3	...	x_n
$P(x = x)$	p_1	p_2	p_3	...	p_n

and that μ_X is the mean of *X*, then the variance of *X* is

$$\sigma_x^2 = (x_1 - \mu_x)^2 p_1 + (x_2 - \mu_x)^2 p_2 + \cdots + (x_n - \mu_x)^2 p_n = \sum (x_i - \mu_x)^2 p_i.$$

The standard deviation of *X* is the square root of the variance.

Using the probabilities from the first digits above, let's first assume that each digit is equally likely.

Thus, we have

$$\sigma_x^2 = \sum (x_i - \mu_x)^2 p_i = (1-5)^2\left(\frac{1}{9}\right) + (2-5)^2\left(\frac{1}{9}\right) + \cdots + (9-5)^2\left(\frac{1}{9}\right)$$

$$= 6.66667 \text{ and}$$

$$\sigma = 2.58199.$$

If the first digits are to follow Benford's Law, we have

$$\sigma_x^2 = \sum (x_i - \mu_x)^2 p_i$$

$$= (1-3.441)^2(0.301) + (2-3.441)^2(0.176) + \cdots + (9-3.441)^2(0.046)$$

$$= 6.060459 \text{ and}$$

$$\sigma = 2.4618.$$

LAW OF LARGE NUMBERS

Recall flipping a coin. It is only when it is flipped a large number of times that the results get closer and closer to 50% heads and 50% tails.

LAW OF LARGE NUMBERS

The Law of Large Numbers states that the actual mean of many trials approaches the true mean of the distribution as the number of trials increases.

AP EXPERT TIP

Using these symbols, the Law of Large Numbers states that $\bar{x}$ will approach μ as the number of trials increases.

How many trials does this mean? Again, think about tossing a coin 100 times, 1,000 times, 5,000 times, or 100,000 times. How close to the actual mean do you want to be? How variable is our random outcome? *The more variability there is in the outcomes, the more trials are needed to ensure that the mean outcome is close to the distribution mean.*

Notation and Convention: Properties of data are called sample properties, while properties of the probability distribution are called model or population properties. The box below shows the conventional notation for samples and population values.

POPULATION AND SAMPLE

Population	*Sample*
μ	*$\bar{x}$ or expected value*
σ standard deviation	*s standard deviation or standard error*

RULES FOR MEANS AND VARIANCES OF RANDOM VARIABLES

RANDOM VARIABLES

(1) If X is a random variable and a and b are fixed numbers, then
$\mu_{a+bx} = a + b\mu_x$ and $\sigma^2_{a+bX} = a + b^2\sigma^2_X$
(multiply or add the same fixed number to every outcome).

(2) If X and Y are random variables, then
$\mu_{X+Y} = \mu_X + \mu_Y$ and
$\mu_{X-Y} = \mu_X - \mu_Y$,
and if X and Y are independent random variables, then $\sigma^2_{X+Y} = \sigma^2_X + \sigma^2_Y$ and
$\sigma^2_{X-Y} = \sigma^2_X + \sigma^2_Y$.

Example:

Let $X = \{1,2,3,4,5\}$ with each outcome equally likely to occur. Then $\mu_x = 3$ and $\sigma^2_x = 2$. Consider the set $2 + 5X$, which would be $\{7,12,17,22,27\}$. Again, each outcome is equally likely to occur.

Calculating μ_{2+5x} and σ^2_{2+5x} we get, $\mu_{2+5x} = 7(0.2) + 12(0.2) + 17(0.2) + 22(0.2) + 27(0.2) = 17$ and $\sigma^2_{2+5x} = (7-17)^2(0.2) + (12-17)^2(0.2) + (17-17)^2(0.2) + (22-17)^2(0.2) + (27-17)^2(0.2) = 50$.

Note that $\mu_{2+5X} = 2 + 5\mu_x = 2 + 5(3) = 17$ and $(\sigma_{2+5X})^2 = 5^2\sigma^2_X = 25(2) = 50$.

Example:

Let X = the random variable of the outcome of rolling one fair die. Therefore,

$X = \{1,2,3,4,5,6\}$, each with probability $\frac{1}{6}$. $\mu_x = 3.5$ and $\sigma^2_X = 2.9167$.

Let Y = the discrete random variable of rolling another fair die. The mean and variance of its probability distribution should be the same as that of the first die: $\mu_Y = 3.5$ and $\sigma^2_Y = 2.9167$.

Consider the sum of the outcomes on the dice when rolled together:
$X + Y = \{2,3,4,5,6,7,8,9,10,11,12\}$.

The outcomes and their probabilities are in the table below.

$X + Y$	2	3	4	5	6	7	8	9	10	11	12
p	$\frac{1}{36}$	$\frac{2}{36}$	$\frac{3}{36}$	$\frac{4}{36}$	$\frac{5}{36}$	$\frac{6}{36}$	$\frac{5}{36}$	$\frac{4}{36}$	$\frac{3}{36}$	$\frac{2}{36}$	$\frac{1}{36}$

Calculating the mean and variance, we see that $\mu_{X+Y} = 7$, which is $\mu_X + \mu_Y = 3.5 + 3.5 = 7$, which is $\sigma^2_{X+Y} = 5.8334$, which is $\sigma^2_X + \sigma^2_Y = 2.9167 + 2.9167 = 5.8334$.

BINOMIAL DISTRIBUTION

Many situations that we are interested in have two possible outcomes: playing a game (win or lose); having a baby (boy or girl); applying for a job (hired or not hired); tossing a coin to see how a soccer game starts (kick off or defend); or manufacturing a product in a factory (defective or not). One of the theoretical probabilities distributions associated with modeling these types of situations is the **binomial distribution**.

BINOMIAL MODEL

1. Each observation falls into one of just two categories (success or failure).
2. The number of observations is the fixed number *n*.
3. The *n* observations are all independent.
4. The probability of success, *p*, is the same for each observation.

If data are produced following the binomial model, then the random variable X = number of successes is called a **binomial random variable**, and the probability distribution of X is a binomial distribution.

The distribution of X number of successes in the binomial model has parameters n and p, where n is the number of observations and p is the probability of success of any one observation.
For $X \in \{0, 1, 2, \dots n\}$, the notation for the binomial distribution is $B(n, p)$.

Remember, a random variable is a variable whose value is a numerical outcome of a random phenomenon. Thus, $\mu_X = \sum x_i p_i$ and $\sigma_X^2 = \sum (x_i - \mu_X)^2 p_i$.

When finding binomial probabilities, remember that we are finding the probability of obtaining k successes in n trials.

Example:

A quality control engineer for a company that distributes field hockey sticks selects an SRS of 20 sticks from a large shipment. The engineer is unaware of the fact that 5% of the sticks in the shipment fail to meet specifications. What is the probability that no more than 1 of the 20 of the sticks in the sample fail inspection?

> X = count of sticks in the sample that do not meet specification
> $B(20, 0.05)$

Remember that the formula for $\binom{n}{k} = \frac{n!}{k!(n-k)!}$ gives the number of ways of arranging k successes among n observations. This is also referred to as the **binomial coefficient**.

Binomial probability is given by: $P(X = k) = \binom{n}{k} p^k (1-p)^{n-k}$.

Back to our hockey stick example:

$$P(X \le 1) = P(X = 1) + P(X = 0) = \binom{20}{1}(0.05)^1(1-0.05)^{20-1} + \binom{20}{0}(0.05)^0(1-0.05)^{20-0}$$

$$= \frac{20!}{1!(20-1)!}(0.05)^1(0.95)^{19} + \frac{20!}{0!(20-0)!}(0.05)^0(0.95)^{20} = 0.3774 + 0.3585 = 0.7359.$$

Mean and Standard Deviation of Binomial Distributions

A softball player gets a hit 82% of her times at bat. What is the mean number of hits made in 9 times at bat? Logically, $(9)(0.82) = 7.38$. This is $\mu = np$. For the standard deviation, $\sigma = \sqrt{np(1-p)}$. In our example, $\sigma = \sqrt{9(0.82)(1-0.82)} = 1.5256$.

MEAN AND STANDARD DEVIATION OF BINOMIAL DISTRIBUTIONS

> $\mu = np$
> $\sigma = \sqrt{np(1-p)}$

Note that q is often used for $(1-p)$.

NORMAL APPROXIMATION FOR THE BINOMIAL DISTRIBUTION

As the number of trials in a binomial distribution gets larger, the binomial distribution gets closer to a normal distribution $N\left(np, \sqrt{np(1-p)}\right)$.

How large? Generally, we will use the normal approximation when n and p satisfy $np \geq 10$ and $n(1-p) \geq 10$.

The accuracy of the normal approximation improves as the sample size increases. It is most accurate for any fixed n when p is close to 0.5, and least accurate when p is close to 0 or 1. Of course, any concern about the degree of accuracy can be eliminated by calculating the exact value for the binomial distribution, rather than applying a normal approximation.

GEOMETRIC DISTRIBUTION

The binomial model requires that the number of trials be fixed, and the binomial variable X counts the number of successes in that fixed number of trials. If the goal is to determine the number of trials necessary for the *first* success, we use the geometric model. The number of trials required to get the first success is called a **geometric random variable**, and its distribution is the **geometric distribution**.

GEOMETRIC MODEL

1. Each observation falls into one of just two categories, success or failure.
2. The variable of interest is the number of trials required to obtain the first success.
3. The n observations are all independent.
4. The probability of success, p, is the same for each observation.

Example:

Suppose we plan to roll a die until we get a 5. The random variable X = the number of trials until the first 5 occurs. Now, verify that the geometric model is met: the observations fall into just two categories; we are interested in the number of throws needed to get the first 5; each roll of the die is independent from any other roll; and the probability of getting a 5 is $\frac{1}{6}$ for each trial.

$P(X = 1) = P(\text{success on the first roll}) = \frac{1}{6}$.

$P(X = 2) = P(\text{success on the second roll}) = P(\text{failure on the first roll and success on the second roll}) = \left(\frac{5}{6}\right)\left(\frac{1}{6}\right) = \frac{5}{36}$.

$P(X = 3) = P(\text{success on the third roll}) = P(\text{failure on the first roll, failure on the second roll and success on the third roll}) = \left(\frac{5}{6}\right)\left(\frac{5}{6}\right)\left(\frac{1}{6}\right) = \frac{25}{216}$.

Remember that the trials are independent, so $P(A \text{ and } B) = P(A)P(B)$.

Notice the pattern: $P(X = n) = \left(\frac{5}{6}\right)^{n-1}\left(\frac{1}{6}\right) = (1 - p)^{n-1}p$.

CALCULATING GEOMETRIC PROBABILITY

If X has a geometric distribution with probability p of success and $(1 - p)$ of failure of each observation, then $P(X = n) = (1 - p)^{n-1}p$.

The probability distribution for rolling a 5 would look like the following table.

X	1	2	3	4	5	6	7	8
$P(X)$	0.1667	0.1389	0.1157	0.0965	0.0804	0.0670	0.0558	0.0465

From this, we can draw a probability histogram, as shown in Figure 13.3.

Figure 13.3

Probability Histogram for Rolling a Five

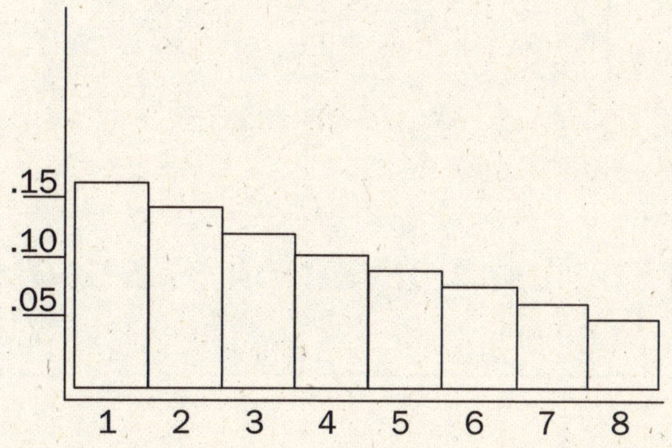

Note that in this example, the distribution is only provided for up to 8 trials, $\sum p_i = 0.7675$. Since we have not considered all possible outcomes in this exercise, the sum of the probabilities does not add to 1.

Mean and Standard Deviation of Geometric Distributions

EXPECTED VALUE OF GEOMETRIC DISTRIBUTIONS

If X is a geometric random variable with probability of success p on each trial, then the mean or expected value of the random variable—that is, the expected number of trials required to get the first success—is $\mu = \frac{1}{p}$.

This result can be derived as follows. We know for any random variable $\mu = np$. With the geometric random variable, we saw $P(X = n) = (1 - p)^{n-1}p = pq^{n-1}$. $P(X)$ for some values of X is summarized below.

X	1	2	3	4	...	n
$P(X)$	p	pq	pq^2	pq^3	...	pq^{n-1}

Thus, we have

$$\mu_X = 1(p) + 2(p)(1-p) + 3(p)(1-p)^2 + \cdots + n(p)(1-p)^{n-1}$$
$$(1)\mu_X = p(1 + 2q + 3q^2 + \cdots + nq^{n-1}), \text{ where } q = 1 - p$$

Multiply both sides by q.

$$(2)q\mu_x = p(q + 2q^2 + 3q^3 + \cdots + nq^n)$$

Now subtract equation (2) from equation (1).

$$\mu_X - q\mu_X = p(1 + q + q^2 + \cdots + q^{n-1})$$
$$\mu_X(1-q) = p\left(\frac{1}{1-q}\right)$$
$$\mu_X = \frac{p}{(1-q)^2} = \frac{p}{p^2} = \frac{1}{p}$$

VARIANCE AND STANDARD DEVIATION OF A GEOMETRIC RANDOM VARIABLE

$\sigma_X^2 = \frac{1-p}{p^2}$ and $\sigma_X = \frac{\sqrt{1-p}}{p}$.

Example:

Andrea is playing a game of chance and skill at the local county fair. The object of the game is to pitch a baseball at a target painted on a wall 90 feet away. Andrea has played this game many times, and has determined that on average, she hits the target 1 out of every 8 times she tries. She believes that her chances of hitting the target are the same for each throw and she has no reason to think that her throws are not independent. Let *X* be the number of throws until she hits the target.

$$E(X) = 8 \text{ since } p = \frac{1}{8} \text{ and } \mu_X = E(X) = \frac{1}{p} = \frac{1}{\frac{1}{8}} = 8.$$

$$\sigma_X^2 = \frac{1-p}{p^2} = \frac{1-0.125}{0.125^2} \text{ or } \frac{\frac{7}{8}}{\frac{1}{64}} = 56 \text{ and thus } \sigma_X = 7.5.$$

The probability that it takes *more* than n trials to see the first success is $P(X > n) = (1 - p)^n$.

What is the probability that it takes Andrea more than 8 throws to hit the target?

$$P(X > 8) = (1 - p)^n = \left(1 - \frac{1}{8}\right)^8 = \left(\frac{7}{8}\right)^8 = 0.3436.$$

SIMULATING RANDOM BEHAVIOR

Often it is not feasible, cost-effective, or time-effective to carry out experiments or test theories and what-ifs. In these cases, simulation is a viable alternative. Large-scale simulations can be run for businesses or research activities using computers. To understand the mechanics of simulation, we can use random number tables or random number generators on calculators.

The concepts behind simulation are best described through an example. Let's say Erin runs a bakery. Her special of the day for Wednesday is Irish soda bread. From past experience, she knows that 27% of the customers buy the special on Wednesday. Erin wants the bread to be fresh, and doesn't want to bake more than she will sell in the first hour of business. She can bake the bread all day, but needs to have some ready when the bakery opens in the morning. She feels she could easily bake three loaves of the soda bread before opening. She wants to know the chances that of the first five customers, more than three will want to purchase the bread.

Solution: Since experience shows Erin that 27% of the customers will buy the soda bread, you let the digits 00–26 represent these customers. Thus, the digits 27–99 would represent the customers who would not purchase the soda bread. To simulate the first five customers, select five sets of two consecutive digits from a random number table. From the following table, following across the first row would yield 85, 65, 26, 24, and 93. Now, note how many of these numbers fall in the 00–26 range. Continue this process and record the number of times in each group of five that you get numbers in the range of 00–26. In the first group 8 5 6 5 2 6 2 4 9 3, we got 2 out of 5 numbers in the range of 00–26. Running through ten such simulations, we get 2, 1, 3, 0, 4, 2, 1, 0, 1, and 2. These simulate the number of customers (out of five) who would purchase the soda bread. Thus, if Erin bakes only three loaves to start the day, there is only a one in ten chance that she would need more than the three loaves.

85652	62493	43891	83493	39170	95823	45597	03555	20061	52735
84023	70144	71519	96013	50776	18328	11658	35690	04159	14831
43793	41070	20165	66354	04238	41890	69184	94017	67353	44280
09769	50850								

Here we have simulated ten samples of five customers. For business and researcher personnel, simulation is a relatively quick, inexpensive, and easy means of estimating customer activity or answering a probability-based research question. With computers, such simulations can be repeated thousands of times in a matter of minutes.

SIMULATING GEOMETRIC MODELS

Now, suppose you and your friend start a lawn service company. You know from past experience that when you approach a new house that you have a probability of 0.10 that they will hire you. Use simulation to find the probability that you will get hired at the first house you approach, then the second, the third, the fourth, and the fifth.

We will again use the random number table from above. This time 0 will represent you getting hired and 1–9 not getting hired. Reading the random number table now, we want to note how many digits it takes before we get a 0.

85652	62493	43891	83493	39170	95823	45597	03555	20061	52735
84023	70144	71519	96013	50776	18328	11658	35690	04159	14831
43793	41070	20165	66354	04238	41890	69184	94017	67353	44280
09769	50850								

Beginning with the first row, it takes us 25 digits before we get the first 0. That is, 85652 62493 43891 83493 39170. Continuing in this same manner, we get the following results.

Number of Tries to Get First Success	Frequency
1	3
2	1
3	1
4	2
5	0
Over 5	10

From this, we can estimate the probabilities of getting hired on the first try, second try, and so on.

Number of Tries to Get First Success	Estimated Probability
1	$\frac{3}{17} = 0.18$
2	$\frac{1}{17} = 0.06$
3	$\frac{1}{17} = 0.06$
4	$\frac{2}{17} = 0.12$
5	$\frac{0}{17} = 0.0$
Over 5	$\frac{10}{17} = 0.59$

Note that this simulation simulates a geometric distribution. The geometric distribution shows the number of trials required to get the first success.

REVIEW QUESTIONS

1. Based on his past performance, the probability that Ben will make a free throw is 0.6. What is the probability that he will make three out of his next five free throws?

   (A) 0.6630
   (B) 0.0960
   (C) 0.3456
   (D) 0.9360
   (E) 0.01536

2. Based on his past performance, the probability that Ben will make a free throw is 0.6. What is the probability that he will miss his first three free throws, and then make his fourth one?

   (A) 0.9744
   (B) 0.1536
   (C) 0.8704
   (D) 0.096
   (E) 0.0384

3. Robin owns a bookstore. She is working on a presentation to convince her partner to spend \$500 on a catchy window display. Robin has data to support the fact that if people come in to browse, 62% will make a purchase. Given that the average purchase is \$12.38, what is the expected amount of sales from the next 20 customers who enter the store?

   (A) \$ 7.68
   (B) \$153.51
   (C) \$247.60
   (D) \$ 94.60
   (E) \$ 58.34

4. A radio station is running a lottery to raise money for a local charity. The prizes are \$10, \$50, \$100, and a grand prize of \$1,000. The chances of winning these amounts are 0.05, 0.015, 0.009, and 0.001, respectively. What are your total expected winnings (minus costs) if you pay \$5 for a ticket?

   (A) \$ 3.15
   (B) \$ 8.15
   (C) \$ 90
   (D) \$290
   (E) –\$ 1.85

5. The scores for the top three golfers on a high school golf team are used to determine which high schools advance to the regional level. The Central High team's top three players have mean scores and standard deviations the following

	Player 1	Player 2	Player 3
μ_x	89.5	94.4	97.2
σ_x	2.3	4.5	3.9

   What are the mean score and standard deviation for the top three golfers on the Central High team?

   (A) $\mu_T = 281.1,\ \sigma_T = 6.38$
   (B) $\mu_T = 93.7,\ \sigma_T = 6.38$
   (C) $\mu_T = 93.7,\ \sigma_T = 3.57$
   (D) $\mu_T = 281.1,\ \sigma_T = 3.57$
   (E) $\mu_T = 281.1,\ \sigma_T = 10.7$

Free-Response Question

A manufacturer of batteries for hearing aids claims that only 4% of their batteries are defective. A consumer watch group is doubtful of the claim and wants to check it. They have a shipment of 500 batteries.

(a) Describe an appropriate model for the number of defective batteries in the shipment.

(b) What is the mean and standard deviation of the distribution?

(c) The consumer group has reason to believe that the rate of defective batteries is at least 5%. Based on your findings in (b), what is the probability that more than 5% of this shipment would be defective if 4% of the manufacturer's batteries are defective?

ANSWERS AND EXPLANATIONS

1. C

The question is one that can be addressed by the binomial model. It meets the four criteria: (1) Each observation falls into one of two categories: success or failure. Ben either makes the free throw or not. (2) There are a fixed number of observations. We are concerned with what happens in five free throws. (3) The n observations are independent. Ben's success on any free throw is not affected by what he did on his previous free throws. (4) The probability of success is the same for each observation. The probability of Ben making a free throw is 0.6. So,

$$P(X = k) = \binom{n}{k} p^k (1 - p)^{n-k}$$

$$P(X = 3) = \binom{5}{3} (0.6)^3 (1 - 0.6)^{(5-3)}$$

$$= \frac{5!}{3!(5 - 3!)}(0.6)^3(0.4)^2 = 0.3456.$$

2. E

The question is one that can be addressed by the geometric model. It meets the four criteria: (1) Each observation falls into one of two categories: success or failure. Ben either makes the free throw or not. (2) The variable of interest here is the number of trials required for the first success. Don't let the wording about "missing three free throws" confuse you—we are interested in the probability that Ben's first success is his fourth throw. (3) The n observations are independent. Ben's success on any free throw is not affected by what he did on his previous free throws. (4) The probability of success is the same for each observation. The probability of Ben making a free throw is 0.6. So,

$$P(X = n) = (1 - p)^{n-1} p$$

$$P(X = 4) = (1 - 0.6)^{(4-1)}(0.6) = (0.4)^3(0.6) = 0.0384.$$

3. B

The mean for a random variable is given by $\mu = np$. Thus, the average number of purchases Robin and her partner can expect is $\mu = np = 20(0.62) = 12.4$. Since the average purchase is $12.38, the expected amount of sales is ($12.38 per purchase)(12.4 purchases) = $153.51.

4. E

Recall that the expected value or mean of a random variable $\mu_X = \Sigma x_i p_i$. We can calculate the probability distribution as follows.

$	0	10	50	100	1,000
Probability	0.925	0.05	0.015	0.009	0.001

$\mu_X = \Sigma x_i p_i = 0(0.925) + 10(0.05) + 50(0.015) + 100(0.009) + 1{,}000(0.001) = \$3.15$. After subtracting out the $5 you paid for the ticket, your expected winnings would be –$1.85.

5. A

Each of the high school players' scores are random variables, and the players' scores are independent from one another. The mean of the sum of random variables is the sum of the means of the individual random variables: $\mu_{X+Y} = \mu_X + \mu_Y$. So, $\mu_{X+Y+Z} = \mu_X + \mu_Y + \mu_Z = \mu_T = 89.5 + 94.4 + 97.2 = 281.1$. For the standard deviation, we work first with the variance, since the variance of the sum is the sum of the $\sigma^2_{X+Y} = \sigma^2_X + \sigma^2_Y$. For the golfers, that would be $\sigma^2_{X+Y+Z} = \sigma^2_X + \sigma^2_Y + \sigma^2_Z = (2.3)^2 + (4.5)^2 + (3.9)2 = 40.75$, and thus $\sigma_{X+Y+Z} = \sqrt{\sigma^2_{X+Y+Z}} = \sqrt{40.75} = 6.38$.

Free-Response Answer

(a) Let X = the number of defective batteries in the shipment. Each battery falls into one of two categories, defective (success) or not defective (failure). There is a fixed number of observations, so $n = 500$. The batteries are independent of each other, and the probability that a battery is defective is the same for each battery. Therefore, the binomial model is an appropriate model.

(b) As the number of trials in a binomial distribution gets larger, the binomial distribution gets closer to a normal distribution, $N\left(np, \sqrt{np(1-p)}\right)$.

Check conditions: $np \geq 10$ and $n(1-p) \geq 10$. Checking, we have $500(0.04) = 20$ and $500(0.96) = 480$. Therefore, both conditions are met.

$\mu = np = 500(0.04) = 20$ and $\sigma = \sqrt{np(1-p)} = \sqrt{500(0.04)(0.96)} = 4.38$.

(c) Since 5% of 500 is 25, this question asks what is the probability that there will be more than 25 defective batteries in the shipment.

$P(X > 25) = P\left(z > \dfrac{25-20}{4.38}\right) = P(z > 1.14) \approx 0.1271$. For a shipment of 500 batteries, the probability of getting at least 25 defective ones is approximately 12.71%.

CHAPTER 14: THE NORMAL DISTRIBUTION

IF YOU LEARN ONLY THREE THINGS IN THIS CHAPTER . . .

1. A probability density function is an expression that gives the frequencies of continuous random variables.
2. The Empirical Rule (68%/95%/99.7%) states that in a normal distribution approximately 68% of the data values fall within one standard deviation of the mean, 95% of the data values fall within two standard deviations of the mean, and approximately 99.7% of the data values fall within three standard deviations of the mean.
3. The *z*-score measures the distance an observation is from the mean in standard deviations.

PROBABILITY DENSITY FUNCTIONS

Mathematical models are formalized expressions of a theory or the causal relationship that exists between variables.

To find probabilities for discrete random variables, we used probability distribution functions. But to find probabilities for continuous random variables, it is not possible to enumerate every possible value the random variable can take on, as well as the corresponding probability. In order to find the probabilities for continuous random variables, we need to use a probability density function.

A **probability density function** is defined as an expression giving the frequencies (and thus the probabilities) of continuous random variables. The area under the graph of the probability density function over all possible values of the random variable must equal one. Think about why this is true.

If we take all possible outcomes together, we would have a total of 100%, and the decimal equivalent 1.0. The graph of the probability density function is always on or above the horizontal axis.

The probability density function describes the overall pattern of the distribution. Generally, you can start with a histogram and draw a smooth curve through the tops of the histogram bars as shown in Figure 14.1. Thus, a density curve can take on any shape. The area under the density curve represents the probabilities of the continuous random variable.

Figure 14.1
Probablility Density Function

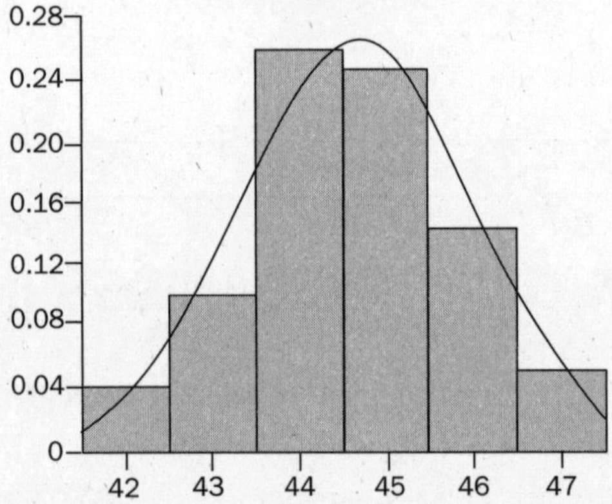

Consider a situation in which the probability of any given value of a continuous random variable is p. The graph of the function would be a rectangle as shown in Figure 14.2. Since the area under the density curve must equal 1, it follows $x(p) = 1$. So if the maximum value of your continuous random variable is 8, then $p = \frac{1}{8}$.

Figure 14.2
Function for Probability of a Continuous Random Variable

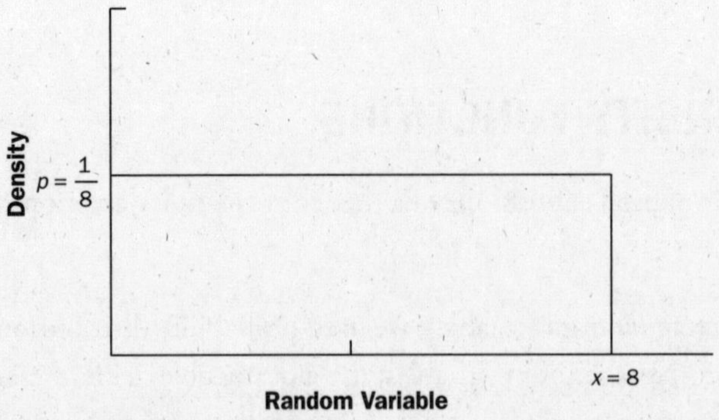

Now let's consider a slightly different situation. Say you are interested in the area on the graph in Figure 14.3 between 3 and 7. We could calculate it simply by multiplying the length of the rectangle by its height. That is $(7 - 3)(0.1) = 0.4$. Thus, $P(3 \leq x \leq 7) = 0.4$. This is also the area from 0 to 7 minus the area from 0 to 3. (7)(0.1) – (3)(0.1) = 0.7 – 0.3 = 0.4. This concept is important in calculating the probabilities of continuous random variables.

Figure 14.3
Area between 3 and 7

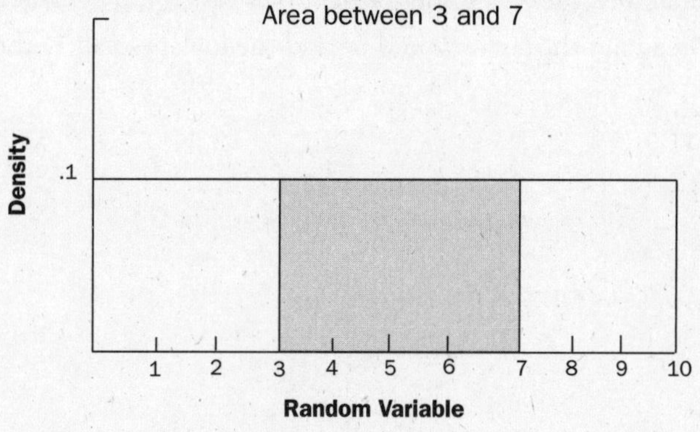

What about $P(X = 3)$? With a continuous random variable, the probability of getting any specific value of X is zero. Returning to our graph, you can see that the width of the rectangle at any specific value of X would be 0, and thus the area is 0. Another way to look at this is that a continuous random variable has an infinite number of possible values it can take on, so you are looking at the $P(X = 3) = \frac{3}{\infty} = 0$.

THE NORMAL DENSITY CURVE

A symmetric, unimodal (bell-shaped) density curve is called the normal curve or normal **distribution**. It is the basis for much of the statistical analysis we perform. Each normal curve is uniquely defined by μ and σ and denoted by $N(\mu, \sigma)$, where μ is the population mean and σ is the population standard deviation. As shown in Figure 14.4, the curve is symmetric about the mean, μ.

Figure 14.4
Normal Curve

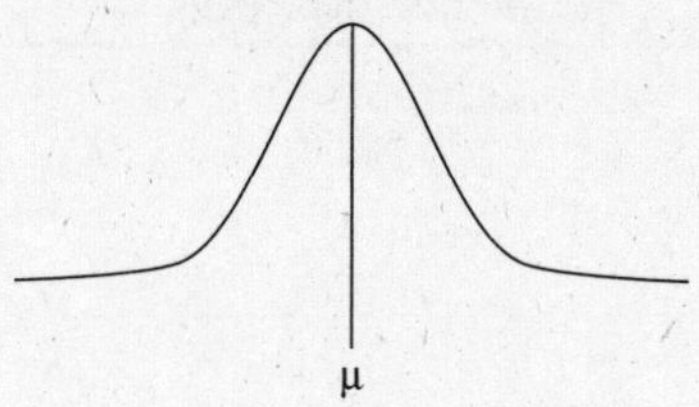

symmetric: mean = median
median = equal areas
mean = balance point

Understanding the Normal Curve

The two normal curves depicted in Figure 14.5 are $N(0,1)$, and $N(0,2)$. Although both curves have a mean of 0, the curve with the standard deviation of 2 is more spread out and has a smaller maximum y value. Remember, the area under both curves is still 1. The larger the standard deviation, the more spread out the curve would be and the lower (smaller) the maximum value of the height.

Figure 14.5
Normal Curves for $N(0, 1)$ and $N(0, 2)$

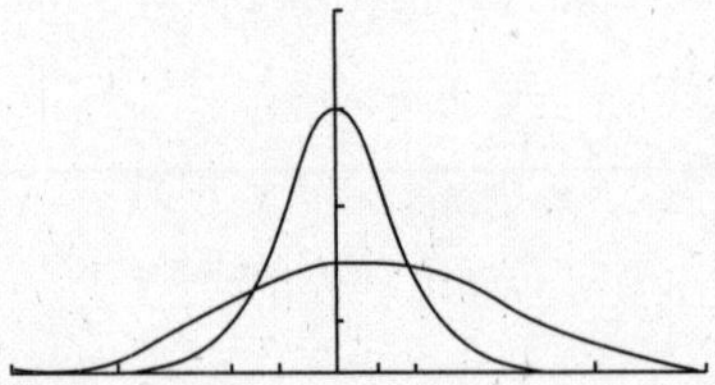

The Empirical Rule (68%/95%/99.7%) states that in a normal distribution approximately 68% of the data values fall within one standard deviation of the mean, 95% of the data values fall within two standard deviations of the mean, and approximately 99.7% of the data values fall within three standard deviations of the mean. This is depicted in Figure 14.6.

Figure 14.6
Empirical Rule (68–95–99.7)

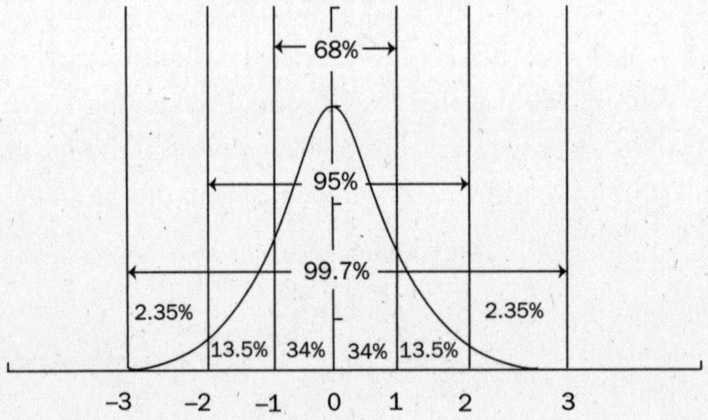

PROPERTIES OF THE NORMAL DISTRIBUTION

- It is symmetric about its mean, μ
- The highest point (maximum value) of the distribution is at $x = \mu$.
- The area under the curve is 1.
- The area on either side of $x = \mu$ is 0.5.
- The Empirical Rule: $\mu \pm 1\sigma = .68$ of the area under the curve
  - $\mu \pm 2\sigma = .95$ of the area under the curve
  - $\mu \pm 3\sigma = .997$ of the area under the curve

Standardizing a Normal Random Variable

It is often necessary to compare data from distributions with different means and standard deviations. This can be accomplished by putting numerical values on the same scale by using the standard deviation. Normalized data points are referred to as *z*-scores. The *z*-score measures the distance an observation is from the mean in standard deviations. This is also called standardizing the data.

$Z = \dfrac{X - \mu}{\sigma}$. If $z = 1.6$, then X is 1.6 standard deviations above the mean.

If $z = -1.8$, then X is 1.8 standard deviations below the mean.

STANDARDIZE DATA

When we standardize data, we transform the data (a random variable with a normal distribution) with a mean μ and a standard deviation σ to a standard normal random variable Z with a mean of 0 and a standard deviation of 1.

Recall that the normal curve is a density curve, and the area under the curve is 1. The normal curve is defined by the function $y = \dfrac{1}{\sigma\sqrt{2\pi}} e^{\frac{-(x-\mu)^2}{2\sigma^2}}$. We can calculate the area under the curve using calculus techniques, or by using the *z*-scores. The Standard Normal Probabilities Table on page 433 provides the area under the curve to the left of the *z*-score. To calculate the area to the right, simply subtract the area to the left from 1.

Figure 14.7
Calculating the Area Under the Curve

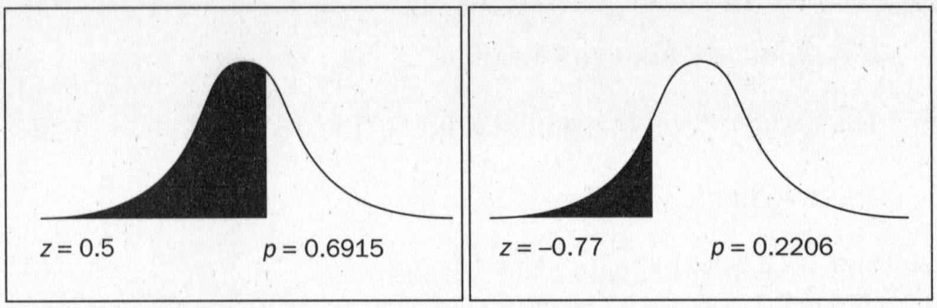

In Figure 14.7, the diagram on the left, $z = 0.5$, indicates that the value is $(0.5)\sigma$ above the mean, and the area to the left or less than this value is 0.6915. Thus, the area to the right or greater than this value is $(1 - p) = (1 - 0.6915) = 0.3085$. These areas also represent the probability of getting a value either less than or greater than the given value. The probability of getting a value less than $z = (0.5)\sigma$ is 0.6915, and the probability of getting a value larger than $z = (0.5)\sigma$ is 0.3085. Similarly, the diagram on the right shows $z = -0.77$. This value is below the mean with the area or probability less than this value is to 0.2206. The area or probability of a value greater than this would be $(1 - 0.2206)$, or 0.7794.

THE NORMAL DISTRIBUTION AS A MODEL FOR MEASUREMENT

AP EXPERT TIP

Use your calculator to find percentages from a normal distribution. You should be familiar with all the statistical functions of your calculator (see chapter 19 for more information on using the calculator).

The area below the normal curve represents a proportion or probability. Thus, we can use the normal curve to model many real-life situations. The first rule when working with normal models is to *draw a diagram*. A picture can help you keep each element straight when you're working. On the horizontal axis, show the raw scores (the actual values) and the corresponding *z*-scores.

Following is a portion of the Standard Normal Probabilities Table for the normal distribution. The table provides the area under the curve to the left of a given *z*-score. Suppose you had a *z*-score of 1.93. Follow the left column of the table down until you find the *z*-score of 1.9. Now move across that row until you locate the cell that intersects with the column 0.03. This cell gives the area 0.9732, which is the area to the left of the *z*-score 1.93. Recall that the area under the curve is equal to 1, so we simply subtract 0.9732 from 1 in order to get the area to the right of this value. Note that the first column gives you the units and tenths place values of the *z*-score, and the remaining columns provide the hundredths place. More precise values for *z*-scores can be obtained using a graphing calculator or statistical software package.

Standard Normal Probabilities Table

z	0.00	0.01	0.02	0.03	0.04	0.05	0.06	0.07	0.08	0.09
0.0	0.5000	0.5040	0.5080	0.5120	0.5160	0.5199	0.5239	0.5279	0.5319	0.5359
0.1	0.5398	0.5438	0.5478	0.5517	0.5557	0.5596	0.5636	0.5675	0.5714	0.5753
1.7	0.9554	0.9564	0.9573	0.9582	0.9591	0.9599	0.9608	0.9616	0.9625	0.9633
1.8	0.9641	0.9649	0.9656	0.9664	0.9671	0.9678	0.9686	0.9693	0.9699	0.9706
1.9	0.9713	0.9719	0.9726	0.9732	0.9738	0.9744	0.9750	0.9756	0.9761	0.9767
2.0	0.9772	0.9778	0.9783	0.9788	0.9793	0.9798	0.9803	0.9808	0.9812	0.9817
2.1	0.9821	0.9826	0.9830	0.9834	0.9838	0.9842	0.9846	0.9850	0.9854	0.9857
2.2	0.9861	0.9864	0.9868	0.9871	0.9875	0.9878	0.9881	0.9884	0.9887	0.9890

Let's say we know that the scores on an exam were roughly normally distributed, with a mean of 82.3 and a standard deviation of 4.7. If we know that Frank scored in the 80th percentile, what grade did he receive?

The next table shows another portion of the Standard Normal Properties Table. Since we know that Frank scored better than 80% of the other students, we know the area to the left of his score would be 0.80. Locate that area in the table and then find the *z*-score that would give you that area. Note that 0.80 doesn't appear exactly in the table. It is between the values 0.7995 and 0.8023. These values are on the row of the *z*-score 0.8, and the value falls between the columns 0.04 and 0.05. Thus we know that the *z*-score for 0.8 is between 0.84 and 0.85. Thus, Frank's score is between 0.84 and 0.85 standard deviations above the mean of 82.3. Since 0.8 is closer to 0.7995 than it is to 0.8023, we will use 0.84 in our calculations. Using the formula for converting values to *z*-scores we have: $z = \frac{x - \mu}{\sigma}$ and $0.84 = \frac{x - 82.3}{4.7}$ giving $x = 86.25$. Frank scored approximately 86 on the exam.

z	0.00	0.01	0.02	0.03	0.04	0.05	0.06	0.07	0.08	0.09
0.7	0.7580	0.7611	0.7642	0.7673	0.7704	0.7734	0.7764	0.7794	0.7823	0.7852
0.8	0.7881	0.7910	0.7939	0.7967	0.7995	0.8023	0.8051	0.8078	0.8106	0.8133
0.9	0.8159	0.8186	0.8212	0.8238	0.8264	0.8289	0.8315	0.8340	0.8365	0.8389

z-SCORE

The *z*-score tells you how many standard deviations above or below the mean the value is.

Example:

The college you are interested in gives the distribution of the SAT math scores for their students as $N(580, 100)$. How high must a student score to be in the upper 10% of all students?

Solution: Picture the normal curve. Recall that the table of the normal distribution gives the area (probability) of a value equal to or less than the one given. Specifically, it gives the area to the left of the value. Since you want to know what the top 10% of scores would look like, the area to the left would be 0.90. That is a *z*-score of $z = 1.2816$.

$$Z = \frac{x - \mu}{\sigma} \text{ So, } \frac{x - 580}{100} = 1.2816, \text{ giving } x = 708.$$

As depicted in Figure 14.8, a student would have to score 708 or better to be in the top 10% of the student body.

Figure 14.8
Calculating the *z*-Score

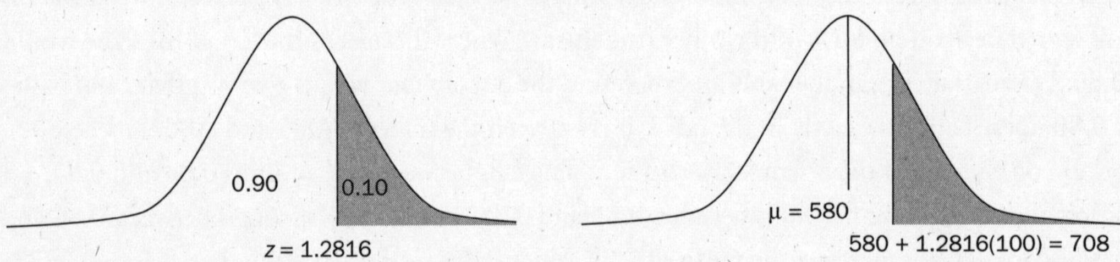

ASSESSING NORMALITY

The properties of the normal distribution allow us to do many tests and procedures in statistics. So often it is important to ascertain that the distribution of the data we are using is approximately normal. This can be done simply by examining one of the common visual displays for univariate data, such as a histogram, boxplot, or stem and leaf plot. You are looking for the display to show a unimodal, symmetric, and bell-shaped curve. Figure 14.9 shows displays of the ages of the presidents of the United States at the time of their inaugurations. In each case, you can easily see that the data has a distribution that is approximately normal.

Figure 14.9
Normal Distributions of Data
(histogram, boxplot, and stem and leaf plot)

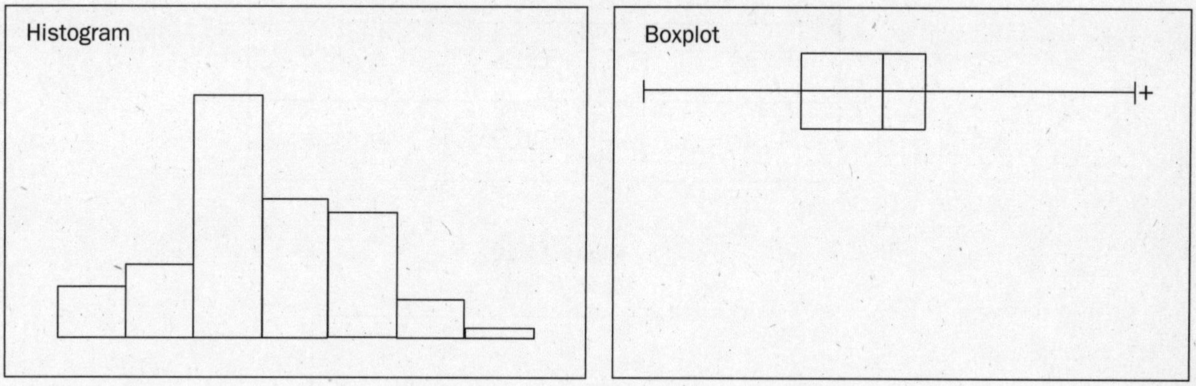

Stem and Leaf

```
4 | 2 3
4 | 6 7 8 9 9
5 | 0 1 1 1 1 1 2 2 4 4 4 4 4
5 | 5 5 5 5 6 6 6 7 7 7 7 8
6 | 0 1 1 1 2 4 4
6 | 5 8 9
```

Another method for assessing normality is the **normal probability plot**. This is done almost exclusively with the aid of technology—that is, with either a graphing calculator or a statistical software package. The normal probability plot is constructed by first ordering all of the observed data from smallest to largest, and then dividing the interval [0, 1] into n intervals where n is the number of observations. Determine the midpoint of each of these subintervals. In our example with the ages of the U.S. presidents, $n = 42$. So our initial interval starting at 0 would be at $0, \frac{1}{42}, \frac{2}{42}, \frac{3}{42}, \ldots, \frac{41}{42}, 1$. Then the midpoint of each of those subintervals would be $\frac{1}{84}, \frac{3}{84}, \frac{5}{84}, \ldots$ You can easily verify that these midpoints are $\frac{1}{2n}, \frac{3}{2n}, \frac{5}{2n}, \ldots, \frac{(2n-1)}{2n}$. Now determine the z-score that has the area $\frac{1}{2n}$, or in our case $\frac{1}{84}$, to the left of it.

The following table provides the first three z-scores. Notice the column headings in the table. The actual ages are the x-values and corresponding the z-scores are the y-values. The coordinates for all of these points are plotted as shown in Figure 14.10.

x	Midpoint	y
42	$\frac{1}{84} = 0.0119$	−2.2602
43	$\frac{3}{84} = 0.0357$	−1.8027
46	$\frac{5}{84} = 0.0595$	−1.5588

Figure 14.10
Plot of *x* and *y*

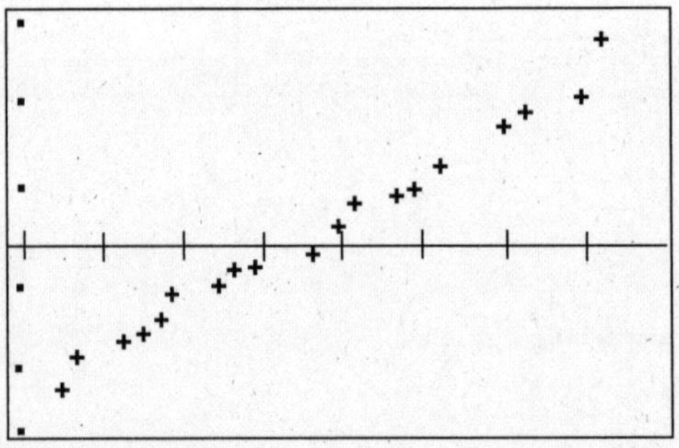

If the data distribution is close to a normal distribution, the plotted points will lie close to a straight line. Outliers in the data will appear as points that are far away from the overall pattern of the plot. Non-normal data will show a nonlinear trend as shown in Figure 14.11.

Figure 14.11
Non-Normal Data for Plot

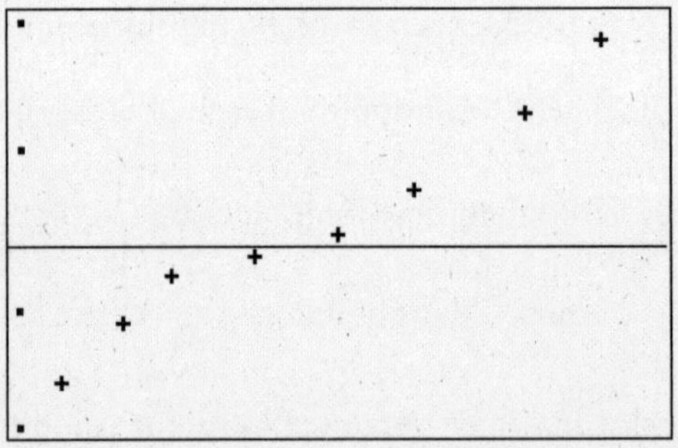

Notice the curved pattern to these points. This is actually a normal probability plot for the probability of first digits as defined by Benford's Law.

REVIEW QUESTIONS

1. The grading at Central High gives a B for grades between 86 and 93. On the English final for seniors, what proportion of the class would get a B if the grades were normally distributed with a mean grade of 86.34 and standard deviation of 14.23?

   (A) 0.07
   (B) 0.4905
   (C) 0.6801
   (D) 0.1896
   (E) 0.0280

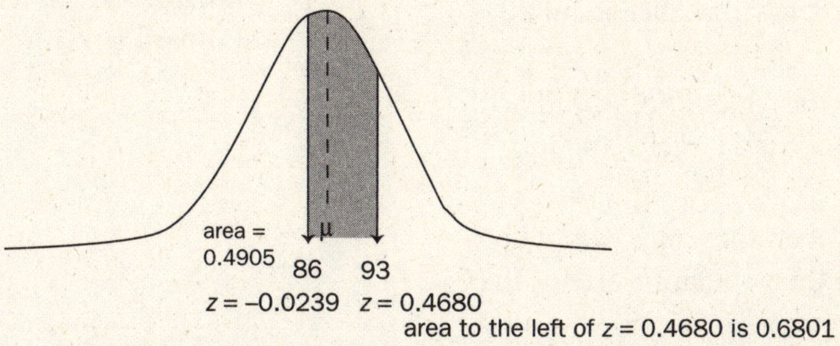

2. The mean GPA for Central High is 2.9, with the standard deviation of 0.5. Assuming the GPAs are normally distributed, what GPA score will place a student in the top 5% of the class?

   (A) 3.72
   (B) 3.43
   (C) 2.08
   (D) 2.90
   (E) 3.38

3. On Sarah's last two biology exams, she scored an 87. The class mean on the first exam was 75, with a standard deviation of 8.9. The class average on the second exam was 73, with a standard deviation of 9.7. Assuming the scores on the exam were approximately normally distributed, on which exam did Sarah score better relative the rest of her class?

   (A) She scored better on the first exam.
   (B) She scored better on the second exam.
   (C) She scored equally on both exams.
   (D) It is impossible to determine because the class sizes are unknown.
   (E) It is impossible to determine because the correlation between the two sets of exam scores is not provided.

4. A researcher notes that two populations of lab mice—one consisting of mice with white fur, and one of mice with gray fur—have the same mean weight, and both have approximately normal distributions. However, the population of white mice has a larger standard deviation than the population of gray mice. If the weights for both of these populations were plotted, how would the curves compare to each other?

   (A) The curves would be identical.
   (B) The curve for the gray mice would be taller because it has a smaller standard deviation.
   (C) The curve for the white mice would be taller because it has a larger standard deviation.
   (D) The curve for the white mice would be taller because the population size of the white mice is larger.
   (E) The curve for the gray mice would be taller because its variance is larger.

5. Which of the following statements is NOT true for normally distributed data?

   (A) The mean and median are equal.
   (B) The area under the curve is dependent upon the mean and standard deviation.
   (C) Almost all of the data lie within three standard deviations of the mean.
   (D) Approximately 68% of all of the data lies within one standard deviation of the median.
   (E) When the data are normalized, the distribution has a mean $\mu = 0$ and a standard deviation $\sigma = 1$

FREE-RESPONSE QUESTION

Webb is a baseball fanatic. He keeps his own statistics on the major league teams and individual players. For the 350 regular starters, Webb has found their mean batting average is 0.229, with a standard deviation of 0.024. His sister is appalled that baseball players get paid the salaries they do and get a hit in fewer than 25% of their attempts at bat. To further her argument, she asks for the following information:

(a) What proportion of players hit more than 25% of the times they are at bat?

(b) Since the players with the top ten batting averages get cash bonuses, what is the lowest batting average that will receive a bonus?

ANSWERS AND EXPLANATIONS

1. D

Here we are looking for the area under the normal curve between the values of 86 and 93. First, we standardize the scores: $z = \frac{x - \mu}{\sigma}$. Then, using the standard normal table, we can find the area under the curve to the left of these values. To find the area in between these two values, we can subtract the area to the left of the lower value from the area to the left of the upper value: $z_{93} = \frac{93 - 86.34}{14.23} = 0.4680$ and $z_{86} = \frac{86 - 86.34}{14.23} = -0.0239$. The corresponding areas are 0.6801 and 0.4905, respectively. The difference between these two areas is 0.1896.

2. A

If a student is in the top 5%, then 95% of the students have GPAs below him or her. The corresponding z-score for having an area of 0.95 to the left is 1.644. Using this, we can substitute into the formula

$z = \frac{x - \mu}{\sigma}$, and solve for x: $1.644 = \frac{x - 2.9}{0.5}$, giving

$x = 3.72$.

3. B

Determine the standardized score for each test grade using the formula $z = \frac{x - \mu}{\sigma}$. For the first test, Sarah would have a standardized test score of $\frac{87 - 75}{8.9} = 1.3483$, and for the second test, $\frac{87 - 73}{9.7} = 1.4433$. Thus, her second test score was 0.095 standard deviations greater than the mean than the first test score.

4. B

A smaller standard deviation means the distribution is less spread out. The data is clustered more closely around the mean, as in Figure 14.5. If the distribution were plotted as a histogram, there would be more frequency to the values around the mean.

5. B

The normal curve is a probability density curve, and the area under a probability density curve is always 1.

FREE-RESPONSE ANSWER

(a) Hitting 25% of the time they are at bat would give players a batting average of 0.25 or higher: $z = \frac{x - \mu}{\sigma} = \frac{0.25 - 0.229}{0.024} = 0.875$. Since we were asked for those whose batting average is at least 25% of their times at bat, we want the area to the right of this z-score.

So, 1 – (area to the left of $z = 0.87$) = 1 – 0.8078 = 0.1922, and 1 – (area to the left of z) = 0.88 = 1 – 0.8106 = 0.1894. So about 19% of the players get hits on average over 25% of the times they are at bat.

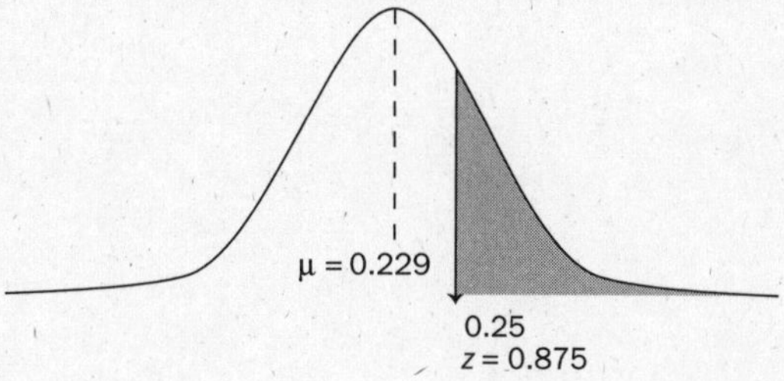

(b) The top ten players would represent $\frac{10}{350} = 0.02857$, or approximately 2.86% of the players. Using the Standard Normal Probabilities Table, the z-score corresponding to the top 2.86% of the players is the z-score that has 1 – 0.2857 = 0.97143 area to the left of it. That z-score is approximately 1.9. Knowing this, we can calculate the actual batting average that is 1.9 standard deviations above the mean. $z = \frac{x - \mu}{\sigma}$ and $1.9 = \frac{x - 0.229}{0.024}$, giving $x = 0.2746$. Specifically, this shows that the top ten players have batting averages of 0.2746 or higher.

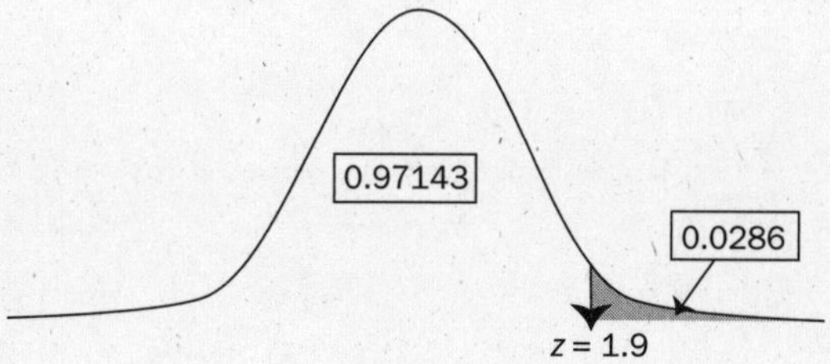

CHAPTER 15: SAMPLING DISTRIBUTIONS

IF YOU LEARN ONLY TWO THINGS IN THIS CHAPTER . . .

1. **Central Limit Theorem for the Mean:** If an SRS of size n is drawn from a population with mean μ and standard σ, then when n is large, the sampling distribution of the sample mean $\bar{x}$ has a normal distribution or approximately normal distribution defined $N\left(\mu, \frac{\sigma}{\sqrt{n}}\right)$.
2. **Central Limit Theorem for a Proportion:** If an SRS of size n is drawn from a large population with a proportion of p, then the sampling distribution of the sample proportion $\hat{p}$ is approximately normal and defined $N\left(p, \sqrt{\frac{p(1-p)}{n}}\right)$.

SAMPLING DISTRIBUTIONS

Since it is generally not feasible to examine the entire population, researchers, scientists, and investigators rely on samples to answer their questions and test their theories. Being able to infer information from a sample back to the entire population with some level of confidence is critical. Statistical inference addresses how often a method would give us a correct result if we used it many times. Not every sample, no matter how carefully designed, will truly represent the population. Being able to quantify this is essential. To do this, we use probability.

As depicted in Figure 15.1, the concept is simple. From a population, we draw a sample. From the sample, we collect statistics. From the statistics, we infer or estimate population parameters. From that, we can infer information about the population.

We examine samples in order to gain reasonable conclusions about a population. It is imperative to be able to understand what the sample results tell you, and, equally as important, what they do *not* tell you.

Figure 15.1
Population Sample

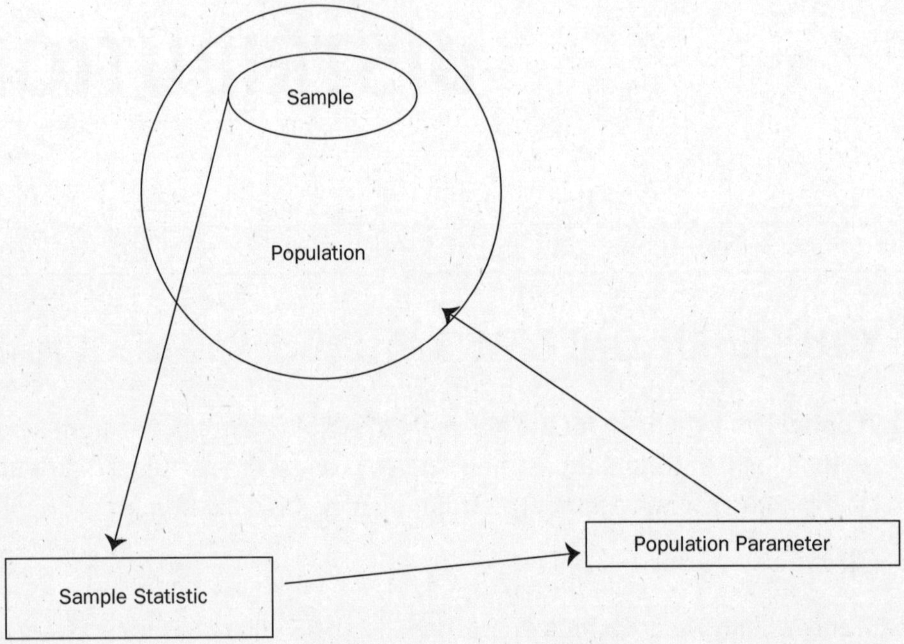

Based on data from the 2000 Census, the mean income for households in the United States was \$57,045. This was calculated based on a sample of only 50,000 households. Political polls give a picture of the support among the population for each candidate. Polls generally only include a few thousand people and yet their results are inferred to the population as a whole.

It is important to keep the symbols and terminology regarding samples and populations straight. Study the table below.

	Population Parameter	Sample Statistic
Mean	μ	$\bar{x}$
Standard Deviation	σ	s
Proportion	p	$\hat{p}$

To understand the relationship between samples and the population, we must clearly understand sampling distributions.

SAMPLING DISTRIBUTION

The distribution of values of a statistic taken from all possible samples of a specific size is the sampling distribution.

Consider the population consisting of the elements $P = \{1,2,3\}$:

$$\mu_p = 2, \sigma_p = 0.8165, \sigma_p^2 = 0.6667.$$

Now consider all possible samples of size 2 with replacement. There would be $3^2 = 9$ possible samples.

Sample	Sample Mean x	s^2	s
1,1	1	0	0
1,2	1.5	0.5	0.707107
1,3	2	2	1.4142
2,1	1.5	0.5	0.707107
2,2	2	0	0
2,3	2.5	0.5	0.707107
3,1	2	2	1.4142
3,2	2.5	0.5	0.707107
3,3	3	0	0
Mean of the column	2.0	0.6667	0.628539

If you had drawn any one of these samples, you would have a basis for estimating the population parameter. Here we are simply talking about the mean of a set of numbers. If the sample were taken of students, we could look at any number of measurable characteristics, such as GPA, family income, height, weight, cholesterol level, blood pressure, or number of siblings.

The table above gives the sampling distribution for all possible samples of size 2 from the population. Note that μ_p = mean of the sample distribution, σ_p^2 = mean of the distribution of sample variances, but $\sigma_p \neq$ mean of the distribution of standard deviation.

A statistic is said to be unbiased if the mean of the sampling distribution is equal to the true value of the parameter being estimated. Thus, $\bar{x}$ and s_x^2 are unbiased estimates of μ and σ_p^2, but s_x is a biased estimator of σ_p.

Figure 15.2 shows is a histogram of the sample means from our example above. This is the sampling distribution of the sample means. Remember, it is over all possible samples of a given size. From this example, we can see that the sampling distribution is symmetric and is approximately normal. This is a key element of statistical inference.

Figure 15.2

Histogram of the Sample Means

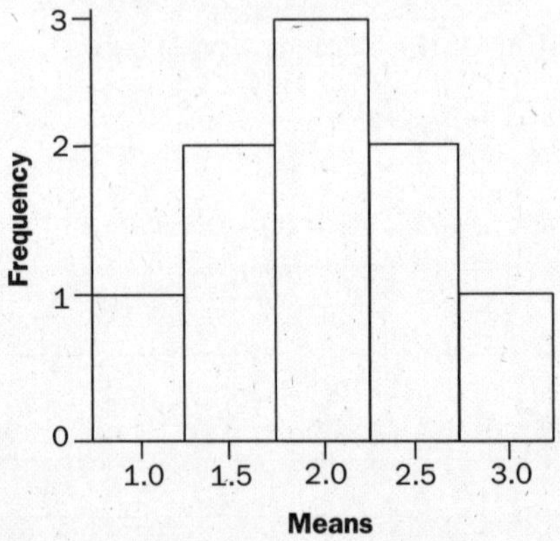

The variability of a statistic is defined by the spread of its sampling distribution. This spread is determined by the sampling design and by the size of the sample. The larger the sample, the smaller the spread will be.

THE DISTRIBUTION OF THE SAMPLE PROPORTIONS

What proportion (or %) of the population watches the local nightly news on TV? Using a simple random sample of 1,500 people, we find that 985 watch the local news, so

$$\hat{p} = \frac{985}{1{,}500} = 0.6566.$$

From this, we want to infer information about the population p. We can do this based on the following facts.

THE SAMPLING DISTRIBUTION SAMPLE PROPORTIONS

The sampling distribution of $\hat{p}$ is approximately normal.

The mean of the sampling distribution is p the population parameter.

The standard deviation of the sampling distribution is $\sqrt{\dfrac{p(1-p)}{n}}$.

Note: The standard deviation gets smaller as the size of the sample gets larger.

How good is $\hat{p}$ as an estimate of the parameter p? To determine this, we consider what would happen if we took a large number of samples. What would the sampling distribution of $\hat{p}$ look like?

The normal distribution provides an approximation to the sampling distribution of $\hat{p}$ when $np > 10$ and $n(1-p) > 10$. If either of these conditions is not met, the normal distribution is not an appropriate model for the probability distribution since the distribution of the data exhibits significant skewness. A population is considered large relative to sample size n if the population is greater than $10n$ ($N > 10n$). This allows the assumption of the constant probability of success required by the binomial criterion.

These conditions must be verified. If these conditions are not met, the normal distribution is not an appropriate choice for the probability distribution $\hat{p}$.

Example:

A polling organization asks an SRS of 1,500 employees if they took things from their place of work for personal use. It is believed that on a national level, 35% of employees engage in the white-collar crime of taking things from their place of employment. What is the probability that a random sample of 1,500 will give a result within 2 percentage points of the true population proportion of 35%?

Solution: $n = 1{,}500$ and $p = 0.35$. We are interested in $P(0.33 \le \hat{p} \le 0.37)$.

Can we use the normal distribution to approximate the sampling distribution of $\hat{p}$?

Thus, $np = 1{,}500(0.35) = 525$ and $n(1-p) = 1{,}500(1-0.35) = 975$. Both are much larger than 10.

$10n \le N = 15{,}000$. It is reasonable to assume that there are more than 15,000 employees in the United States.

Since the conditions are met, the normal approximation is an appropriate model.

The sampling distribution of $\hat{p}$ has a mean $\mu_{\hat{p}} = 0.35$. Calculate its standard deviation:

$$\sigma_{\hat{p}} = \sqrt{\frac{p(1-p)}{n}} = \sqrt{\frac{0.35(1-0.35)}{1{,}500}} = 0.0123.$$

Standardize $\hat{p}$ and use the z-scores to calculate the desired area.

$$z = \frac{\hat{p} - \mu_{\hat{p}}}{\sigma_{\hat{p}}}\ z = \frac{0.33 - 0.35}{0.0123} = -1.63 \text{ and } z = \frac{0.37 - 0.35}{0.0123} = 1.63.$$

Figure 15.3
Mean for the Sampling Distribution of p

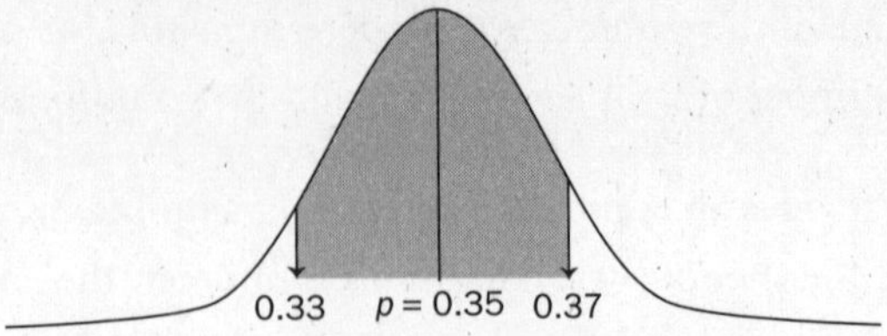

$$P(0.33 \leq \hat{p} \leq 0.37) = P(-1.63 \leq z \leq 1.63) = 0.9484 - 0.0516 = 0.8968$$

Thus, approximately 90% of all samples will give a result within 2 percentage points of true mean (the mean of the population). Figure 15.3 shows the sampling distribution.

Sampling Distributions of the Difference of Two Proportions from Independent Samples

Often the point of interest for researchers or investigators is the comparison between two population proportions. For example, do males and females feel differently about social issues? We have all of the techniques we need to do this, we just need to recall a few facts. Remember that we are dealing with random variables. The key facts regarding random variables that we need are: Given X and Y are random variables, then $\mu_{X+Y} = \mu_X + \mu_Y$, and if X and Y are independent random variables, then $\sigma^2_{X+Y} = \sigma^2_X + \sigma^2_Y$ and $\sigma^2_{X-Y} = \sigma^2_X + \sigma^2_Y$. Remember that variances add, while standard deviations do not.

If two independent SRSs of sizes n_1 and n_2 are drawn from large populations whose parameters are p_1 and p_2, then the sampling distribution of the difference of the sample proportions is approximately normal, with the mean difference equal to $(p_1 - p_2)$ and the standard deviation equal to $\sqrt{\frac{p_1(1-p_1)}{n_1} + \frac{p_2(1-p_2)}{n_2}}$.

Key points to remember:

- $(\hat{p}_1 - \hat{p}_2)$ is an unbiased estimator $(p_1 - p_2)$.
- Both samples must satisfy the conditions $np > 10$ and $n(1 - p) > 10$.
- Both populations should be large relative to their respective samples satisfying the conditions $N > 10n$.

THE DISTRIBUTION OF THE SAMPLE MEANS

Sample proportions arise most often with categorical data. *What percent of…? What proportion of…?* The sampling distribution of $\bar{x}$ is the distribution of the value of $\bar{x}$ in all possible samples of the same size from the population.

Two characteristics of averages that should be kept in mind are the following:

1. Averages are less variable than individual observations.
2. Averages are more normally distributed than individual observations.

As we demonstrated by taking all possible samples of size 2 with respect to the population of $\{1,2,3\}$, $\bar{x}$ is an unbiased estimate of μ.

MEAN AND STANDARD DEVIATION

Let $\bar{x}$ equal the mean of an SRS of size n drawn from a large population with mean μ and standard deviation σ.

The mean of the sampling distribution of $\bar{x}$ is μ and its standard deviation is $\frac{\sigma}{\sqrt{n}}$.

Key concepts regarding the sampling distribution of a sample mean are the following:

- Values of $\bar{x}$ are less spread out for large samples.
- The standard deviation decreases at a rate $\sqrt{n}$.
- $\frac{\sigma}{\sqrt{n}}$ should only be used for standard deviation when the population is at least ten times greater than the sample.
- The shape of the distribution of $\bar{x}$ depends on the shape of the distribution of the population. If the distribution of the population is normal, then so is the distribution of the sample mean.

THE DISTRIBUTION OF A SAMPLE MEAN

If an SRS of size n is drawn from a population that has a normal distribution with mean μ and standard σ, then the sample mean $\bar{x}$ has the normal distribution $N\left(\mu, \frac{\sigma}{\sqrt{n}}\right)$.

For the population $P = \{2,4,6\}$, with $\mu = 4$ and $\sigma = 1.63299$, consider the means of all possible samples of size $n = 2$ (with replacement). Calculate the mean and standard deviation of these sample means.

Sample	Mean
2,2	2
2,4	3
2,6	4
4,2	3
4,4	4
4,6	5
6,2	4
6,4	5
6,6	6
	$\frac{\sum x_i}{n} = \frac{36}{9} = 4$

The mean of the means $\mu_{\bar{x}}$ is 4 and the standard deviation is

$$\sigma_{\bar{x}} = \sqrt{\frac{(2-4)^2 + (3-4)^2 + (4-4)^2 + (3-4)^2 + (4-4)^2 + (5-4)^2 + (4-4)^2 + (5-4)^2 + (6-4)^2}{9}}$$

$$= \sqrt{\frac{12}{9}}$$

$$= 1.1547.$$

Notice that these are exactly $\mu_{\bar{x}} = \mu = 4$ and $\sigma_{\bar{x}} = \frac{\sigma}{\sqrt{n}} = \frac{1.63299}{\sqrt{2}} = 1.1547$.

Sampling Distributions of the Difference of Sample Means

Comparing the means of some measured value from two samples follows the same general rules as we saw for the sampling distribution of a sample mean. Here, however, we are interested in comparing two groups on a particular attribute. We may be interested in the average wages of males and females in similar jobs; we may be interested in whether males or females have higher cholesterol levels. Remember that these are random variables, and the rules for random variables apply. They are worth repeating here. Given X and Y are random variables, then $\mu_{X+Y} = \mu_X + \mu_Y$ and $\mu_{X-Y} = \mu_X - \mu_Y$, and if X and Y are independent random variables, then $\sigma^2_{X+Y} = \sigma^2_X + \sigma^2_Y$ and $\sigma^2_{X-Y} = \sigma^2_X + \sigma^2_Y$. Remember that variances add, while standard deviations do not.

Given the sample means $\bar{x}_1$ and $\bar{x}_2$ for two independent SRSs , then $(\bar{x}_1 - \bar{x}_2)$ is an unbiased estimator of $(\mu_1 - \mu_2)$, and the standard deviation of the sampling distribution of $(\bar{x}_1 - \bar{x}_2)$ is $\sqrt{\frac{\sigma_1^2}{n_1} + \frac{\sigma_2^2}{n_2}}$.

THE CENTRAL LIMIT THEOREM

Not all populations have a normal distribution. In cases where the population is not normally distributed, as the sample size increases, the sampling distribution of the sample mean gets closer and closer to a normal distribution. This is true no matter what the distribution is for the population.

CENTRAL LIMIT THEOREM FOR THE MEAN

If an SRS of size n is drawn from a population with mean μ and standard σ, then when n is large, the sampling distribution of the sample mean x has a normal distribution or approximately normal distribution defined by $N\left(\mu, \frac{\sigma}{\sqrt{n}}\right)$.

There are several important implications that follow from the Central Limit Theorem. First, since the mean of the sampling distribution of $\bar{x}$ is μ, then $\bar{x}$ is an unbiased estimator of μ. Secondly, if the population distribution is *not normal*, then the sampling distribution approaches a normal distribution as n increases. Therefore, if it cannot be determined whether the population is normal, n should be large. In general, $n \geq 30$.

NOTE OF CAUTION: It is easy to confuse the distribution of a single sample with the sampling distribution for the sample mean. Remember: When we talk about sampling distribution, we are referring to the sample statistic for all possible samples of a given size. Think about our example of the population of {1,2,3} and all possible samples of size 2.

Example:

A tire manufacturer advertised that a new brand of tires has a mean life of 40,000 miles, with a standard deviation of 2,000 miles. A research team examined a random sample of 100 of these tires and determined that the tires in the sample had a mean life of 39,000. If the mean life is indeed 40,000 miles, how likely is it that a random sample of 100 would have a mean life of 39,000?

Solution: By the Central Limit Theorem, the set of all samples of size 100 has a mean $\bar{x} = \mu$, so $\bar{x} = 40{,}000$, and a standard $s = \frac{\sigma}{\sqrt{n}}$, so $s = \frac{2{,}000}{\sqrt{100}} = 200$.
$z = \frac{x - \mu}{\sigma} = \frac{39{,}000 - 40{,}000}{200} = -5$. Since this is off the charts for standard normal distribution, the area under the curve or the probability would be close to zero. Therefore, it is highly unlikely to get a sample of 100 tires with a mean of 39,000 if the manufacturer's claim is true.

CENTRAL LIMIT THEOREM FOR A PROPORTION

If an SRS of size n is drawn from a large population with a proportion of p, the sampling distribution of the sample proportion $\hat{p}$ is approximately normal and defined by $N\left(p, \sqrt{\frac{p(1-p)}{n}}\right)$.

Remember:

$\hat{p}$ is an unbiased estimator of p:

- The normal distribution provides an accurate approximation to the sample distribution provided $np > 10$ and $n(1 - p) > 10$.
- A population is considered large relative to a sample if $N > 10n$.

SIMULATION OF SAMPLING DISTRIBUTIONS

Simulation can provide a viable option when cost or logistics make it unfeasible to sample. One of the most common types of simulation is the use of random number tables or technology that generates random numbers. The methodology is identical to what we did previously in chapter 13. Let's look at an example to reinforce the basics.

Remember our example about people watching the nightly news? Well, Mr. Tuff, the owner of a local business, is interested in buying advertising to be aired during the news program. He would like to know with some certainty that his money would be well spent.

Let's say the TV station reports that 66% of the population watches the nightly news. Mr. Tuff is skeptical and believes the percentage to be lower—more likely 40%—based on a survey he commissioned of 10 randomly selected people. How likely would it be for Mr. Tuff's survey to come up with 40% if the true population proportion was 66%?

Solution: Using simulation, we can approximate the proportion of times we would obtain a sample of 10 people in which 4 or fewer watch the nightly news by taking repeated samples with a true proportion of 66%. Use two-digit numbers from 00 to 65 to represent those people in the population who watch the nightly news. The two-digit numbers from 66 to 99 will represent those who do not watch the nightly news. Using a random number table, we will consider 20 consecutive digits as representing 10 two-digit numbers. This represents our sample of 10 people. We will look at each two-digit number and count the number of values that are between 00 and 65, inclusive. In order to answer the question, we need to repeat this process a number of times. For our purposes, we will repeat the process 10 times, keeping track of the number of samples out of 25 that had 4 or fewer (40% or under) numbers between 00 and 65. To facilitate this example, the random number table was broken down into groups of 10 two-digit numbers. The underlined number at the end of each row is the number of times numbers between 00 and 65 appeared.

85 65 26 24 93 43 89 13 49 75 5
39 17 09 58 23 45 59 70 35 55 9
20 06 15 27 35 84 02 37 01 44 9
71 51 99 60 13 50 77 61 83 28 6
11 65 88 35 69 00 41 59 14 83 7
14 37 93 41 07 02 01 56 63 53 9
40 42 38 41 89 06 91 84 94 01 6
76 73 53 44 28 00 97 69 50 85 5
04 28 73 92 09 63 02 29 34 65 8
84 23 88 54 77 66 89 36 50 17 5
49 47 33 56 34 21 43 61 08 23 10
73 64 53 38 04 85 99 00 64 50 7

Out of 10 samples of size 10, there were none with 4 or fewer numbers representing those people who watch the evening news. Based on this simulation, it seems somewhat doubtful that Mr. Tuff's feelings that only 40% of the people watch the nightly news is correct. Obviously, we would want to do a lot more repetitions of this process to get a more accurate estimate for the probability.

OTHER SAMPLING DISTRIBUTIONS

There are several key sampling distributions that readily model the real world. We will review two specific and very important ones here. They are the ***t*-distribution** and the **chi-squared distribution**.

t-DISTRIBUTIONS

The Central Limit Theorem says that if you take random samples of size n from a population of mean μ and standard σ, then as n gets large, $\bar{x}$ approaches the normal distribution with mean μ and standard $\frac{\sigma}{\sqrt{n}}$. This depends on large sample size and knowing σ. But sample sizes are often small, and σ is rarely if ever known.

When σ was known, we relied on the normal distribution to answer many of our questions. If we do not know σ, we have to use s and the standard deviation of $\bar{x}$ becomes $\frac{s}{\sqrt{n}}$. This is referred to as the **standard error**. The resulting statistic does not have a normal distribution. It has the t-distribution, and as n increases, the t-distribution approaches the normal distribution. The t-distribution is actually a family of distributions. The actual distribution is based on the number of degrees of freedom. The **degrees of freedom** (*df*) is the number of independent observations $\bar{x}$. Further, the t-distribution is a probability density function. The t-distributions are designated $t(k)$, which represents the t-distribution with k degrees of freedom. All t-distributions are mound-shaped and symmetric, and all t values are positive. The spread of a t-distribution is slightly larger than the normal distribution. This results in a greater probability in the tails of the t-distribution and a lesser probability around the mean.

The graphs in Figure 15.4 show a comparison of the normal distribution with various t-distributions. As the degrees of freedom increase, t-distributions approach the normal distribution.

Figure 15.4
Normal Distribution (dashed) with Various *t*-Distributions

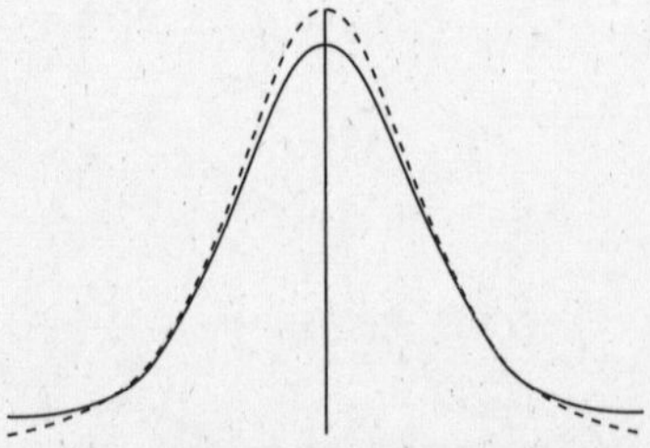

Standard Normal Distribution
t-Distribution with 3 *df*

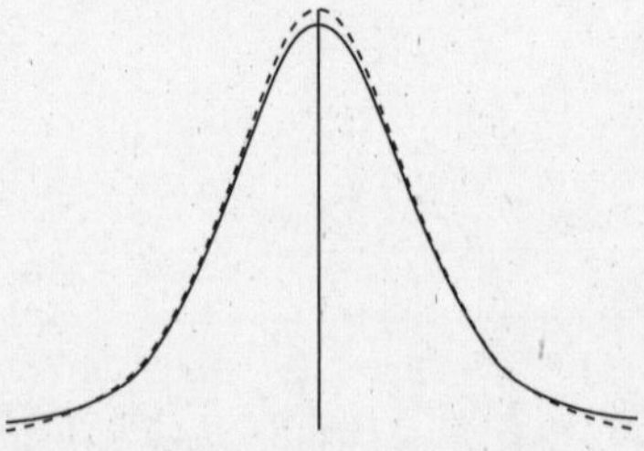

Standard Normal Distribution
t-Distribution with 7 *df*

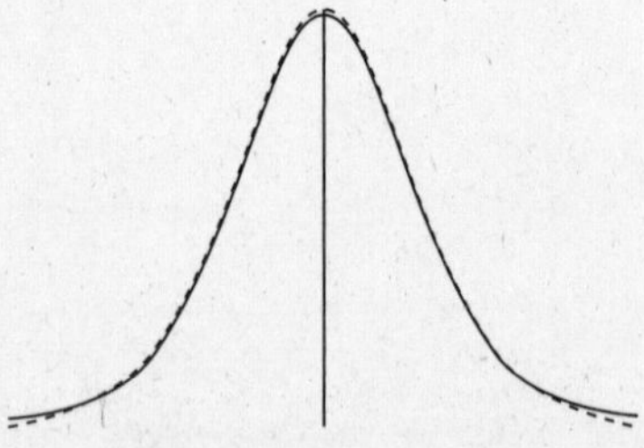

Standard Normal Distribution
t-Distribution with 20 *df*

Chi-squared Distributions

Chi-squared distributions are a family of distributions that take only positive values and are skewed to the right. It has one parameter, which is degrees of freedom. Since it is a probability density curve, the area under the curve equals 1.

The curve begins at 0 on the horizontal axis, increases to a peak, and approaches the horizontal axis asymptotically. The extent of its skewness is determined by its degrees of freedom. As the number of degrees of freedom increases, the curve becomes more symmetrical and looks more like the normal curve.

The three density curves shown in Figure 15.5 are all chi-square distributions. They are $\chi^2(x,2)$, $\chi^2(x,6)$, and $\chi^2(x,18)$. Notice how as the degree of freedom increases the curve becomes more symmetric and more normal in appearance.

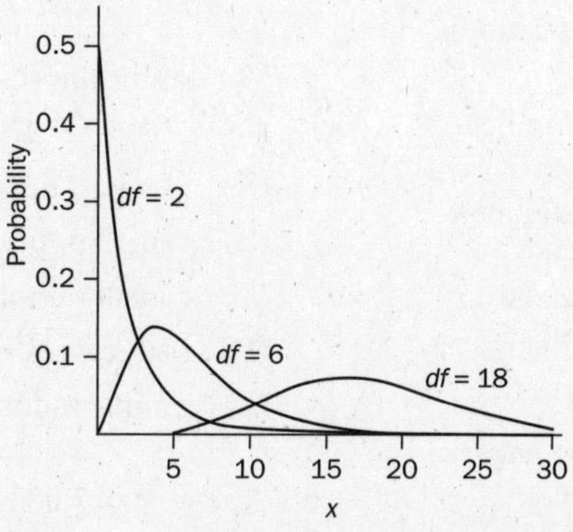

Figure 5.5 Chi-squared Distributions

REVIEW QUESTIONS

1. Which of the following statements about the *t*-distribution is true?

    (A) The *t*-distribution has a mean of 0 and a standard deviation of 1.

    (B) The *t*-distribution has a larger variance than the standard normal distribution.

    (C) The smaller the degrees of freedom, the smaller the variance for the *t*-distribution.

    (D) The *t*-distribution is a skewed distribution.

    (E) The normal distribution is flatter and more spread out than the *t*-distribution.

2. Two samples of corn were taken from a field to test the percent of corn plants infested with worms. The USDA states that approximately 28% of all corn plants are infested. One sample contains 100 ears of corn and the second sample contains 500 ears. Which sample has the larger standard deviation?

    (A) The sample of 500 will have the larger standard deviation.

    (B) It is impossible to determine which sample would have the larger standard deviation without additional information.

    (C) Both samples will have the same standard deviation.

    (D) The sample of 100 ears will have a smaller standard deviation.

    (E) The sample of 100 ears will have the larger standard deviation.

3. Which of the following statements best describes a sampling distribution of a sample mean?

    (A) It is $\bar{x}$.

    (B) It is the distribution of all possible values of a population parameter.

    (C) It is the distribution of values of a statistic taken from all possible samples of a specific size.

    (D) It is an unbiased estimator μ.

    (E) It is the normal distribution with $\bar{x} = 0$ and $s = 1$.

4. The conditions that $np > 10$ and $n(1-p) > 10$ are necessary to guard against

    (A) a skewed distribution.

    (B) a small population size.

    (C) a small sample size.

    (D) a large standard deviation.

    (E) a non-randomly selected sample.

5. A sample of 5,000 female adults was randomly drawn from the United States. It is known that the diastolic blood pressure for adult women in the United States is $N(80, 12)$. What is the mean and standard deviation of the distribution of the sample means?

    (A) $\bar{x} = 0.16$, $s = 0.1697$

    (B) $\bar{x} = 80$, $s = 12$

    (C) $\bar{x} = 80$, $s = 0.1697$

    (D) $\bar{x} = 3.58$, $s = 0.024$

    (E) Cannot be determined from the information given.

FREE-RESPONSE QUESTION

In selling the produce from her garden, Mrs. Clark knows that 35% of the women that stop will buy squash or green beans, but only 23% of men will. The County Agricultural Department that runs the local Farmer's Market, conducted a random sample of county residents to determine what they should offer at the Farmer's Market. They took two independent surveys of 100 men and 150 women. What is the probability that the difference in the percentages of men and women who would buy squash and green beans is more than 10%?

ANSWERS AND EXPLANATIONS

1. B

The *t*-distribution is actually a family of distributions. The *t*-distributions are mound shaped, symmetric, and centered at 0. The actual distribution is based on the number of degrees of freedom. The degrees of freedom (*df*) is the number of independent observations $\bar{x}$. As the number of degrees of freedom increases, the *t*-distribution approaches the normal distribution. The spread of a *t*-distribution is slightly larger than the normal distribution. This results in a greater probability in the tails of the *t*-distribution.

2. E

The standard deviation of the sampling distribution is $\sqrt{\frac{p(1-p)}{n}}$. The standard deviation gets smaller as the size of the sample increases. In this problem, $p = 0.28$ and the standard deviation for each sample would be $\sqrt{\frac{0.28(0.72)}{100}} = 0.045$ and $\sqrt{\frac{0.28(0.72)}{500}} = 0.020$. Therefore, the sample of 100 would have the larger standard deviation. Remember that, all other things being equal, the larger sample will have a smaller standard deviation.

3. C

The distribution of values of a statistic taken from *all possible samples* of a specific size is the sampling distribution.

4. A

The normal distribution provides an approximation to the sampling distribution of $\hat{p}$ when $np > 10$ and $n(1-p) > 10$. If either of these conditions is not met, the normal distribution is not an appropriate model for the probability distribution since the distribution of the data exhibits significant skewness.

5. C

If an SRS of size n is drawn from a population that has a normal distribution with mean μ and standard deviation σ, then the sample mean $\bar{x}$ has the normal distibution $N\left(\mu, \frac{\sigma}{\sqrt{n}}\right)$. Thus, we have $\bar{x} = 80$ and $s = \frac{\sigma}{\sqrt{n}} = \frac{12}{\sqrt{5,000}} = 0.1697$.

Free-Response Answer

If two independent SRSs of sizes n_1 and n_2 are drawn from large populations whose parameters are p_1 and p_2, the sampling distribution of the difference of the sample proportions is approximately normal with the mean difference equal to $(p_1 - p_2)$ and the standard deviation equal to $\sqrt{\frac{p_1(1-p_1)}{n_1}+\frac{p_2(1-p_2)}{n_2}}$.

Both samples must satisfy the conditions $np > 10$ and $n(1-p) > 10$. Both populations should be large relative to their respective samples satisfying the conditions $N > 10n$.

$$n_1 p_1 = 150(0.35) = 52.5 \quad n_1(1-p_1) = 150(0.65) = 97.5 \quad 10n_1 = 10(150) = 1{,}500$$

$$n_2 p_2 = 100(0.23) = 23 \quad n_2(1-p_2) = 100(0.77) = 77 \quad 10n_2 = 10(100) = 1{,}000$$

It is reasonable to assume that the population of women in the county is greater than 1,500 and that of men is greater than 1,000. Thus, all conditions are met:

$$s = \sqrt{\frac{p_1(1-p_1)}{n_1}+\frac{p_2(1-p_2)}{n_2}} = \sqrt{\frac{0.35(1-0.35)}{150}+\frac{0.23(1-0.23)}{100}} \approx 0.05733.$$

We are interested in the difference being more than 10%, so we use 0.10 for $(\hat{p}_1 - \hat{p}_2)$:

$$z = \frac{(\hat{p}_1-\hat{p}_2)-(p_1-p_2)}{\sqrt{\frac{p_1(1-p_1)}{n_1}+\frac{p_2(1-p_2)}{n_2}}} = \frac{0.10-(0.35-0.23)}{0.05733} \approx -0.3489.$$

The probability that is associated with $z \approx -0.3489$ is 0.3632. The probability is that the difference between women and men in their purchasing of squash or green beans is greater than $10\% \approx 1 - 0.3632 \approx 0.6368$.

CHAPTER 16: ESTIMATION

IF YOU LEARN ONLY FOUR THINGS IN THIS CHAPTER . . .

1. Finding confidence intervals for large samples ($n \geq 30$) can be broken down into five easy steps:

   Step 1. Identify $\overline{x}$, s, n, and the desired confidence level (C) in the problem.
   Step 2. Verify that all assumptions/conditions for the procedures are met.
   Step 3. Find the appropriate z-value from the bottom of the line of the t-distribution table.
   Step 4. Calculate the ME and create the confidence interval by adding and subtracting ME from x $(x - ME, x + ME)$.
   Step 5. State what the confidence interval means within the context of the problem.

2. The sampling distributions for the means of small samples are distributed according to the t-distribution.

3. The mean $\mu_{\hat{p}}$ and standard deviation $\sigma_{\hat{p}}$ of the sampling distribution of the sample proportions $\hat{p}$ can be found using the mean and standard deviation for a binomial variable: $\mu_{\hat{p}} = \frac{np}{n} = p$ $\sigma_{\hat{p}} = \frac{\sqrt{npq}}{n} = \sqrt{\frac{npq}{n^2}} = \sqrt{\frac{pq}{n}}$.

4. The standard deviation of the difference between the two means is $\sqrt{\frac{\sigma_1^2}{n_1} + \frac{\sigma_2^2}{n_2}}$ for $\sigma_1^2 \neq \sigma_2^2$ and $\sigma\sqrt{\frac{1}{n_1} + \frac{1}{n_2}}$ for $\sigma_1^2 = \sigma_2^2$.

   If σ is unknown, we substitute s for σ as follows:

   $\sqrt{\frac{s_1^2}{n_1} + \frac{s_2^2}{n_2}}$ for $s_1^2 \neq s_2^2$ and $s\sqrt{\frac{1}{n_1} + \frac{1}{n_2}}$ for $s_1^2 = s_2^2$.

POINT ESTIMATES AND MARGIN OF ERROR

Often a researcher is interested in some measurement of the population, and finds it impossible to measure each and every individual. The researcher will then draw a **sample**, or a subset, of a workable size from the population and measure each individual in the sample. The researcher is now ready to estimate the population **parameter**, a measured characteristic about the population, using the measurements calculated from the sample. These measurements are referred to as **sample statistics** (see Figure 16.1).

Figure 16.1
Sample Statistics

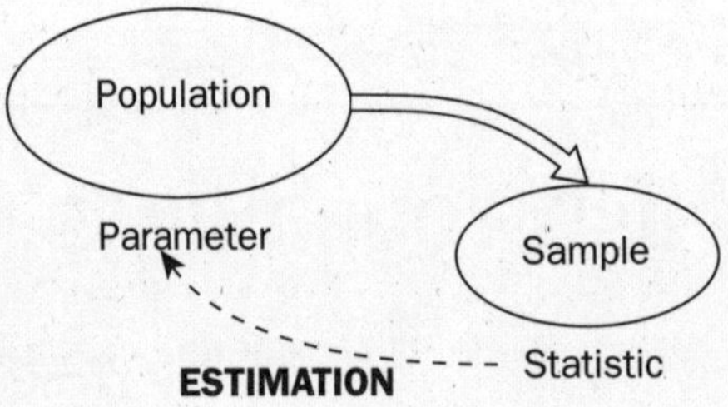

POINT ESTIMATES

The easiest way to estimate the unknown parameter is to use the value obtained from the sample. For instance, if a politician is interested in estimating the percentage of people who will vote for him in an upcoming election, he could draw a sample of eligible voters from the population. After finding the voting preference of each person in the sample, the sample proportion ($\hat{p}$) can be calculated. Using these statistics, the politician can then estimate the population proportion (p). This is known as a **point estimate**, a single value that estimates a population parameter.

Point estimates are very useful if they are unbiased, or if they do not systematically tend to overestimate or underestimate the value of the true parameter. For estimating the proportion of a population, the best unbiased estimate is the proportion of the sample.

To estimate the mean (μ) of a population, the best unbiased estimator is the sample mean ($\bar{x}$). Careful study and analysis of the distributions for mean, median, midrange, and mode has shown that $\bar{x}$ is the best estimate of μ. Besides being unbiased, $\bar{x}$ is a better estimator than the other measures of central tendency because the sampling distribution for $\bar{x}$ has less variation than those for median, midrange, and mode.

MARGIN OF ERROR

When we use ($\hat{p}$) to estimate the population proportion, we know that we may not be completely accurate. Since there is some doubt as to the accuracy of the estimate, it is important to provide a measurement that indicates the maximum amount the point estimate may deviate from the parameter. This is known as the **margin of error (ME)**. The margin of error will provide us with a range of values ($\hat{p} \pm ME$) in which we are *likely* to find the true population parameter.

Since this range of values may or may not contain the parameter, it is necessary to indicate how *likely* it is that the parameter will indeed fall in this range. The *likelihood* of this happening is a probability, and is usually given as a **confidence level**. Thus, the range of values, or interval, is called a **confidence interval**. The confidence level becomes vital to calculating the margin of error, since $ME =$ (*critical value based on confidence level*) × (*standard deviation of the estimate*). The standard deviation of the estimate is also referred to as the standard error of the estimate.

Suppose our politician mentioned in the last example found that 57% of the sample would vote for him. If the margin of error for a 95% confidence level was 4%, then he could be 95% confident that the true population proportion that would vote for him is contained in the interval 0.57 ± 0.04 or (0.53, 0.061). Written in percentage form, we see that the 95% confidence interval is between 53% and 61% of the population.

Widths of Confidence Intervals

It would be wonderful to be 99% confident of every estimate we make. (We can never be 100% sure, unless we measure the entire population.) However, the more confident we wish to be, the wider the range of values that would most likely contain the population parameter. If the width is too wide, it becomes meaningless. How helpful is it to know that the true proportion of voters that favor a certain candidate is between 20% and 60%? For confidence intervals to be helpful as an estimate of the population parameter, they must be highly likely (or reliable) while also being useful (not too wide).

Another factor that will affect the width of a confidence is the sample size. It only makes sense that the more people we sample, the more accurate we would expect our estimate to be. This concept of "more accurate" relates to narrower ranges of values for the confidence interval. In all our measures for standard deviation of the estimate, we will find each of them contains a division by n or $\sqrt{n}$. Therefore, increasing n will decrease the margin of error and thus decrease the width of the interval.

In summary, the best point estimate for the population proportion is the sample proportion. The best point estimate for the population mean is the sample mean. Point estimates may not be completely accurate. The margin of error indicates the maximum amount we expect a point estimate to be off with a certain amount of confidence. A confidence interval is an interval of the form (point estimate) + (margin of error). The margin of error depends on both the sample size and the confidence level, C, chosen. Larger sample sizes lead to narrower intervals while higher confidence levels create wider intervals. The confidence level C tells us that C% of the intervals constructed in this manner will capture the true population parameter we are trying to estimate. Note, while the parameter is a fixed value, the interval will change from sample to sample because the point estimates change.

ASSUMPTIONS/CONDITIONS

Using samples to estimate population parameters is, in essence, modeling the population. In order to ensure that a model is a good representation of the population, certain conditions or assumptions must be met. This is true in all statistical modeling. If the conditions or assumptions are not met, the model may not be an appropriate representation of the population, causing the estimates to be inaccurate. It is important to ensure that appropriate assumptions and conditions are checked prior to carrying out any statistical procedure.

GENERAL FORMAT OF A CONFIDENCE INTERVAL

The AP exam provides a formula sheet that appears in the Appendix of this book. The sheet gives a general form of a confidence interval as statistic ± (critical value) (standard deviation of statistic). The critical value is found from either a *z*-table or *t*-table. Formulas for the standard deviations of various statistics are also found on the formula sheet. This information will help you remember how to find the various types of confidence intervals that are explained in the following section.

In addition, your calculator should be able to quickly calculate confidence intervals for you. The calculator function should only be used for multiple-choice questions or as a check. You are expected to show all work on a free-response question.

CONFIDENCE INTERVALS FOR LARGE SAMPLES

The methods we use for calculating confidence intervals are valid only if the sample data is from a randomized sample design. The Central Limit Theorem, as depicted in Figure 16.2, shows that the sampling distribution of the sample means for any population when the sample size is large enough, say $n \geq 30$, is distributed with a mean ($\mu_{\bar{x}}$) equal to the population mean (μ) and a standard deviation ($s_{\bar{x}}$) equal to the population standard deviation (σ) divided by the square root of the sample size (n). In other words: $\mu_{\bar{x}} = \mu$, and $s_{\bar{x}} = \dfrac{\sigma}{\sqrt{n}}$.

Figure 16.2
Illustration of the Central Limit Theorem

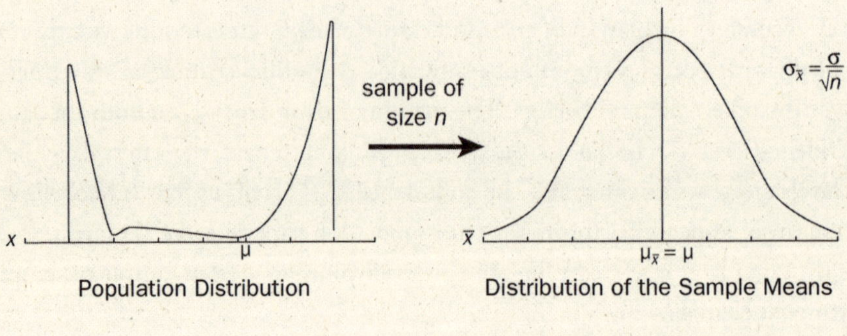

No matter what the population distribution is, for large sample sizes, the distribution of the means of all samples (the sampling distribution) will be normally distributed with a mean of μ and a standard deviation of $\frac{\sigma}{\sqrt{n}}$.

Thus, the sample mean we found may be located anywhere along the horizontal axis of the sampling distribution. However, there is much more likelihood of it being located near the population mean than far from it. Using this fact, along with the Central Limit Theorem, we can now calculate the *ME* for the mean of a sample. To do this we must multiply our standard error of the estimate by the appropriate *z*-value from the standard normal distribution for the level of confidence that we desire. The most common *z*-values along with their corresponding confidence levels are given in Figure 16.3.

Figure 16.3
Common *z*-Values and Confidence Levels

Confidence Level	z
90%	±1.645
95%	±1.960
99%	±2.576

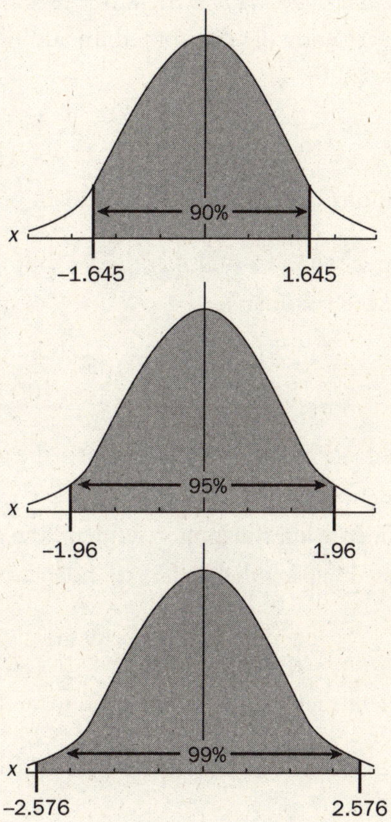

The most common *z*-values can be found on the bottom row (∞) of the *t*-distribution table on page 435. You will use the confidence levels (C) that are on the bottom of the table. Of course, these values can also be obtained from the Standard Normal Probabilities Table; however, the *t*-distribution table is very straightforward to use for the common confidence levels.

The *ME* can now be found by multiplying the appropriate *z*-value by the standard error of the mean $\frac{\sigma}{\sqrt{n}}$. In other words, $ME = z_c \frac{\sigma}{\sqrt{n}}$. Unfortunately, rarely do we know the standard deviation of the population (σ). Therefore, we must estimate σ by using S (the sample standard deviation).

Example:

A simple random sample of 100 students was taken by a researcher to estimate the mean GPA for all students in a school. The sample mean was found to be 3.12, and the standard deviation was 0.86. Find the margin of error for the sample mean for a 95% confidence interval.

First, we need to verify that the assumptions/conditions our procedures are based on are met. In this example, they are:

(1) *The sample was a simple random sample.* Given.

(2) *The population is normal or* $n \geq 30$. It is reasonable to assume that the GPAs of students in a school are normally distributed. In addition, our sample size in this example is larger than 30 ($n = 100$).

(3) σ is unknown.

With these assumptions/conditions verified, we can proceed.

The margin of error is found by multiplying the appropriate *z*-value (1.96 for 95% confidence) by the standard error of the estimate:

$$ME = 1.96\frac{0.86}{\sqrt{100}} = 0.169.$$

Now that we have calculated *ME*, we can complete the process of showing our estimate as an interval, known as the confidence interval. To do this, we merely add and subtract *ME* from the mean (i.e., $\bar{x} \pm ME$). The confidence interval is often written as an algebraic interval, or two numbers separated by a common enclosed in parentheses. The confidence interval for our example is

$$3.12 \pm 0.169 \text{ or } (2.951, 3.289).$$

We are 95% confident that the true population mean GPA is between 2.951 and 3.298.

Finding confidence intervals for large samples ($n \geq 30$) can be broken down into five easy steps:

(1) Identify $\bar{x}$, s, n, and the desired confidence level (C) in the problem.

(2) Verify that all assumptions/conditions for the procedures are met.

(3) Find the appropriate *z*-value from the bottom of the line of the *t*-distribution table. (Note: For confidence levels not shown in the table, see the section, "Finding *z*-Values from the Standard Normal Probabilities Table.")

(4) Calculate the *ME* and create the confidence interval by adding and subtracting *ME* from $\bar{x}$.

$$(\bar{x} - ME, \bar{x} + ME).$$

(5) State what the confidence interval means within the context of the problem.

Example:

A quality control inspector tested the volume of 36 randomly selected bottles of soda from the assembly line. He obtained a mean volume of 1.98 liters with a standard deviation of 0.076 liters. Find the 98% confidence interval for this problem.

Step 1: Identify $\bar{x}$, σ (or s), n, and C.

$\bar{x} = 1.98$, $s = 0.076$, $n = 36$, and $C = 98\%$.

Step 2: Verify all assumptions/conditions for the procedures are met.

(1) *The sample was a simple random sample.* Given

(2) *The population is normal or* $n \geq 30$. It is reasonable to assume that the volume of soda bottles is normally distributed. In addition, our sample size in this example is larger than 30 ($n = 36$).

(3) σ is unknown.

Step 3: Find the appropriate *z*-value.

$z^* = 2.326$. The *z* value for a confidence interval is denoted z^*.

Step 4: Calculate *ME*.

$$ME = z^* \frac{s}{\sqrt{n}}.$$

$$ME = 2.326 \frac{0.076}{\sqrt{36}} = 0.029.$$

Create the confidence interval.

$\bar{x} + ME = 1.98 \pm 0.029$ or $(1.951, 2.009)$.

Step 5: State what the confidence interval means within the context of the problem.

We are 98% confident that the true population mean volume of the soda bottles is between 1.951 and 2.009 liters.

AP EXPERT TIP

We already know the exact value of our sample statistic from our data, so it doesn't make sense to estimate it!

Interpreting the Confidence Interval

In the last example, we found the 98% confidence interval to be (1.951, 2.009). Remember that the confidence interval is an estimate for the population parameter and not the sample statistic. From our sample data, we are estimating the true population value.

Keep in mind that the 98% confidence indicates that if confidence intervals for all possible samples of size *n* were constructed for the given population, 98% of those intervals would contain the population parameter.

AP EXPERT TIP

Remember that on an AP problem, you will be expected to explain the meaning of your confidence interval in context of the problem.

Finding *z*-Values from the Standard Normal Distribution Table

Occasionally, you may be asked to find a confidence interval for a confidence level (C) that is not listed in the *t*-distribution table. In these cases, you must find the *z*-values that relate to the probability of $\frac{(1-C)}{2}$ in both tails of the standard normal distribution. For example, suppose you are to find a 97% confidence interval. Since each tail would have to contain $\frac{(1-0.97)}{2} = 0.015$, you would look this up in the Standard Normal Probabilities Table (Table A in the test booklet) found on page 433 of the Appendix. You will not be finding a *z*-value, but rather looking up the area in the main part of the table. You will find this value on the 14th line, 8th column of the table. This relates to a *z*-value of −2.17. Since the sign is not important, we would use a *z*-value of 2.17 for a 97% confidence interval.

CONFIDENCE INTERVALS FOR SMALL SAMPLES

In finding confidence intervals thus far, we have relied on the Central Limit Theorem. This theorem holds for samples that are large ($n \geq 30$). When a sample is small—which is often the case—the sampling means are not normally distributed. However, the sampling distributions for the means of small samples are distributed according to the *t*-distribution.

The *t*-distribution is similar to the *z*-distribution, but has greater variability. Therefore, the bell-shaped curve is wider and flatter than the standard normal distribution. There are many *t*-distribution curves. Many of the properties of the standard normal distribution hold for the *t*-distribution, such as symmetry around the mean, and the mean being equal to zero. The defining characteristic of this family of curves is a parameter known as the degrees of freedom (df). Degrees of freedom is equal to one less than the sample size ($df = n - 1$). The larger the df, the smaller the variability in the curve. In fact, as df increases, the *t*-distribution curve approaches the standard normal distribution curve. When $df = 29$, the difference between the two curves is so small that the z-distribution can be used. Thus, we determine the definition of the "large sample" to be $n \geq 30$.

Degrees of Freedom (df): The number of independent observations given $\bar{x}$. There is a different t-distribution for each sample size. $t(k)$ = t-distribution with k degrees of freedom.

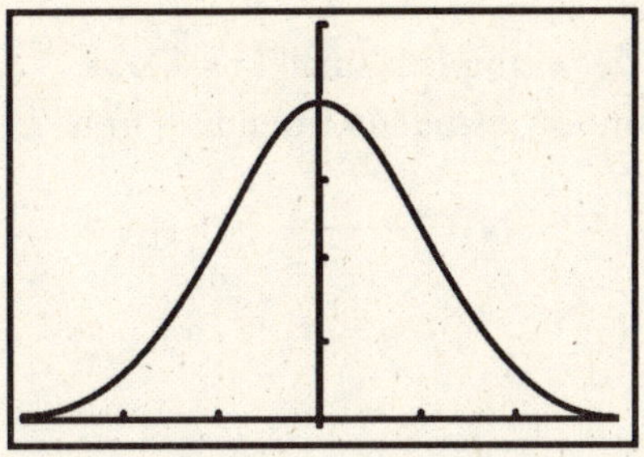

Standard Normal Distribution

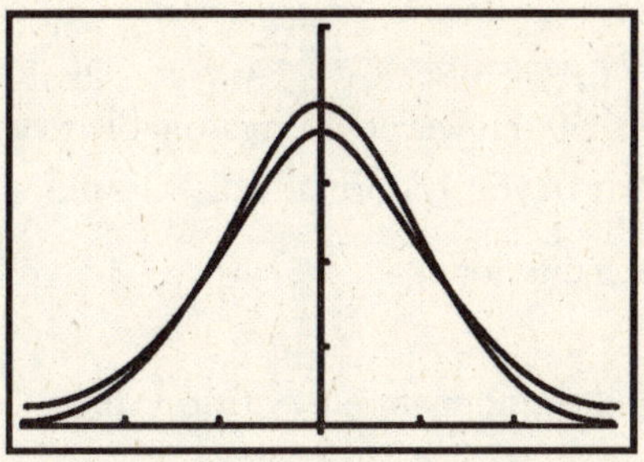

Distribution with 3 df

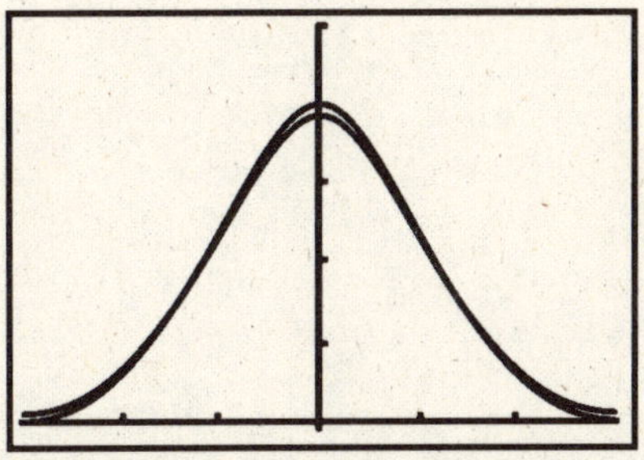

t-Distributions with 7 df

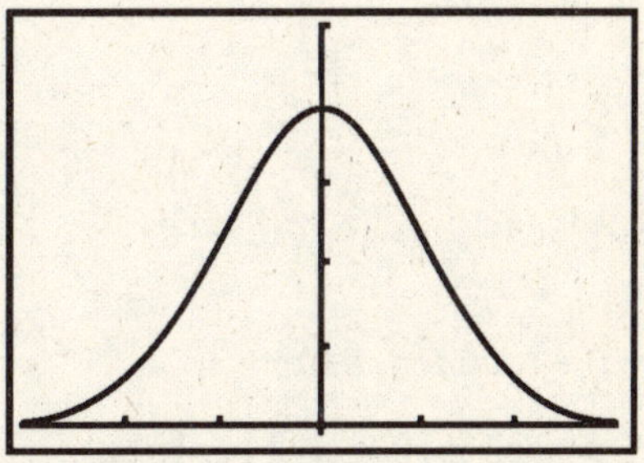

Normal Distribution and
t-Distribution with 20 df

Critical values for creating various confidence levels are provided in the t-distribution table in Table B in the Appendix. To find the critical value you will need to know the degrees of freedom (df). This number is just $n - 1$. You *must* look on the line that corresponds to the degrees of freedom, not the sample size, to find the appropriate t-critical values for your problem. The steps for creating a confidence interval for small samples are the same as for large samples, except that you will need to find the t-critical value rather than the z-value.

Example:

You must estimate the number of students who will attend the annual Math Meet at your university this year. Over the past 8 years, the attendance has been rather consistent, with a mean of 122 and a standard deviation of 12. Estimate the mean number of attendees with 95% confidence. Assume that each year's attendance is independent of previous years.

Step 1: Identify $\bar{x}$, σ (or S), n, and C.

$\bar{x} = 122$, $S = 12$, $n = 8$, and $C = 95\%$.

Step 2:

(1) *The sample is a simple random sample.* You have actual population counts for the past 8 years.

(2) *The population is normal or $n \geq 30$.* Our sample size in this example is small ($n = 8$), so $n \leq 30$. However, it is reasonable to assume that the number of students attending from year to year is normally distributed.

(3) σ is unknown.

Step 3: Find the appropriate t-critical value.

For $df = 8 - 1$ or 7, the t-critical value for 95% confidence level is 2.365:

Step 4: Calculate ME.

$$ME = t^* \frac{s}{\sqrt{n}}.$$

$$ME = 2.365 \frac{12}{\sqrt{8}} = 10.03.$$

Create the confidence interval.

$\bar{x} \pm ME = 122 \pm 10.0$ or (112, 132).

Step 5: State what the confidence interval means within the context of the problem.

We are 95% confident that the number of students who will attend this year's annual Math Meet will be between 112 and 132.

CONFIDENCE INTERVALS FOR A PROPORTION

We now turn our attention to creating confidence intervals for proportions. Recall that the normal distribution can be used to estimate the binomial distribution for large samples. A large sample is defined as $n > 30$, $np > 10$, and $nq > 10$, where n is the sample size (some texts use np and $nq > 5$), p represents the population proportion and $q = 1 - p$. When p and q are not known, use the sample statistics $\hat{p}$ and $\hat{q}$.

This commonly accepted rule allows us to use the standard normal distribution to create confidence intervals about the population proportion. The mean ($\mu_{\hat{p}}$) and standard deviation

($\sigma_{\hat{p}}$) of the sampling distribution of the sample proportions ($\hat{p}$) can be found using the mean and standard deviation for a binomial variable.

$$\mu_{\hat{p}} = \frac{np}{n} = p.$$

$$\sigma_{\hat{p}} = \frac{\sqrt{npq}}{n} = \sqrt{\frac{npq}{n^2}} = \sqrt{\frac{pq}{n}}.$$

Recall that the standard deviation of a sampling distribution is called the standard error of the estimate. Therefore, $\sigma_{\hat{p}}$ is the standard error of the proportion. The *ME* for the estimate now becomes $ME = z_c\sqrt{\frac{\hat{p}\hat{q}}{n}}$, where z_c is the appropriate value from the standard normal distribution for the level of confidence desired.

As with the confidence intervals for the mean, there are assumptions and conditions that must be verified. They are as follows:

(1) *The sample is a simple random sample.*

(2) *The population is large relative to the sample, $10n < N$.*

(3) *$np > 10$ and $nq > 10$.*

We can now find the confidence interval for a proportion by noting that the basic outline of the confidence interval is the *point estimate* ± *ME.* In this case, the confidence interval becomes

$$\hat{p} \pm z_c\sqrt{\frac{\hat{p}\hat{q}}{n}}.$$

Remember to always give your answer within the context of the problem. Don't just give the numerical range of the interval.

The steps for finding the confidence interval for a proportion are similar to those for means of a large sample, with different values being identified in the first step and the appropriate new formula being used to find *ME.*

Example:

An insurance company wishes to estimate the number of claims that are settled within 30 days. A random sample of 75 of the company's claims found that 60 were settled within 30 days. Find the 95% confidence interval for the true population proportion.

Step 1: Identify $\hat{p}$, n, and C. Calculate q. Verify that the normal distribution approximation can be used for the sampling means (i.e., $n\hat{p} > 10$ and $n\hat{q} > 10$).

$\hat{p} = \frac{60}{75} = 0.8$, $n = 75$, $C = 95\%$.
$q = 1 - 0.8 = 0.2.$

Step 2: Verify the assumptions/conditions.

(1) *The sample is a simple random sample.* Given.

(2) *The population is large relative to the sample,* $10n < N$. It is reasonable to assume that $10n$, which is 750, is less than the total number of claims for the insurance company.

(3) $np > 10$ and $nq > 10$.
$n\hat{p} = 75(0.8) = 60 > 10$, $n\hat{q} = 75(0.2) = 15 > 10 \rightarrow$ valid to use normal approximation.

Step 3: Find the appropriate *z*-critical value.

For 95% confidence, $z^* = 1.96$.

Calculate *ME:*

$$ME = z^* \sqrt{\frac{\hat{p}(1-\hat{p})}{n}} = z^* \sqrt{\frac{\hat{p}\hat{q}}{n}}.$$

$$ME = 1.96\sqrt{\frac{(0.8)(0.2)}{75}} = 0.091.$$

Step 4: Create the confidence interval:

$\hat{p} + ME = 0.8 \pm 0.091$ or $(0.709, 0.891)$.

Step 5: State what the confidence interval means within the context of the problem.

We are 95% confident that the percent of claims settled by the insurance company within 30 days is between 0.709 and 0.891.

CONFIDENCE INTERVALS FOR THE DIFFERENCE BETWEEN TWO MEANS

When finding confidence intervals for the difference between two means, we will continue to use the basic outline for finding confidence intervals, which is *point estimate* ± *margin of error*, where the *ME* is *(critical value)(standard deviation of the statistic)*. However, in the case of comparing two means, the point estimate will be the difference between the two means $(\bar{x}_1 - \bar{x}_2)$. The standard deviation of the difference between the two means is

$$\sqrt{\frac{\sigma_1^2}{n_1} + \frac{\sigma_2^2}{n_2}}, \quad \text{for} \quad \sigma_1^2 \neq \sigma_2^2 \qquad \text{and} \qquad \sigma\sqrt{\frac{1}{n_1} + \frac{1}{n_2}}, \quad \text{for} \quad \sigma_1^2 = \sigma_2^2.$$

If σ is unknown, we substitute s for σ as follows.

$$\sqrt{\frac{s_1^2}{n_1} + \frac{s_2^2}{n_2}}, \quad \text{for} \quad s_1^2 \neq s_2^2 \qquad \text{and} \qquad s\sqrt{\frac{1}{n_1} + \frac{1}{n_2}}, \quad \text{for} \quad s_1^2 = s_2^2.$$

The sampling distribution for the test statistics $(\bar{x}_1 - \bar{x}_2)$ will be normally distributed if the degrees of freedom (df) is greater than 29. Otherwise, the sampling distribution will be distributed according to the t-distribution. The degrees of freedom will be the smaller of $n_1 - 1$ or $n_2 - 1$.

The assumptions/conditions that must be checked when calculating the confidence interval for the difference between two means are the following:

(1) The data are from simple random samples.

(2) The two samples are independent.

(3) The populations are normally distributed, or $n_1 + n_2 \geq 40$.

Example:

A college administrator wishes to estimate the difference in the mean GPA between students who do not belong to a fraternity or sorority and those who do. He takes a random sample of 40 students from each group. The means obtained are 2.14 for members of a fraternity or sorority, and 2.25 for nonmembers. Assume the standard deviation (σ) for both populations is 0.6. Find the 90% confidence interval for the difference in the population means.

Step 1: Identify $\bar{x}_1$, $\bar{x}_2$, n_1, n_2, σ_1^2, σ_2^2 and C. Determine if σ_1^2, σ_2^2 are equal or unequal. Calculate the test $\bar{x}_1 - \bar{x}_2$.

$\bar{x}_1 = 2.25$, $\bar{x}_2 = 2.14$, $n_1 = n_2 = 40$, $C = 90\%$

assuming $\sigma_1^2 = \sigma_2^2 = 0.36$

$\bar{x}_1 - \bar{x}_2 = 2.25 - 2.14 = 0.11$.

Step 2: Verify the assumptions/conditions.

(1) *The data are from simple random samples.* Given.

(2) *The two samples are independent.* Based on the nature of the groups, those who are members of fraternities or sororities and those who are not, it is reasonable to assume that one sample has no impact on the other, and that they are therefore independent.

(3) *The populations are normally distributed,* or $n_1 + n_2 \geq$. Each sample included 40 students, so the sum of the two samples is 80. Thus $n_1 + n_2 \geq$.

Step 3: Find the appropriate z-critical value.

For 90% confidence, $z_c = 1.645$.

Calculate ME.

$$ME = z^*\sigma\sqrt{\frac{1}{n_1} + \frac{1}{n_2}}.$$

$$ME = 1.645 \cdot 0.6\sqrt{\frac{1}{40} + \frac{1}{40}} = 0.22.$$

Step 4: Create the confidence interval.

$(\bar{x}_1 - \bar{x}_2) \pm ME = 0.11 \pm 0.22$ or $(-0.11, 0.33)$.

Step 5: State what the confidence interval means in words in context with the problem.

We are 90% confident that the difference between the mean GPA for those not in fraternities or sororities and those who are is between –0.11 and 0.22.

Don't be confused by the negative values. Remember: This is an interval around the estimate of the difference between two proportions. Here, a negative value would indicate that the second proportion was larger than the first.

Example:

Two independent random samples were taken to compare the means of the two populations. The sample statistics are summarized in the table below. Construct a 99% confidence interval for the estimation of the difference in the means of the two samples.

Sample	n	$\bar{x}$	s
A	35	63.5	5.2
B	50	50.4	9.6

Step 1: Identify $\bar{x}_1$, $\bar{x}_2$, n_1, n_2, s_1^2, s_2^2, and C. Determine if s_1^2 and s_2^2 are equal or unequal. Calculate the test $\bar{x}_2 - \bar{x}_1$.

$\bar{x}_1 = 63.5$, $\bar{x}_2 = 50.4$, $n_1 = 35$, $n_2 = 50$, $C = 99\%$.

$s_1^2 = 5.2$, $s_2^2 = 9.6$.

$\bar{x}_1 - \bar{x}_2 = 63.5 - 50.4 = 13.1$.

Step 2: Verify the assumptions/conditions.

(1) *The data are from simple random samples.* Given.

(2) *The two samples are independent.* Given.

(3) *The populations are normally distributed,* or $n_1 + n_2 \geq 40$. $n_1 + n_2 = 35 + 50 = 85$. Thus, $n_1 + n_2 \geq 40$.

Step 3: Find the appropriate t-critical value with $df = \min\{n_1 - 1, n_2 - 1\} = \min\{34, 49\} = 35$.

Since our table does not have $df = 34$, we err on the conservative side and take the next lowest value the table contains. In this case, it would be $df = 30$. For 99% confidence, $t_{34} = 2.750$.

Calculate *ME*.

$$ME = t^* \sqrt{\frac{s_1^2}{n_1} + \frac{s_2^2}{n_2}}.$$

$$ME = 2.750\sqrt{\frac{5.2}{35} + \frac{9.6}{50}} = 1.605.$$

Step 4: Create the confidence interval.

$(\bar{x}_1 - \bar{x}_2) \pm ME = 13.1 \pm 1.605$ or (11.495, 14.705).

Step 5: State what the confidence interval means within the context of the problem.

We are 99% confident that the difference between means of these two populations is between 11.495 and 14.705.

CONFIDENCE INTERVALS FOR THE DIFFERENCE BETWEEN TWO PROPORTIONS

Again we turn to the basic form for finding a confidence interval (*sample statistic* ± *margin of error*) to find the confidence interval for the difference between two proportions. The sample statistics will, of course, be $\hat{p}_1 - \hat{p}_2$. The standard deviation for the estimate, or standard error of the estimate, is given by the formula

$$\sqrt{\frac{\hat{p}_1(1-\hat{p}_1)}{n_1} + \frac{\hat{p}_2(1-\hat{p}_2)}{n_2}}, \text{ when } \hat{p}_1 \neq \hat{p}_2;$$

$$\sqrt{\hat{p}(1-\hat{p})}\sqrt{\frac{1}{n_1} + \frac{1}{n_2}}, \text{ when } \hat{p}_1 = \hat{p}_2.$$

The assumptions/conditions that must be verified are the following.

(1) The data are collected from simple random samples.

(2) The populations are large, or $10n_1 < N_1$ and $10n_2 < N_2$.

(3) $n_1\hat{p}_1 > 10$ and $n_1(1-\hat{p}_1) > 10$.
$n_2\hat{p}_2 > 10$ and $n_2(1-\hat{p}_2) > 10$.
(Note: Some texts use 5.)

(4) The samples are independent.

Example:

In studying her campaign plans, a candidate wishes to know the difference between men's and women's views regarding her appeal as a candidate. She asks her campaign manager to take two random samples and find the 98% confidence interval for the estimated difference in proportions.

A sample of 500 voters is taken from each population. The proportion that finds the candidate appealing is 38.2% for men and 46.1% for women.

Step 1: Identify $\hat{p}_1, \hat{p}_2$, n_1, n_2, and C. Calculate the test statistic $\hat{p}_1 - \hat{p}_2$.

We'll let sample 1 be the women since it has a larger proportion.

$\hat{p}_1 = 0.461, \hat{p}_2 = 0.382$, $n_1 = 500$, $n_2 = 500$, $C = 98\%$.

$\hat{p}_1 - \hat{p}_2 = 0.461 - 0.382 = 0.079$.

Step 2: Verify assumptions/conditions

(1) *The data are from simple random samples.* Given.

(2) *The populations are large, or* $10n_1 < N_1$ *and* $10n_2 < N_2$. The sample of men and the sample of women were both 500. It is reasonable to assume that the population of male voters is more than 5,000 and the population of women voters is more than 5,000.

(3) $n_1\hat{p}_1 > 10$ *and* $n_1(1 - \hat{p}_1) > 10$.

$n_1\hat{p}_1 = 500(0.461) = 230.5 > 10$ and $n_1(1 - \hat{p}_1) = 500(0.539) = 269.5 > 10$.

$n_2\hat{p}_2 > 10$ *and* $n_2(1 - \hat{p}_2) > 10$.

$n_2\hat{p}_2 = 500(0.382) = 191 > 10$ and $n_2(1 - \hat{p}_2) = 500(0.618) = 309 > 10$.

(4) *The samples are independent.* It is reasonable to assume that the samples of men and women are independent.

Step 3: Find the appropriate *z*-critical value.

For 98% confidence, $z_c = 2.326$.

Calculate *ME:*

$$ME = z^* \sqrt{\frac{\hat{p}_1(1 - \hat{p}_1)}{n_1} + \frac{\hat{p}_2(1 - \hat{p}_2)}{n_2}}.$$

$$ME = 2.326\sqrt{\frac{(0.461)(0.539)}{500} + \frac{(0.382)(0.618)}{500}} = 0.072.$$

Step 4: Create the confidence interval.

$(p_1 - p_2) \pm ME = 0.079 \pm 0.072$ or $(0.007, 0.151)$.

Step 5: State what the confidence interval means within the context of the problem:

We are 98% confident that the difference that in the proportion of female and male voters who find the candidate appealing is between 0.007 and 0.151, with females in greater proportion throughout the confidence interval.

CONFIDENCE INTERVALS FOR THE SLOPE OF A LEAST-SQUARES REGRESSION LINE

The least-squares regression line is the line that best fits the data based on minimizing the sum of the squares of the residuals. The slope of the least-squares regression line is very important, since it indicates the rate of change of one variable in relation to the other. For this reason, we may want to calculate a confidence interval around our estimate of the slope of the least-squares regression line.

The confidence interval for the slope of the least-squares regression line is calculated in the same manner as our other confidence intervals. In this case, it is: (estimate of the slope of the regression line) ± (critical value)(standard error of the slope). The calculations for the standard error of the slope are rather intensive, so usually a computer program is called upon to calculate these values.

Most programs will provide various values in the output of a least-squares regression analysis. Among these will be two very important values: the coefficient of the constant and the coefficient of the independent variable. The slope of the line is the coefficient of the independent variable. Usually, these coefficients are reported with their respective standard errors. These values, multiplied by the appropriate critical value, would be used as the maximum error that is added and subtracted from the slope:

$$\textit{coefficient of variable} \pm (\textit{critical value}) \times (\textit{standard error of the estimate}).$$

As we have seen before, there are assumptions/conditions that must be met in order for this confidence interval to be meaningful. They are the following:

(1) The scatterplot should indicate a reasonable linear relationship.

(2) The set of observations represent the population and are randomly selected.

(3) The residual plot should not show any curved pattern and is reasonably scattered. It shows little skewness and no extreme outliers.

(4) The errors around the regression line at each value of x follow a normal distribution.

The degrees of freedom (*df*) for the confidence interval for the slope of a least-squares regression line is $df = n - 2$. The 2 comes from the fact that there are two things we don't know, the slope and the y-intercept, so there are two things that can vary.

Example:

In a physical fitness course, many variables are measured to ascertain a student's level of fitness. Among these are the number of push-ups and the number of sit-ups a student can complete in a certain time period. Ten students were randomly chosen for this test, and the number of push-ups and sit-ups were recorded. A sample of a portion of the computer output for these data is shown in the table. Find the 95% confidence interval for the slope of the least-squares regression line.

Variable	Coefficient	SE Coef
Constant	14.9	0.248
Push-ups	0.66	0.113

We can see from the output that the least-squares regression line is $y = 14.9 + 0.66x$. We are interested in the confidence interval to estimate the slope of the population. Therefore, our test statistic is 0.66, the estimated slope, and its respective standard error is 0.113. The $df = 10 - 2$, or 8. In practice, it is essential that the assumptions/conditions be verified. For this example, we do not have the necessary data to check them. In cases like this, we simply make a statement that the information was not available to validate these assumptions/conditions for calculating the confidence interval, and thus the confidence interval must be viewed with that in mind.

We must look up the *t*-critical value for 8 degrees of freedom and a 95% confidence level in the *t*-distribution table. We find the *t*-critical value is 2.306. Using these numbers, we construct the confidence interval:

> coefficient of variable ± (critical value) × (standard error of the estimate) = 0.66 ± 2.306 × 0.113 or (0.399, 0.921).

We are 95% confident that the slope of the true regression line for the population is between 0.399 and 0.921, assuming that the assumptions/conditions for developing the confidence interval are satisfied.

While computer outputs for regression vary from program to program, each will contain a table that looks like the following.

```
Dependent variable: MPG
No Selector
R squared = 75.6% R squared (adjusted) = 75.1%
s = 2.413 with 50 - 2 = 48 degrees of freedom
Source       Sum of Squares      df       Mean Square    F-ratio
Regression   865.41               1       865.41         149
Residual     279.57              48         5.82436

Variable     Coefficient       s.e. of Coeff   t-ratio     Prob
Constant     48.7393             1.976           24.7      ≤ 0.0001
Weight       -0.00821362       673.8e-6         -12.2      ≤ 0.0001
```

The regression equation for this output is $MPG = 48.7393 - 0.00821362 \cdot \text{weight}$. $s = 2.413$ is the standard error of the residuals and 673.8×10^{-6} is the standard error of the slope.

CALCULATING SAMPLE SIZES

A topic related to creating confidence intervals is calculating the size of a sample required to obtain usable information for making statistical inferences. You will recall that in creating confidence intervals, we calculate the margin of error of the estimate (ME), which determines the width of the intervals. Two factors that are under the control of the researcher determine the size of ME: the confidence level and the sample size. Remember that a confidence interval estimate is useless if it is too wide. Therefore, determining the sample size to allow a confidence interval of reasonable width for a desired confidence level is an important part of the study design. In addition, since there are generally resource and time constraints in conducting samples, researchers are interested in knowing the smallest sample size that will give them the accuracy and precision they want.

Recall the equation for calculating ME: $ME = z_c \frac{\sigma}{\sqrt{n}}$. Solving this equation for n, we get

$$ME = z_c \frac{\sigma}{\sqrt{n}}$$

$$ME\sqrt{n} = z_c\sigma$$

$$\sqrt{n} = \frac{z_c\sigma}{ME}$$

$$n = \left(\frac{z_c\sigma}{ME}\right)^2.$$

This provides us with the formula for calculating the sample size.

From the formula, we see that to determine the sample size, we must know the level of confidence desired, the margin of error we are willing to accept (or the width of the interval we desire), and the population standard deviation. It is the last of these that poses the most difficulty, since if the population standard deviation was known, we would have no need to study the population to estimate the mean. Therefore, researchers will often estimate the population standard deviation from previous studies, using a conservative estimate of $p = 0.5$. When $p = 0.5$, the product $p(1 - p)$ is the maximum. Thus, $\sqrt{\frac{p(1-p)}{n}}$ is the largest and the interval will err on the side of being a bit larger than possibly necessary. Another possibility is to use the rule of thumb that σ can be estimated by $\frac{1}{4}R$ where R represents the range of values expected to be found in the population.

AP EXPERT TIP

You could also just use the original margin of error (ME) formula, plug in the given values, and solve for n.

Finally, when calculating the sample size, it is important to realize that the resulting value provides you with the minimum number of individuals required to meet the parameters set in the equation. If the answer is a decimal, which is often the case, and you round using standard rounding protocol, you will specify a sample size that is too small whenever you round off the decimal. For this reason, it is very important that you *always round up* when dealing with decimal answers.

Similarly, we can find the formula for calculating the sample size required for estimating the population proportion, p:

$$ME = z_c\sqrt{\frac{pq}{n}}$$

$$ME\sqrt{n} = z_c\sqrt{pq}$$

$$\sqrt{n} = \frac{z_c\sqrt{pq}}{ME}$$

$$n = \left(\frac{z_c}{ME}\right)^2 pq.$$

Again, we see that to calculate the sample size we must determine and estimate some parameters. In particular, we must estimate the population proportion, which happens to be the parameter we wish to estimate. Consequently, we will again refer to previous studies, or in the absence of such use $p = 0.5$. A quick look at the function $f(p) = p(1 - p)$, shows that it reaches a maximum value of 0.25 when $p = 0.5$. Thus, using $p = 0.5$ for the estimating n will yield the most conservative sample size possible, (as depicted below).

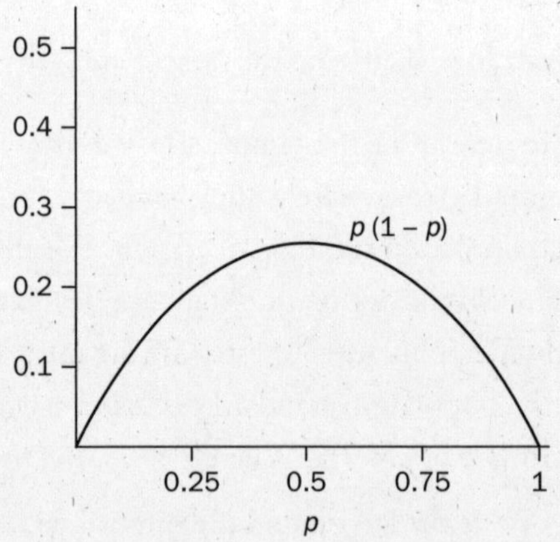

REVIEW QUESTIONS

1. An ecologist would like to estimate the mean carbon monoxide level of the air in a particular city. The carbon monoxide levels are measured on 14 days during a month and recorded. A histogram of the 14 readings is roughly symmetrical, with no outlying values. The mean and standard deviation of these values are 5.4 and 2.2, respectively. Assume the 14 days can be considered a simple random sample of all days. Which of the following is a correct statement?

   (A) A 95% confidence interval for μ is $5.4 \pm 2.145 \times \frac{2.2}{\sqrt{14}}$.

   (B) A 95% confidence interval for μ is $5.4 \pm 2.145 \times \frac{2.2}{\sqrt{13}}$.

   (C) A 95% confidence interval for μ is $5.4 \pm 2.160 \times \frac{2.2}{\sqrt{14}}$.

   (D) A 95% confidence interval for μ is $5.4 \pm 2.160 \times \frac{2.2}{\sqrt{13}}$.

   (E) The sample is too small to trust the results.

2. Campaign managers conduct regular polls to estimate the proportion of people who will vote for their candidate in an upcoming election. Shortly before the actual election, the campaign manager doubles the sample size of the poll. What effect does this have on the estimate?

   (A) It increases the reliability of the estimate.

   (B) It decreases the standard deviation of the sampling distribution of the sample proportion.

   (C) It decreases the variability in the population.

   (D) It will reduce the effect of confounding variables.

   (E) It reduces the bias that comes from interviewer effect.

3. When comparing a 95% confidence interval with a 99% confidence interval created from the same data, how will the intervals differ?

   (A) The sample size must be known to determine the difference.

   (B) The mean of the sample must be known to determine the difference.

   (C) The use of the *t*-distribution or the *z*-distribution will determine how the two intervals differ.

   (D) The 95% interval will be wider than the 99% interval.

   (E) The 95% interval will be narrower than the 99% interval.

4. Which of the following is true about the *t*-distribution?

   (A) The *t*-distribution is symmetrical about the mean.

   (B) The *t*-distribution has more variation than the standard normal distribution.

   (C) The *t*-distribution with k degrees of freedom has a smaller variance than the *t*-distribution with $k + 1$ degrees of freedom.

   (D) Both *a* and *b* are true.

   (E) All of the above are true.

5. In a random sample of 22 college women, each was asked for her height (to the nearest inch) and her weight (to the nearest 5 pounds). A linear regression was a satisfactory description of the relationship between the height and weight. The results of the regression analysis are shown below.

Dependent Variable: Weight				
Source	**Sum of Squares**	***df***	**Mean Square**	**F-Ratio**
Regression	230.0	1	230.0	94.0791
Residual	48.875	20	2.44475	
Variable	**Coefficient**	**SE of Coeff**	***t*-ratio**	**Prob**
Constant	−168.5	0.3153	37	< 0.0001
Height	4.71	0.294	11.8	< 0.0001
R squared = 63.7%		R adjusted = 63.1%		
s = 0.7145 with 22 − 2 = 20 degrees of freedom				

Which of the following should be used to calculate the 95% confidence interval for the slope of the regression line?

(A) $4.71 \pm 2.086 \times 0.294$

(B) $4.71 \pm 1.96 \times 0.6174$

(C) $4.71 \pm 2.074 \times 0.294$

(D) $-168.5 \pm 2.086 \times 0.3153$

(E) $-168.5 \pm 1.96 \times 0.3153$

6. To estimate the proportion of TV viewers watching a certain special, how large of a random sample is required so that the margin of error is 0.04 with 99.6% confidence?

(A) 18

(B) 36

(C) 96

(D) 1,296

(E) 1,492

7. A quality control engineer at a steel mill must estimate the mean tensile strength of a new machine using a random sample of 12 beams. The actual population distribution for this machine is unknown, but graphical displays of the sample indicate that the assumption of normality is reasonable. Since there are no historical data for this prototype machine, the variability of the process is completely unknown. The engineer determines a t-distribution rather than a z-distribution because

(A) he has a small sample, making the z-distribution inappropriate.

(B) he is using data rather than theoretical methods to determine the mean.

(C) the data comes from only one machine.

(D) the variability of the machine is unknown.

(E) the t-distribution results in a narrower confidence interval.

8. A random sample of 25 tourists who visited Hawaii this summer spent an average of \$1,420 on this trip with a standard deviation of \$285. The 95% confidence interval for the mean money spent by all tourists who visit Hawaii is

(A) (\$1,302, \$1,538).

(B) (\$1,308, \$1,531).

(C) (\$1,397, \$1,443).

(D) (\$1,363, \$1,477).

(E) (\$1,385, \$1,465).

9. A sample of 1,000 adults showed that 31% of them are smokers. To estimate the proportion of people in the entire population who smoke, what additional information would you need?

(A) The size of the population

(B) The amount of confidence you desire in your estimate

(C) The standard deviation for the number of smokers

(D) The length of time the people smoked

(E) All the information you need is contained in the problem.

10. A company wants to estimate the mean net weight of all 32-ounce packages of its Yummy Taste cookies at 95% confidence. It is known that the standard deviation of net weights is 0.1 ounce. The sample size that will yield the margin of error within 0.02 ounces of the population mean is

(A) 9.

(B) 10.

(C) 96.

(D) 97.

(E) More information is needed.

11. Which of the following statements is false?

(A) Larger random errors lead to larger confidence intervals.

(B) The size of the random errors has no effect on the width of the confidence interval.

(C) A systematic error leads to incorrect answers.

(D) Systematic errors have no effect on the size of the confidence interval.

(E) Random errors can be minimized by averaging over a larger number of observations.

12. Does the 95% confidence interval change if the reliability of the measurements decreases?
    (A) No, the confidence interval remains the same.
    (B) Yes, the interval becomes narrower because the standard deviation of the measurements increases.
    (C) Yes, the interval becomes wider because the standard deviation of the measurements increases.
    (D) Yes, if the standard deviation of the population is known, but no if the standard deviation of the population is estimated.
    (E) Yes or no, depending on the sample size.

13. Increasing the sample size by a factor of 4 will have what effect on the margin of error?
    (A) It will increase the margin of error by a factor of 4.
    (B) It will decrease the margin of error by a factor of 4.
    (C) It will increase the margin of error by a factor of 2.
    (D) It will decrease the margin of error by a factor of 2.
    (E) It will decrease the margin of error by a factor of 16.

14. The principal of Southside High School, a large urban school of 4,252 students, took a simple random sample of 250 Southside students and found that 43% of them were involved in extracurricular activities. The 90% confidence interval for the estimate of students involved in extracurricular activities at Southside High School is
    (A) (0.3899, 0.4701).
    (B) (0.3780, 0.4820).
    (C) (0.3778, 0.4822).
    (D) (0.3785, 0.4815).
    (E) (0.1327, 0.2112).

15. The owner of a large local restaurant noticed that, when diners paid by credit card, women generally left cash tips and men tended to include the tip on the credit card. To test his observation, he took a simple random sample of 85 women paying by credit card during a two-month period; he noted that only 15 put the tip on the credit card. During the same two-month period, the restaurant owner took a simple random sample of 95 men paying by credit card; he noted that 42 included the tip on the credit card. The 95% confidence interval for the difference between the proportion of men and women who include the tip on the credit card is
    (A) (0.1370, 0.3943).
    (B) (0.1577, 0.3736).
    (C) (0.1603, 0.4081).
    (D) (0.0762, 0.2238).
    (E) Cannot be determined from the information provided.

16. Two independent samples of high school students were taken from a large metropolitan school district. Each group was asked if they would be in favor of reinstituting the draft for military duty. The first sample consisted of 50 students from a high school in an affluent neighborhood, and the results indicated that 23% supported the proposal. The second sample consisted of 75 students from a high school in a low-income area, and the results indicated that 17% supported the proposal. Find the 90% confidence interval for the difference in proportion between these two groups.

    (A) (−0.0843, 0.2042)
    (B) (−0.0611, 0.1811)
    (C) (0.0, 0.1811)
    (D) (0.0, 0.2042)
    (E) (−0.0230, 0.1430)

17. A local politician wants to estimate the percentage of voters who plan to support a referendum to curb development in the county. How large of a sample will be needed to ensure a margin of error of no more than 3%, with 95% confidence?

    (A) 896
    (B) 752
    (C) 632
    (D) 1,068
    (E) More information is needed to compute the appropriate sample size.

18. Ridout County, a large suburban county, conducted a two-year study of teenage drivers 16–18 years of age to determine if males or females were involved in more accidents. A random sample of 100 females showed that they had an average of 15.3 accidents per month, with a standard deviation of 3.5. A second independent sample of 150 males showed that they had an average of 18.7 accidents per month, with a standard deviation of 4.1. Determine a 95% confidence interval for the difference in the number of accidents per month between female and male drivers.

    (A) (−3.8968, −2.9032)
    (B) (−4.3490, −2.4510)
    (C) (17.0, 17.4844)
    (D) (16.527, 17.473)
    (E) (−4.3639, −2.4363)

19. Hoover High School examines a random sample of 35 boys and 30 girls to determine if the scores on the state's English Standards examination differ for the two groups. The 99% confidence interval for the difference in the mean scores between boys and girls is (−10.67, −6.89). What can you conclude about the scores for boys and girls at Hoover High School?

    (A) There is no difference in the scores between girls and boys.
    (B) Girls score better than boys do on the English Standards examination.
    (C) Boys score better than girls do on the English Standards examination.
    (D) Who scored better cannot be determined without knowing the mean scores and standard deviations.
    (E) Who scored better cannot be determined since the sample sizes are different.

20. A national news magazine surveyed 1,500 adults in the United States, and found that 37% disapproved of the Administration's handling of domestic issues. The magazine reported the results as 37% ± 3%. What degree of confidence is reported in these results?

(A) 97%

(B) 56%

(C) 94%

(D) 99%

(E) There is not sufficient information to determine the degree of confidence.

21. Professor Graham wants to reduce the width of the confidence interval around his estimate of the proportion of adults who are carriers of a certain bacteria. What can he do to accomplish this?

(A) Decrease his sample size.

(B) Increase the confidence level.

(C) Change his estimate of $\hat{p}$.

(D) Increase his sample size.

(E) None of these will result in a smaller confidence interval.

22. A random sample of adult male physicians at Memorial Hospital was taken, and the mean cholesterol level was found to be 183 mg/dL. A 95% confidence interval for the corresponding population mean is 183 ± 17 mg/dL. Which of the following statements must be true?

(A) Ninety-five percent of the population measurements fall between 166 and 200.

(B) Ninety-five percent of the sample measurements fall between 166 and 200.

(C) If 100 samples were taken, you could predict that 95 of the sample means would fall between 166 and 200.

(D) $P(166 \leq \bar{x} \leq 200) = 0.95$

(E) If $\mu = 160$ then this $\bar{x}$ of 183 mg/dL would be unlikely to happen.

23. A recent survey of 500 people reported that 67% of American adults believe that high gasoline prices are caused by the greed of oil companies. The margin of error was reported as 3%. What does this margin of error mean?

(A) No more than 70% of the population believes that high gasoline prices are caused by the greed of oil companies.

(B) The actual parameter is between 64% and 70%.

(C) It is unlikely that the reported statistic would be 67%, unless the true value was between 64% and 70%.

(D) Three percent of the people were not surveyed.

(E) Three percent of the time, the value obtained would be different from 67%.

24. A random sample of five snack foods available in the vending machines in the school cafeteria contained the following amounts of sodium (in mg): 310, 350, 320, 280, and 340. What is the 90% confidence interval for the amount of sodium in mg per snack food?

(A) 320 ± 26.1

(B) 320 ± 20.2

(C) 320 ± 18.78

(D) 320 ± 24.7

(E) 320 ± 21.78

25. A marketing expert is hired to estimate the relationship between the amount of money spent on advertising and the total sales in dollars for an electronics company. The prediction equation developed is: total sales = 8.941 (advertising dollars) + 73.842 (amounts in thousands of dollars).

    Given this information, the 95% confidence interval around the estimate 8.941

    (A) is a range estimate for the total sales that the store can expect.
    (B) indicates there is a 95% probability that the ratio of advertising dollars to total sales dollars is 8.941:1.
    (C) indicates there is a 95% probability that the ratio of total sales dollars to advertising dollars is 8.941:1.
    (D) indicates a 95% confidence that the true slope of the regression line is within the interval.
    (E) is a meaningless estimate.

FREE-RESPONSE QUESTIONS

1. In a random sample of college students, one question asked students which brand of bottled water they prefer. They were given a choice of brand A or brand B in that order. Out of 300 students surveyed, 180 chose brand A and 120 chose brand B.

   (a) Based on the results of this survey, construct and interpret a 95% confidence interval proportion of students who chose brand A.

   (b) What is the meaning of 95% confidence?

2. A company is considering installing a facsimile machine at one of its offices. As part of the decision process, the company's manager wants to estimate the average number of documents that would be transmitted daily if the machine were installed. From experience at other offices, the company manager believes the standard deviation of the number of documents sent daily is 32. The manager also believes the number of documents transmitted daily is a normally distributed random variable. The machine is tested over a random sample of 15 days, and the resulting sample mean is 267. Give a 99% confidence interval for the average number of documents that would be transmitted daily if the machine were installed.

3. An economist for the government needs to estimate the mean income for households not covered by health insurance in the city of Albany. She collects a random sample of 1,500 families, and finds the mean income for the sampled households is $18,870 with a standard deviation of $7,240.

   (a) Define the population of interest.

   (b) Calculate a reasonable confidence interval around this estimate of the population mean, and interpret your response.

4. A random sample of 150 seniors at State University were asked if they had cheated on an exam or major paper at any time during their college career. A total of 93 seniors reported that they had cheated.

   (a) Calculate a 95% confidence interval for the proportion of all seniors who had cheated during their college careers.

   (b) Interpret the meaning of the 95% confidence interval.

   (c) How many students should be surveyed to obtain a 95% confidence interval that is within 1% of the correct percent of seniors who cheated?

   (d) How would the length of the confidence interval be affected if the confidence level were changed to 80%? Justify your answer.

5. The gender pay gap has long been debated. Erin decided to study the topic for her senior thesis. One aspect of her thesis was to examine the differences in the perceptions of men and women regarding the gender pay gap. She conducted a random sample of 95 men, of which 12 responded that they feel a pay gap exists between men and women. Erin conducted a second independent random sample of 115 women, and 28 responded that they feel a pay gap exists between men and women.
    (a) Find the margin of error for the difference in the proportion of men and proportion of women that feel that there is a pay gap between men and women at the 95% confidence level.
    (b) Calculate the 95% confidence interval for this difference.
    (c) Interpret this interval.

ANSWERS AND EXPLANATIONS

1. C

Choice (A) is incorrect because the 2.145 is the t-critical value for 14 degrees of freedom. Remember, degrees of freedom is $n - 1$. In this case, the degrees of freedom value is 13. Therefore, the t-critical value must come from the line for 13 in the t-distribution table.

Choice (B) is incorrect since not only was n used to find the t-critical value instead of $n - 1$, but the degrees of freedom were used in the denominator of the standard error instead of n. This is typically a problem when a student gets confused about when to use n and $n - 1$.

For choice (C), the proper value of 13 is used to find the t-critical value, and 14 (n) is used in the formula for the standard error of the estimate. This is the correct way to calculate a confidence interval for a small sample.

Choice (D) used the correct critical value from the t-distribution, but forgot to use the entire sample size (n) not the degrees of freedom (df) in the denominator of the standard error of the mean.

Choice (E) is incorrect since the t-distribution allows you to use very small samples to find estimates. The biggest problem you would encounter with such a small sample is that the width in the resulting interval could be too large to be usable. That, however, is a different issue than the one addressed in this answer, which is your ability to "trust" the estimate. The confidence level of 95% provides you with 95% certainty about the estimate.

Note: This problem is typical of the type of problem you can expect to see if you know the mechanics of calculating a confidence interval, although this particular problem focuses more on utilizing the correct numbers for df and for n. Other erroneous answers could utilize the z-critical value in place of the t-critical value to determine if you can distinguish the proper distribution to use. Remember, you *must* use the t-distribution for samples smaller than 30 when σ is not known and $df = n - 1$. As for using the correct denominator, substitute exactly what is asked for into the formula.

2. B

Choice (A) has to do with the reliability of the estimate. Reliability is based on the level of confidence. This is determined by the researcher and has nothing to do with sample size. Although smaller samples may force the researcher to utilize a lower confidence level, it is reading too much into this problem to assume that the doubling of the sample size will cause the researcher to increase the confidence level, thus increasing reliability. This choice would be incorrect in light of the other choices.

Choice (B) is correct because when the sample size is increased, you divide by a larger value when calculating the standard deviation of the estimate. This, of course, will lead to a small value for the standard deviation of the estimate. Since the formula calls for the division of $\sqrt{n}$, the amount of the decrease you will expect from a doubling of the sample size will be about 1.4 times.

Choice (C) is incorrect since no sampling technique or procedure will have any effect on the variability of the population. Be careful, since the increased sample size will decrease the variability of the sampling distribution. However, this answer indicates the population, and sampling *never* affects the population parameters.

Choice (D) indicates that increasing the sample size will, in some way, decrease the effect of confounding variables. Unfortunately, increasing the sample size only continues to carry the effect of confounding variables. To decrease these effects, the researcher *must* revise the sampling plan to control for these variables. Sample size may change as a result of that revision. However, increasing the sample size will not guarantee a decrease in the confounding effects. Therefore, this answer is incorrect.

Choice (E) is obviously incorrect. If there is an interviewer bias in this sampling plan, it will continue to be present until the issue is addressed and the interviewers are properly trained to not introduce bias when polling the voters. Again, increasing sample size will only continue the bias, not remove it.

Note: This problem is really a conceptual problem dealing with your understanding of the factors that affect the confidence interval estimate. Although the term "confidence interval" is not included in this problem, you should realize that it involves this concept; it is a problem about estimating a population parameter, which is accomplished through the use of confidence intervals. Remember, the only variables that can affect the confidence interval estimate are: (1) sample size, which, when increased, decreases standard error (or standard deviation of the sampling distribution) and thus the width of the interval; and (2) level of confidence, which, when increased, will increase the width of the interval but not affect the standard error of the estimate. The only other variables used in calculating confidence intervals are $\hat{x}$ and S, neither of which the researcher has any control over.

3. E

Choice (A) indicates that sample size will affect the width of the intervals differently. However, the confidence level merely affects the critical value that is used in calculating the maximum error. Provided the sample size remains constant, which can be assumed since the intervals are being made on the same sample, the effect on each interval will be the same.

Choice (B) is incorrect since the mean of the sample has nothing to do with the width of the interval. The mean anchors the interval, since it is always in the middle of the confidence interval.

Choice (C) indicates that the distribution used to determine the critical value will have differing effects on intervals based on the confidence level. A quick look at the t-distribution table shows that for any df (including the ∞ line) the critical values increase consistently as the interval levels increase. Therefore, although the use of the t-distribution will increase the size of a confidence interval over the z-distribution, that increase would be consistent for both confidence levels. For this reason, this answer is incorrect.

Choice (D) is also incorrect. As the confidence level increases, the z-critical value (or t-critical value) increases. Since the critical value is multiplied by the standard error of the estimate to find the value that is added and subtracted from the mean, its effect will be directly proportional. Therefore, increases in the critical value will increase the maximum error and, thus, increase the interval width.

For those reasons mentioned for (D), choice (E) is the correct answer.

Note: Again, this problem is testing your understanding of the concepts involved in creating confidence intervals—in particular, the factors that affect the sizes of the intervals. Again, remember that confidence level is directly proportional to interval widths, while increases in sample size decrease interval widths. These are the only two factors that affect confidence intervals.

4. D

Choice (A) is true. You will recall that the *t*-distribution shares many of the properties of the standard normal distribution, including the property of symmetry about the mean.

Choice (B) is also true. The major difference between the *t*-distribution and the standard normal distribution is that the former has more variability, and is thus wider and flatter than the latter.

Choice (C) is false. While the degrees of freedom do affect the amount of variability in the *t*-distribution, the effect is in the opposite direction than that stated in the problem. The smaller the *df*, the larger the variability in the resulting *t*-distribution. Logically this makes sense, since a smaller *df* comes from a smaller sample size, which would be expected to have a large amount of variation in the sampling distribution.

Choice (D) is correct since it indicates that both (A) and (B) are true.

Choice (E) is obviously incorrect since (C) is false.

Note: Be careful when presented with this type of question, in which you must find the true statement. Do not rush to fill in the first true statement you find. Be sure to read through all the answers to ensure that there is not more than one correct statement. In this case, you will generally find choices that allow you to choose combinations of the given statements. If you have read *all* the single statements and identified those that you believe to be true, then identifying the correct answer will be much easier for you.

5. A

Choice (A) is correct because the coefficient for the independent variable (height) along with its standard error is used. In addition, the *t*-critical value is found for the proper degrees of freedom (20).

Choice (B) uses the incorrect number for standard error (*s* instead of *s.e. of the coef*), and also uses the *z*-critical value instead of the *t*-critical value for 20 *df*.

Choice (C) is calculated with the wrong *t*-critical value. In this answer, the *t*-value related to 22 degrees of freedom is found. This is a common mistake if one is looking up the sample size instead of *df* in the *t*-distribution table.

Choice (D) uses the wrong information. We are not looking for the confidence interval for the constant, which is often meaningless (as in this example). Therefore, we will not be interested in any information printed on the "constant" line.

Choice (E) also uses the information for the constant and the *z*-critical rather than the *t*-critical value.

Note: Although computer printouts differ in layout, all of them will list two variables, one of which is "constant," and the actual values and standard errors for those variables. You will need to select the value for the variable that is not listed as "constant" and its respective standard error. Most printouts will show the degrees of freedom, as did the one in this example problem. However, if the degrees of freedom are not given to you on the printout, remember that it is $n - 2$. The reason we subtract 2 instead of 1 is that for each subject, we took two measurements.

6. D

This problem requires you to find the z_c from the standard normal distribution table. Remember that you will look up $\frac{1-0.996}{2} = 0.002$ in the table to obtain a z_c of 2.88. Also remember the proper formula for calculating sample size for a proportion is $n = \left(\frac{z_c}{ME}\right)^2 pq$, where $q = 1 - p$.

Since we have not been given any previous information regarding the population proportion (p), we will use 0.5 for p.

In choice (A), the correct formula was used with the correct z-value. Unfortunately, $\frac{z_c}{E}$ was not squared. Forgetting to square is a common mistake made when calculating sample size. It will generally lead to a very small sample. Be aware of this error if a sample size calculation is small.

In choice (B), the $\frac{z_c}{E}$ was doubled instead of squared. Again, this is a common mistake in utilizing this formula. It, too, leads to rather small sample sizes, which should be a reminder to check for correct calculation of the formula.

In choice (C), the z-critical value for 96% was used instead of 99.6%. Usually this type of mistake happens when a problem is read too quickly and an easy answer is being sought. It takes a bit longer to find the proper critical value for 99.6%.

Choice (D) is the correct answer since 2.88 was used for the critical value and the formula was calculated correctly.

In choice (E), the z-critical value for 99.8% was used instead of 99.6%. This also is a common error made when reading or answering a question too quickly.

7. A

Choice (A) is correct since the t-distribution is ideally suited for estimating a population parameter from a small sample with an unknown population variance (σ).

Choice (B) is incorrect. Researchers almost always use samples as opposed to theory to estimate means. In fact, if the engineer were to use a theoretical approach to find the mean, she would have no need for statistical methods, which are all built around inferring information about population parameters from sample statistics.

Choice (C) indicates the engineer's decision is dependent upon the fact that only one machine is being used to collect data. The limited method of collecting data may influence bias in the collection process, but has no influence on the statistical methods used.

Choice (D) is partially correct. If we have no idea about the population variation when using a small sample, we are to use the t-distribution. However, if this was a large sample, the engineer would have used the z-distribution even though σ is unknown. Therefore, the decision to use the t-distribution does not rest solely on this fact.

Choice (E) is obviously wrong since the t-critical values for any given confidence level get larger, not smaller, as the sample size decreases. Since the critical value is directly proportional to the maximum error, it would increase, not decrease, the width of the confidence interval.

8. A

In this problem, the answers are given in an algebraic interval format. Regardless of how the answers are given, the correct answer will be found by calculating

$$1{,}420 \pm 2.060\left(\frac{285}{\sqrt{25}}\right).$$

In choice (A), these calculations are performed correctly yielding the noted interval.

In choice (B), the critical value for the *z*-distribution (1.960) was used in lieu of the proper value from the *t*-distribution (2.060). This is the most common error made on an estimating problem. It is imperative that you remember to double check your sample size before calculating the standard error of the estimate.

In choice (C), the denominator used was 25 instead of $\sqrt{25}$. This, of course, will increase the interval size.

In choice (D), the critical value was omitted in the formula. This is not a common error, but can occur when keying in numbers into a calculator.

9. B

Choice (A) is not correct since the size of the population is not important when doing any inferential statistics. This may seem counterintuitive, but the only numbers we require in creating confidence intervals are sample size, and the sample statistics along with critical value based on the size of the sample.

Choice (B) is correct since a confidence interval is an interval estimate that provides some amount of certainty (or confidence level) that the true parameter is contained in it.

Choice (C) is incorrect. A standard deviation is only needed when finding confidence intervals for the mean. This problem asks for the confidence interval of a proportion.

Choice (D) is not relevant to the problem.

Choice (E) is obviously wrong since we need to know the confidence level to proceed with the problem.

10. D

To complete this problem, you will use the following formula,

$$n = \left(\frac{1.96 \times 0.1}{0.02}\right)^2$$

and round the answer up.

Choice (A) does not complete the problem by squaring the value found when multiplying and dividing the proper numbers.

Choice (B), like the previous answer, also does not include the square, However, the answer obtained is rounded up.

In choice (C), the value for *n* is calculated correctly. However, the answer was rounded off instead of up. Although this follows the conventions of rounding numbers, sample sizes must always be rounded up.

Choice (D) is correct since the formula was calculated correctly and the answer was rounded up to obtain 97.

Choice (E) is obviously incorrect. We do indeed have sufficient information to use the formula and calculate the sample size needed.

11. B

Choice (A) is indeed true. The size of a confidence interval is determined by the standard deviation. Larger random errors lead to a larger standard deviation.

Choice (B) is false, making this the correct answer.

Choice (C) is also true. Any error leads to an incorrect answer, and since we cannot avoid errors, no answer is ever correct. This is the pessimistic view of life. The optimistic view is that some errors are worse than others and that we should try to make our errors as small as possible. But the statement is very true.

Choice (D) is true because the effect of a systematic error is that the mean is different from the true value, but they do not influence the size of the interval as such. The width of the confidence intervals is determined by the random error.

Choice (E) is true. Random errors fluctuate in either direction within the observations and typically result from taking limited measurements in different ways. Taking more measurements should average out some of these fluctuations.

12. C

Choice (A) is incorrect because the standard deviation does affect the size of the interval.

Choice (B) is incorrect because the effect of the standard deviation is a direct relation. Remember that you multiply by the standard deviation to calculate the standard error of the estimate. Therefore, larger standard deviations lead to larger standard errors, which, in turn, lead to wider, not narrower, confidence intervals.

Choice (C) is correct due to the reasons outlined above.

Choice (D) and (E) are wrong, knowing that (C) is correct.

13. D

Recall the formula for the margin of error for a proportion is $z^*\sqrt{\frac{\hat{p}(1-\hat{p})}{n}}$. If the sample size n is increased by a factor of 4, $z^*\sqrt{\frac{\hat{p}(1-\hat{p})}{4n}}$, we can see that

$$z^*\sqrt{\frac{\hat{p}(1-\hat{p})}{4n}} = z^*\frac{\sqrt{\hat{p}(1-\hat{p})}}{\sqrt{4n}} = z^*\frac{\sqrt{\hat{p}(1-\hat{p})}}{2\sqrt{n}}$$
$$= z^*\frac{\sqrt{\hat{p}(1-\hat{p})}}{\sqrt{n}}\cdot\frac{1}{2}.$$

Thus, the margin of error is decreased by a factor of 2 if the sample size is increased by a factor of 4.

14. D

Lay out the information you know:

$n = 250 \quad (1-\hat{p}) = q = 0.57$

$\hat{p} = 0.43 \quad z = 1.645$ for 90% confidence interval

Check assumptions/conditions:

1. *The sample is a simple random sample.* Given.
2. *The population is large relative to the sample,* $10n < N$. $10(250) = 2{,}500 < 4{,}252$
3. $np > 10$ and $nq > 10$: $250(0.43) = 107.5 > 10$; $250(1-0.43) = 142.5 > 10$

$$\hat{p} \pm z^*\sqrt{\frac{\hat{p}q}{n}} = 0.43 \pm 1.645\sqrt{\frac{(0.43)(0.57)}{250}}$$
$$= (0.3785, 0.4815)$$

The 90% confidence interval for the estimate of students involved in extracurricular activities at Southside High School is $(0.3785, 0.4815)$.

15. A

Lay out the information you know:

$n_1 = 95 \quad n_2 = 85$

$\hat{p}_1 = \frac{42}{95} = 0.4421$

$\hat{p}_2 = \frac{15}{85} = 0.1765$

z^* for 95% confidence interval $= 1.96$

1. *The data are collected from simple random samples.* Given.

2. *The populations are large,* or $10n_1 < N_1$ and $10n_2 < N_2$.

$$10n_1 = 10(95) = 950 \quad 10n_2 = 10(85) = 850$$

Since it is given that it is a large restaurant, it is reasonable to conclude that in the period of two months, there were more than 950 men paying with credit cards and more than 850 women paying with credit cards.

3. $n_1\hat{p}_1 > 10$ and $n_1(1-\hat{p}_1) > 10$

$n_2\hat{p}_2 > 10$ and $n_2(1-\hat{p}_2) > 10$

$n_1\hat{p}_1 = 95(0.4421) = 41.99 > 10$
$n_1(1-\hat{p}_1) = 95(0.5579) = 53.00 > 10$
$n_2\hat{p}_2 = 85(0.1765) = 15.00 > 10$
$n_2(1-\hat{p}_2) = 85(0.8235) = 69.99 > 10$

(Note: some texts use 5.)

4. The samples are independent. This is implied in the question.

$$(\hat{p}_1 - \hat{p}_2) \pm z^* \sqrt{\frac{\hat{p}_1(1-\hat{p}_1)}{n_1} + \frac{\hat{p}_2(1-\hat{p}_2)}{n_2}}$$

$$(0.4421 - 0.1765) \pm 1.96\sqrt{\frac{0.4421(1-0.4421)}{95} + \frac{0.1765(1-0.1765)}{85}} = (0.1370, 0.3943).$$

The 95% confidence interval for the difference between the proportion of men and women who include the tip on the credit card is (0.1370,0.3943).

16. **B**

Lay out the information you know:

$n_1 = 50 \quad n_2 = 75 \quad \hat{p}_1 = 0.23 \quad \hat{p}_2 = 0.17$
z^* for 90% confidence interval $= 1.645$

1. *The data are from simple random samples.* Given.

2. *The populations are large,* or $10n_1 < N_1$ and $10n_2 < N_2$.

$$10n_1 = 10(50) = 500 \quad 10n_2 = 10(75) = 750$$

Since this is a large metropolitan school district, it is reasonable to conclude that there are more than 500 students attending school in the affluent neighborhood, and more than 750 students attending school in the low-income neighborhood.

3. $n_1\hat{p}_1 > 10$ and $n_1(1-\hat{p}_1) > 10$

$n_2\hat{p}_2 > 10$ and $n_2(1-\hat{p}_2) > 10$

$n_1\hat{p}_1 = 50(0.23) = 11.5 > 10$
$n_1(1-\hat{p}_1) = 50(0.77) = 38.5 > 10$
$n_2\hat{p}_2 = 75(0.17) = 12.75 > 10$
$n_2(1-\hat{p}_2) = 75(0.83) = 62.25 > 10$

(Note: some texts use 5.)

4. The samples are independent. This is implied in the question.

$$(\hat{p}_1 - \hat{p}_2) \pm z^* \sqrt{\frac{\hat{p}_1(1-\hat{p}_1)}{n_1} + \frac{\hat{p}_2(1-\hat{p}_2)}{n_2}}$$

$$(0.23 - 0.17) \pm 1.645\sqrt{\frac{0.23(1-0.23)}{50} + \frac{0.17(1-0.17)}{75}} = (-0.0611, 0.1811).$$

The 90% confidence interval for the difference in the proportion between students in the affluent high school that support the draft and those in the lower-income high school is (–0.0611,0.1811).

17. D

margin of error = $z^* \sqrt{\frac{p^*(1-p^*)}{n}}$

Since p is unknown, the conservative estimate $p^* = 0.5$ is used. For a 95% confidence interval, $z^* = 1.96$.

The margin of error must be no more than 3%, so:

$$0.03 \geq z^* \sqrt{\frac{p^*(1-p^*)}{n}}$$

$$0.03 \geq 1.96\sqrt{\frac{0.5(1-0.5)}{n}}$$

$$0.01531 \geq \sqrt{\frac{0.25}{n}}$$

$$0.002345 \geq \frac{0.25}{n}$$

$$n \geq 1{,}067.008.$$

Since a sample size must be a whole number greater than 1,067.008, we round up to 1,068. Therefore, the sample size needed by the politician to ensure a margin of error no more than 3% and a 95% confidence level is 1,068.

18. E

$\bar{x}_1 = 15.3,\ \bar{x}_2 = 18.7,\ n_1 = 100,\ n_2 = 150,\ C = 95\%$

$s_1 = 3.5,\ s_2 = 4.1$

Verify the assumptions/conditions.

1. *The data are from simple random samples.* Given.
2. *The two samples are independent.* Given.
3. *The populations are normally distributed,* or $n_1 + n_2 \geq 40$. $n_1 + n_2 = 100 + 150 = 250$. Thus, $n_1 + n_2 \geq 40$.

Find the appropriate t-critical value with df = min $\{n_1 - 1, n_2 - 1\} = \min\{99,149\} = 99$.

Since the t-Distribution Table does not have $df = 99$, we use the next lowest number of degrees of freedom that appears on the table, which is 80. The t-critical value for $df = 80$ is 1.990. So,

$$(\bar{x}_1 - \bar{x}_2) \pm t^* \sqrt{\frac{s_1^2}{n_1} + \frac{s_2^2}{n_2}}$$

$$(15.3 - 18.7) \pm 1.990\sqrt{\frac{3.5^2}{100} + \frac{4.1^2}{150}}$$

$$= -3.4 \pm 1.990\sqrt{0.1225 + 0.1121}$$

$$= -3.4 \pm 0.9639 = (-4.3639, -2.4363).$$

The 95% confidence interval for the difference in the number of accidents per month between female and male drivers is (–4.3639,–2.4363).

19. B

You are given the confidence interval; therefore, you do not need any other information about the samples. Boys are being compared to girls, so the negative values indicate the girls' scores are higher. Since the entire confidence interval is negative, you are 99% confident that girls score higher on the state's English Standards examination.

20. D

Recall that a confidence interval about a proportion is calculated by $\hat{p} \pm z^* \sqrt{\frac{\hat{p}(1-\hat{p})}{n}}$.

We are given that the margin of error is ±3%, so we have:

$$\text{margin of error} = z^*\sqrt{\frac{\hat{p}(1-\hat{p})}{n}}$$
$$0.03 = z^*\sqrt{\frac{0.37(1-0.37)}{1{,}500}}$$
$$0.03 = z^*(0.011434).$$
$$2.624 \approx z^*.$$

From the Standard Normal Probabilities Table (z-Table), we find that this relates to a probability of approximately $1 - 2(1 - 0.9956) = 0.9912$, or 99.12%.

21. D

The confidence interval for a proportion is $\hat{p} \pm z^*\sqrt{\frac{\hat{p}(1-\hat{p})}{n}}$. Since $\hat{p}$ is your estimate, it does not change; therefore, choice (C) is incorrect. The width of the confidence interval is determined by z^* and n. The confidence level determines the z^* value. The higher the confidence level, the more confident you are that the true proportion is contained in the interval; this almost means that the higher the confidence level, the wider the interval. Thus, choice (B) is incorrect. As you decrease the sample size, the quantity $\sqrt{\frac{\hat{p}(1-\hat{p})}{n}}$ gets larger, thus increasing the width of the confidence interval; therefore, choice A would be incorrect. Increasing the sample size will decrease the value of $\sqrt{\frac{\hat{p}(1-\hat{p})}{n}}$ and thus decrease the width of the confidence interval. Choice (D) is the correct choice.

22. C

A confidence interval is not a statement of probability. Therefore, choice (D) is incorrect. Choice (A) is incorrect since it states that the confidence interval is based on the population mean rather than the sample estimate. A confidence interval is an interval around the sample estimate. Choice (B) states that the interval would contain 95% of the sample values rather than being an interval around the sample estimate, so choice (B) is incorrect. Choice (E) is incorrect since the confidence interval is centered on the sample estimate, and does not provide a sense of the probability that the estimate is good or bad. Choice (C) is correct; a confidence interval states that if the samples were taken in the same manner, you could predict that 95 out of 100 of them would have a mean between 166 and 200.

23. C

The margin of error for a confidence interval around a sample proportion is $z^*\sqrt{\frac{\hat{p}(1-\hat{p})}{n}}$. Literally, the margin of error is the distance of the endpoints of a confidence interval from $\hat{p}$. It is determined by the level of confidence. The more confident you want the interval to be, the larger the margin of error will be, and thus the wider your interval will be. Choices (A) and (B) incorrectly relate the margin of error to the population value, and not the sample estimate. Choice (D) is incorrect since it relates the 3% to the portion of the population who were not surveyed. Similarly, choice (E) is incorrect because it incorrectly ties the 3% to the frequency with which an estimate is obtained that is different from 67%. Choice (C) correctly interprets the use of the margin of error in providing a confidence interval around the sample proportion, and the fact that the sampled proportion would be unlikely unless the true value was within the confidence interval.

24. A

A confidence interval about the mean for a small sample is given by $\bar{x} \pm t^*\frac{s}{\sqrt{n}}$. For the data provided, $\bar{x} = 320$ and $s_{\bar{x}} = 27.386$. For a 90% confidence level, $t_4 = 2.132$ and the margin of error is $ME = t^*\frac{s}{\sqrt{n}} = 2.132\frac{27.386}{\sqrt{5}} = 26.1$.

Therefore, the confidence interval would be $\overline{x} \pm ME = 320 = 26.1$, and the correct choice is (A).

25. D

The value 8.941 is an estimate of the slope of the regression line. The confidence interval states that you are 95% confident that the true slope would lie within the calculated interval. Thus, choice (D) is the correct choice. Total sales is the dependent variable, and the confidence interval is around the slope, not the dependent variable; choice (A) is incorrect. Choices (B) and (C) are incorrect because the confidence interval is not a statement of probability. Choice (B) is also incorrect because it defines the slope as the change in the independent variable divided by the dependent variable.

Free-Response Answer

1. (a) $\hat{p} = \dfrac{180}{300} = 0.6,$

$n = 300,$

$C = 95\%,$

$q = 1 - 0.8 = 0.4.$

$n\hat{p} = 180 > 10$, $n\hat{q} = 120 > 10 \rightarrow$ Normal approximation can be used.

Simple random sample $\rightarrow$ bias should be minimal to none.

$10n < N \rightarrow$ It is reasonable to assume that the total number of college students is greater than 10(300) or 3,000.

For 95% confidence, $z_c = 1.96$.

$$ME = 1.96\sqrt{\frac{(0.6)(0.4)}{300}} = 0.055.$$

CI: 0.6 ± 0.055 (0.545,0.655).

We are 95% confident that the true proportion of students who prefer brand A water to brand B is in the interval 54.5% to 65.5%.

(b) The meaning of 95% confidence is that if all the possible samples of size 300 were taken from this population, and confidence intervals were then created around each point estimate for each of these samples, 95% of them would contain the population mean.

It is important to note that assumptions have been met when completing inferences regarding the population. In this case, $n\hat{p}$ and $n\hat{q}$ are both greater than 10, and the sample was chosen in such a way as to reduce or eliminate bias. Be sure you carefully read the problem and include a complete answer. Part (a) asks for a confidence interval *and* interpretation. Leaving off the latter will result in an incomplete answer and a loss of points.

Also notice the difference in the questions posed for part (a) and part (b). At first glance, the second question asks for an interpretation of the interval. However, it does not ask you for the meaning of the 95% confidence *interval*, but rather just for the meaning of 95% confidence. This is asking you to explain the meaning of your confidence, not your interval. By the way, another reasonable explanation for part (b) could be to state that $P(0.545 < \bar{x} < 0.655) = 0.95$. However, the answer provided above is more general and would be preferable to a specific answer.

2. Assume a normally distributed population with known σ.

$n = 15 < 30$, therefore a t-distribution is appropriate.

Simple random sample: Given.

$\bar{x} = 267$

$s = 32$

$n = 15$, $df = 14$

t-critical $= 2.977$

$$ME = 2.977\left(\frac{32}{\sqrt{15}}\right) = 24.6$$

The 99% confidence interval is therefore 267 ± 24.6 or (242.4, 291.6).

We are 99% confident that the number of documents transmitted daily is between 242 and 292.

3. (a) The population of interest is households not covered by health insurance in the city of Albany.

(b) Lay out what you know: $\bar{x} = \$18{,}870$, $s = \$7{,}240$, $n = 1{,}500$

Check your assumptions/conditions.

1. *The sample was a simple random sample.* Given.
2. *The population is normal,* or $n \geq 30$. It is reasonable to assume that the incomes of households not covered by health insurance are normally distributed. In addition, our sample size in this example is larger than 30 ($n = 1{,}500$).
3. *σ is unknown.*

Since a confidence level is not specified, you can choose. Dealing with economic data, a 90% confidence level would be reasonable. The z^* critical value for a 90% confidence level is 1.645.

$$\begin{aligned}\bar{x} \pm z^* \frac{s}{\sqrt{n}} &= \$18{,}870 \pm 1.645\left(\frac{\$7{,}240}{\sqrt{1{,}500}}\right)\\ &= \$18{,}870 \pm \$307.51\\ &= (\$18{,}562.49, \$19{,}177.51).\end{aligned}$$

You are 90% confident that the true mean household income for households without health insurance in Albany is between \$18,562.49 and \$19,177.51.

4. (a) Lay out what you know.

$n = 150$, $\hat{p} = \frac{93}{150} = 0.62$,

z^* critical value for 95% confidence level $= 1.96$.

1. *The sample is a simple random sample.* Given.

2. *The population is large relative to the sample,* $10n < N$ and $10(150) = 1{,}500$. At a large university, it is reasonable to assume that the total population of seniors is greater than 1,500.

3. $np > 10$ and $nq > 10$

$150(0.62) = 93 > 10 \qquad 150(0.38) = 57 > 10$

$$\hat{p} \pm z^*\sqrt{\frac{\hat{p}(1-\hat{p})}{n}} = 0.62 \pm 1.96\sqrt{\frac{0.62(0.38)}{150}}$$
$$= (0.5423, 0.6977)$$

(b) The 95% confidence interval means you are 95% confident that the true population proportion of seniors who cheated at some point in their college career is between 0.5423 and 0.6977.

(c) To ensure a margin of error of at most 1%:

$$\text{margin of error} = z^*\sqrt{\frac{\hat{p}(1-\hat{p})}{n}}$$

$$0.01 \geq 1.96\sqrt{\frac{0.62(0.38)}{n}}$$

$$0.005102 \geq \sqrt{\frac{0.2356}{n}}$$

$$0.00002603 \geq \frac{0.2356}{n}$$

$$n \geq 9{,}050.8$$

$$\bar{x} \pm t^*\frac{s}{\sqrt{n}} = \$612 \pm 1.671\left(\frac{\$156}{\sqrt{61}}\right)$$
$$= (\$578.62, \$645.38).$$

The 90% confidence interval for the mean amount spent by customers paying with a credit card is (\$578.62,\$645.38).

A sample of at least 9,051 seniors would be required to ensure a margin of error no more than 1% with a 95% confidence level.

(d) If the confidence level was changed to 80%, the width of the confidence interval would decrease. This is because the *z*-score needed to construct the interval would be smaller—1.282 as opposed to 1.96 for a 95% confidence level—which would lead to a narrower interval. Also, since you are being less confident with your estimate, the estimate does not need to be as precise.

5. (a) Lay out what you know:

$$n_1 = 95, \quad \hat{p}_1 = \frac{12}{95} = 0.1263,$$

$$n_2 = 115, \quad \hat{p}_2 = \frac{28}{115} = 0.2435,$$

95% confidence level, $z^* = 1.96$.

Check assumptions/conditions.

1. *The data are collected from simple random samples.* Given.

2. *The populations are large,* or $10n_1 < N_1$ and $10n_2 < N_2$.

$10n_1 = 10(95) = 950$

$10n_2 = 10(115) = 1{,}150$

It is reasonable to assume the population of men is greater than 950 and the population of women is greater than 1,150, so this condition is met.

3. $n_1\hat{p}_1 > 10$ and $n_1(1 - \hat{p}_1) > 10$

$n_2\hat{p}_2 > 10$ and $n_2(1 - \hat{p}_2) > 10$

$n_1\hat{p}_1 = 95(0.1263) = 11.998 > 10$

$n_1(1 - \hat{p}_1) = 95(0.8737) = 83 > 10$

$n_2\hat{p}_2 = 115(0.2435) = 28 > 10$

$n_2(1 - \hat{p}_2) = 115(0.7565) = 87 > 10$

4. *The samples are independent.* Given.

$$ME = z^* \sqrt{\frac{\hat{p}_1(1 - \hat{p}_1)}{n_1} + \frac{\hat{p}_2(1 - \hat{p}_2)}{n_2}}$$

$$= 1.96\sqrt{\frac{0.1263(1 - 0.1263)}{95} + \frac{0.2435(1 - 0.2435)}{115}}$$

$$= 1.96\sqrt{0.00276} = 0.10303.$$

(b) The 95% confidence interval for the difference between the proportion of men and the proportion of women who feel that there is a pay gap between the genders is

$$(\hat{p}_1 - \hat{p}_2) \pm ME = (0.1263 - 0.2435) \pm 0.10303$$
$$= (-0.22023, -0.01417).$$

(c) The 95% confidence interval means that you are 95% confident that the true difference in the proportion between the men who feel that there is a pay gap between the genders and the proportion of women who feel there is a pay gap between the genders is between −0.2202 and −0.0142. This implies that the proportion of men who feel there is a pay gap between the genders is less than the proportion of women who do.

CHAPTER 17: HYPOTHESIS TESTING

IF YOU LEARN ONLY FOUR THINGS IN THIS CHAPTER . . .

1. Hypothesis testing is a formal procedure to determine the validity of a claim.
2. When testing claims about the proportion of a population, a normal distribution may be used to approximate the binomial distribution provided $np > 10$ and $nq > 10$ (some texts will use 5), where p is the hypothesized proportion of successes in the population and $q = 1 - p$.
3. The test statistic for the comparing the difference of two proportions is:

$$z = \frac{(\hat{p}_1 - \hat{p}_2) - (p_1 - p_2)}{\sqrt{\frac{p_1 q_1}{n_1} + \frac{p_2 q_2}{n_2}}}$$

where $q_1 = 1 - p_1$ and $q_2 = 1 - p_2$.

4. The z- and t-statistic for one-sample hypothesis tests about the mean are

$$z = \frac{\bar{x} - \mu}{\frac{\sigma}{\sqrt{n}}} \quad and \quad t = \frac{\bar{x} - \mu}{\frac{s}{\sqrt{n}}}$$

where μ is the hypothesized mean.

IT ALL STARTS WITH A CLAIM

In the previous chapter, we were concerned with estimating a population parameter based on the information obtained from a sample. In statistical analysis, another important topic to consider is **hypothesis testing or tests of significance**, which is a formalized procedure to determine the validity of a claim about the population. The claim can be just about anything: "a majority of the people will vote for candidate X"; "the mean age of all students is 22 years"; "the average age of women is higher than that of men," and so on.

Let's say that you are playing a game that involves flipping a coin with a friend. You bet on tails. After 50 tosses, the coin has turned up heads 40 times. At this point, you may make a claim that the coin is weighted and favors heads over tails. To validate this claim, you would have to decide if 40 out of 50 (or 80%) is far enough from the expected value of 25 heads (or 50%) to accept this claim. Would your opinion change if there were 30 out of 50 heads (or 60%)? The question remains as to how far away from the expected value would warrant concluding that the coin is unfair.

To answer this question, you may consider that a fair coin would turn up heads 50% of the time. Intuitively, you may set a range of probabilities for heads that you deem "reasonably likely" to occur by chance when flipping a fair coin 50 times. (See Figure 17.1.) You would then compare the proportion of heads obtained in your game to your range of likely probabilities. If the 80% heads that occurred during the game falls outside your "likely to occur" area, you would conclude that the coin does indeed favor heads. (Of course, this does not confirm that the coin is weighted... only that it favors heads under the given conditions.) If the 80% falls within the "reasonably likely to occur by chance" area, you would be left in a quandary. Although you cannot prove the coin is unfair, neither do you have proof that it is fair. All you can conclude at this point is that the 80% heads could occur by chance, and you cannot prove the coin favors heads.

Figure 17.1

Possible Range of Probabilities for Flipping a Fair Coin

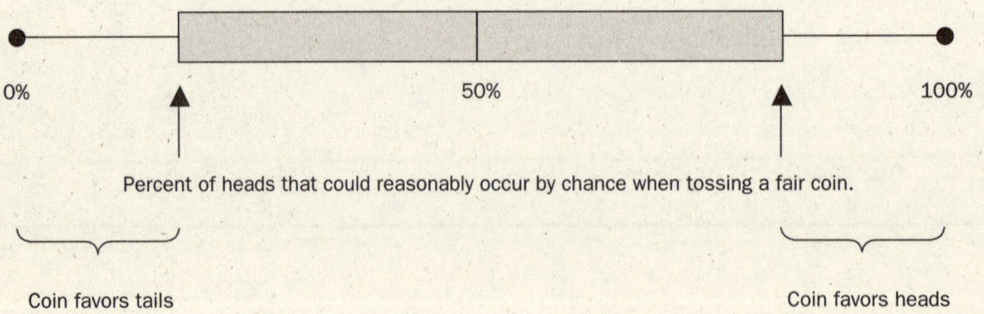

Hypothesis testing is one of the procedures of statistical inference. Statistical inference is an area of statistics that focuses on drawing conclusions from sample data, and uses probability to express the strength of the conclusions. Statistical inference is based on the long-run, regular behavior of an event that is described by probability. For statistical inference techniques to provide a reasonable basis for decision making, the data must come from a random sample or randomized experiment. The techniques of statistical inference cannot overcome basic flaws in the sample design or experiment design.

The Hypothesis

Hypothesis testing addresses the question, "Could these observations really have occurred by chance?" The first step in hypothesis testing is to formulate your hypothesis. This is a statement of the expected value of the parameter and is called the **null hypothesis**. The null hypothesis (H_0) states that equality (=) exists—that is, that the parameter equals some specific value. A null hypothesis could be that the mean family income in Rhode Island is $83,500. The **alternative hypothesis** (H_a) states what you really think is true. The mean family income in Rhode Island is less than $83,500. Or maybe you think it is greater than $83,500. You may not know if it is greater than or less than and just think it is not equal to $83,500. In other words, the alternative hypothesis is a statement of greater than, less than, or not equal to (>, <, or ≠) the value specified in the null hypothesis.

An example of hypotheses using symbols is

$H_0: \mu = \$83{,}500,$

$H_a: \mu \neq \$83{,}500.$

In the example of our coin toss, we made a claim that the coin was weighted or that it favors heads. We based that claim on observing 40 out of 50 heads. In testing our claim, we would have:

$H_0: p = 0.50,$

$H_a: p > 0.50.$

Significance Level or α–Level

On the AP exam, it is also important that you define the variables that are used in your hypotheses. For example, you could write "p is the true probability of heads when tossing a coin." Also, never use $\bar{x}$ or $\hat{p}$ when stating your hypotheses. These are sample statistics symbols, and we are testing a claim about the population parameter.

In order to proceed with the hypothesis test, it is necessary to decide how much error you are willing to accept in validating the hypotheses. We will control for the amount of error we are willing to accept by setting the significance level, or α level, for the hypothesis test. Normally the level of significance, or alpha (α) level, is set at 5%, meaning we are willing to risk a 5% error in rejecting a true null hypothesis. For more critical tests, this may be reduced to 2% or 1%. A significance level of 5% would be the same as saying $\alpha = 0.05$.

The α level is our decision rule. It describes what we feel would be convincing evidence to reject our null hypothesis. If the null hypothesis were true, then what is the probability of observing a test statistic at least as extreme as the one we observed? The α level is always set before you perform your hypothesis test and represents what you feel is the cutoff point for extreme results.

AP EXPERT TIP

The AP exam provides a formula sheet that appears in the Appendix of this book. The sheet gives a general form of a test statistic as:

$$\frac{\text{(statistic} - \text{parameter)}}{\text{(standard deviation of statistic)}}$$

The Test Statistic

The hypothesis test literally compares an observed value to some expected or hypothesized value. In our coin example, we are comparing our 80% heads to the expected 50% value. Essentially what is done in a hypothesis test is that we standardize (normalize) our observed value and determine how far our sample statistic is from the expected or hypothesized value. This distance is measured in standard error units.

AP EXPERT TIP

While your calculator can also perform significance tests (see chapter 19), it is still important that you complete all the steps of a significance test, and that you include the formulas when answering a free-response problem. Your calculator can help make sure all of your work is correct.

The *p*-Value

Once the hypothesis has been stated and an acceptable α level has been determined, then calculate the *p*-value. The *p*-value is a probability statement that answers the question, "If the null hypothesis were true, then what is the probability of observing a test statistic at least as extreme as the one observed?" This is determined by your test statistic. Recall that you can use the *z*-score (how many standard deviations you are from the mean) to determine the area under the curve to the left or to the right of your value, and this area is the probability of obtaining a value within that region.

The Decision

Once calculated, the *p*-value is compared to your fixed significance level α. If the *p*-value $< \alpha$, then you rule out the null hypothesis. When a statistician says that findings were significant, it means that the results were not likely to have happened just by chance. Keep in mind the normal curve and the 68-95-99.7 rule. You are essentially determining how far out in the tails of the curve your observed data falls. Yes, there is a chance a result like that could occur, but the probability is so small that it is highly unlikely. The smaller the *p*-value is, the stronger the evidence against the null hypothesis.

Tests of significance can be one-sided or two-sided. If your alternative hypothesis states "not equal to" ($\neq$), then you will have a two-sided test—one on each side of the "reasonable region." If the alternative hypothesis contains a greater than (>) or less than (<) condition, then you have a one-sided test. For greater than (>), the test is a right-tailed (or right-sided) test. For less than (<), the test is a left-tailed (or left-sided) test. You can consider the inequality symbol as the head of an arrow pointing to the direction of the test.

Since the sampling distribution for the mean and proportion are both distributed normally for $n \geq 30$ and according to the t-distribution for $n < 30$, the boundary of the critical region(s) or rejection regions will be the z-critical value or t-critical value associated with your α level. (See Figure 17.2.) Remember: If your p-value falls within these tail regions, you have strong evidence against the null hypothesis.

Figure 17.2
Boundaries of Critical Regions

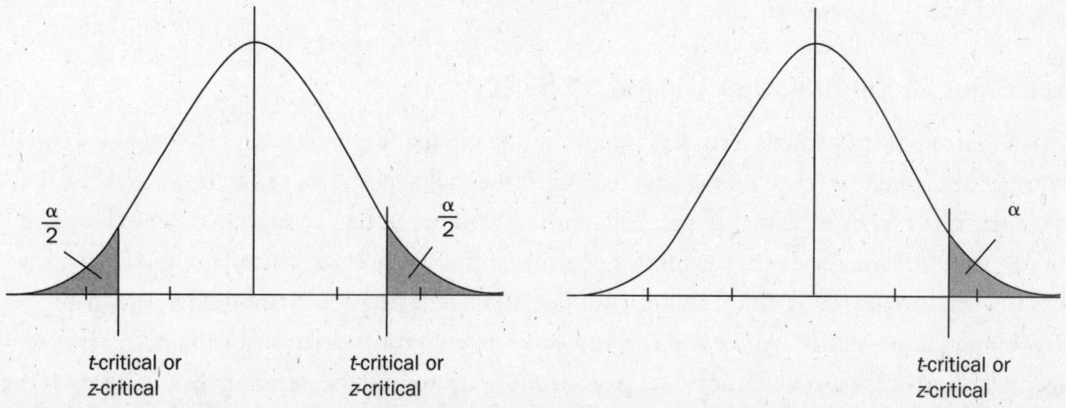

Distribution of sample means under null hypothesis for a two-tailed test (left) and one-tailed test (right). The *t*-distribution is used when $n < 30$ and σ is unknown. Otherwise, use the *z*-critical values obtained from the normal distribution.

Let us return to the example of 50 coin flips. We have decided that we will risk a 5% chance of making an error. This means there is up to a 5% chance you could accuse a friend of cheating when she really is not cheating. Since our alternative hypothesis (H_a) stated $p > 0.50$, we are conducting a right-tailed, or right-sided, test. We will have one critical region as outlined in Figure 17.3. The boundary of the critical region can be found by observing the z-critical value for 95% of the curve to the right. That is, $z = 1.645$ for $\alpha = 0.05$. For a one-tailed test, the entire α value is in the tail. For a two-tailed test, half of the α value is in each tail.

Figure 17.3
Rejection Region

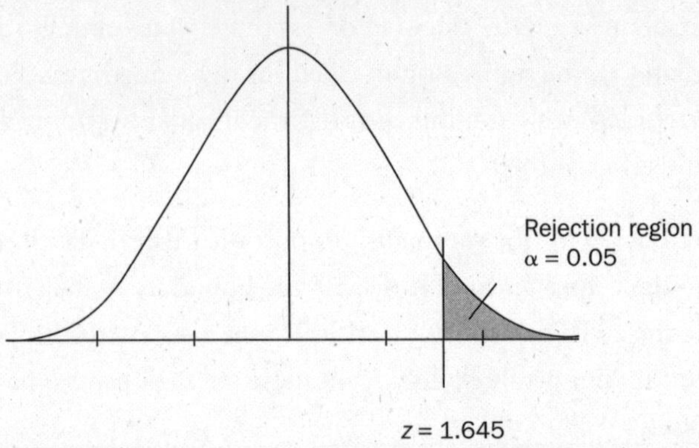

Once you have calculated your p-value, compare it to α. Remember: If the p-value < α, then we rule out the null hypothesis.

Rejecting, Accepting, and Failing to Reject

When performing hypothesis tests, it is important to note that we *never* accept the null hypothesis. We may have evidence based on the data to suggest the null hypothesis is false; however, the evidence can never prove that it is true. This entire situation can be likened to a court of law. The jury is presented with evidence regarding a person's guilt or innocence in reference to a crime. The jury deliberates over the evidence and reaches a decision. The decision is to find the defendant either guilty or not guilty. Notice that the jury *never* states that the defendant is innocent; they can only tell whether the evidence they were given is sufficient to find the person guilty, or whether the evidence was insufficient to find the person guilty.

In the same way, when performing a hypothesis test, you are presented with evidence from a sample regarding a claim being made about a population parameter. Instead of deliberating with 11 other jurors, you perform a formalized procedure to test the validity of the evidence. This will lead you to one of two decisions: reject H_0 (the evidence indicates it is not true), or fail to reject H_0 (the evidence does not support rejecting its validity).

As stated earlier, it is preferable for an alternative hypothesis to include what we think to be true. In this way, a definitive conclusion can be reached regarding the claim. We either reject H_0 and accept the claim that the population is different than stated in H_0, or we fail to reject H_0 and reject the claim of difference from the null. Notice that rejecting H_0 does lead to accepting H_a in much the same way that rejecting a defendant's innocence leads to finding him guilty. However, failing to reject the null hypothesis cannot lead us to claim that the null is true.

Steps in Performing a Hypothesis Test

Before continuing with the various types of hypothesis tests that will be covered on the AP exam, let's review the basic steps that are included in *all* hypothesis tests.

Step 1: State the null and alternative hypothesis (H_0 and H_a, respectively) and define parameters to be used. It is good practice to state both the null and alternative hypotheses symbolically and in words.

Step 2: Identify the test to be used and check appropriate assumptions/conditions.

Step 3: Calculate the test statistic, *p*-value, and compare the *p*-value with your α.

Step 4: Clearly state your decision regarding the null hypothesis based on your *p*-value.

Type I and Type II Errors

Since we are dealing with a random event, there will always be a chance that we could reject our null hypothesis (H_0) when it is really true. We could also make the error of failing to reject H_0 when it is false. The first type of error is known as a Type I error, while the latter is known as a Type II error. Refer to the table below.

	H_0 Is True	H_0 Is False
Reject H_0	Type I error	Correct
Fail to Reject H_0	Correct	Type II error

The probability that a fixed level α hypothesis test will reject H_0 when a particular alternative value of the parameter is true is called the **power of a test** against the alternative. The power of the test is found by calculating 1 – *P*(Type II error). (Standard notation is $\alpha = P(\text{Type I error})$ and $\beta = P(\text{Type II error})$.) However, in order to calculate the probability of a Type II error, one would need to know the true population parameter. Of course, if the true population parameter were known, then there would be no need to perform a hypothesis test. Therefore, it is almost impossible to calculate the true probability of a Type II error of a hypothesis test. There are some algorithms that make certain assumptions regarding the population parameter being tested and can estimate the probability of a Type II error; however, controlling for this error is generally done through setting the sample size and α level. (See Figure 17.4.)

Figure 17.4
Estimation of P (Type II Error)

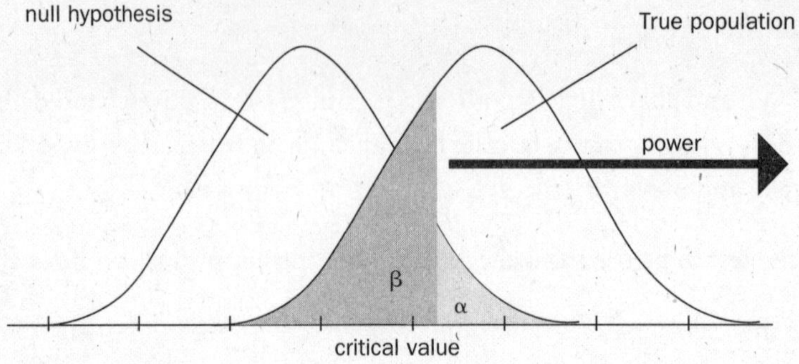

Increasing Power

It would be nice to run all hypothesis tests at a 1% alpha level. However, the lower the P(Type I error), the higher the P(Type II error), and, consequently, the lower the power of the test. Therefore, we must consider that the more accurate our test, the less chance we have of finding a difference from the hypothesized parameter if one does indeed exist. However, it is possible to decrease the P(Type II error) by increasing the sample size. Consider that the larger the sample, the smaller the standard deviation of the sampling distribution. Therefore, the power of our test $(1 - \beta)$ can be increased by increasing the risk we are willing to take in rejecting a true null hypothesis (Type I error), or by increasing the sample size. In summary, an increase in the Type I error risk and/or sample size will decrease the risk of a Type II error, and, consequently, increase the power of the hypothesis test. (See Figure 17.5.)

Figure 17.5
Relationship Between Type I and II Errors and Power of Test

The effects of increasing (darker region) α and n on the Type II error and power can be seen in this figure. Notice that there is an inverse relationship between α and β, while there is a direct relationship between n and β.

LARGE SAMPLE TEST FOR A PROPORTION

When testing claims about the proportion of a population, we can use the normal distribution to approximate the binomial distribution provided $np > 10$ and $nq > 10$ (some texts will use 5), where p is the hypothesized proportion of successes in the population and $q = 1 - p$. Additionally, the inference procedures are reasonably accurate if the population is at least 10 times larger than the sample, or $10n < N$. The test statistic is found by calculating

$$z = \frac{\hat{p} - p}{\sqrt{\frac{pq}{n}}}$$

where

$\hat{p}$ = sample proportion,
p = hypothesized proportion,
$q = 1 - p$, and
n = the sample size.

This test is most accurate when p is close to 0.5, and least accurate when p is close to 0 or 1.

Example:

A magazine editor claims that over 40% of the readers are women. A sample of 70 readers of the magazine is taken, of which 32 are women. Test the editor's claim using $\alpha = 0.05$.

Solution:

Step 1:

H_0: Women make up 40% of the readers of the magazine.

H_a: Women make up more than 40% of the readers of the magazine.

$H_0: p = 0.40$, $H_a: p > 0.40$ $\quad \hat{p} = \frac{32}{70} = 0.457$ and $\alpha = 0.05$

Step 2: This is a right-tailed, one-sample z-test for proportions. (See Figure 17.6.)

In order to perform this test, you must first ensure that the requirements for using the normal approximation have been met, and calculate the sample proportion $\hat{p}$.

$10n = 10(70) = 700$. It is reasonable to assume that the population of readers of the magazine is greater than 700.

$np = 70(0.4) = 28 > 10$, and $nq = 70(0.6) = 42 > 10 \rightarrow$ Normal approximation holds.

Step 3:

$$z = \frac{\hat{p} - p}{\sqrt{\frac{pq}{n}}} = \frac{0.457 - 0.40}{\sqrt{\frac{(0.4)(0.6)}{70}}} = 0.973.$$

Calculate the p-value: $P(z < 0.973) = 0.1646$.

Is the p-value $< \alpha$? $0.1646 > 0.05$

Step 4: Since the p-value is not less than α, we fail to reject H_0 and we do not have sufficient evidence against the statement that the proportion of women readers is over 40%.

Figure 17.6
Failure to Reject H_0

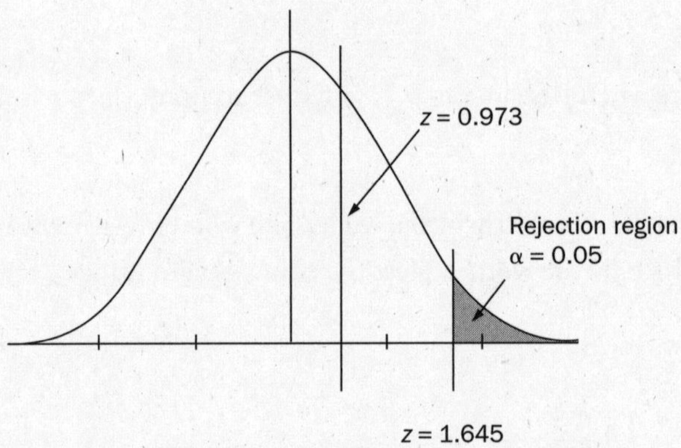

Example:

A recent national survey indicated that 19% of employees in the manufacturing industry are concerned about losing their jobs. A politician would like to use this information in a speech focusing on employee issues. However, he doubts the 19% figure and has his own survey done. A random sample of 1,000 employees in manufacturing jobs is surveyed, and 230 indicate they are concerned with losing their jobs. At a 5% significance level, does this result support the politician's doubt?

Solution:

Step 1:

H_0: $p = 0.19$ (19% of employees in manufacturing jobs are concerned about losing their jobs.)

H_a: $p \neq 0.19$ (The percent of employees in manufacturing jobs that are concerned about losing their jobs is different than 19%.)

$\alpha = 0.05$

Step 2: This is a two-tailed, one-sample *z*-test for proportions.

In order to perform this test, you must first ensure that the requirements for using the normal approximation have been met, and calculate the sample proportion: $\hat{p} = \frac{230}{1,000} = 0.23$.
$10n = 10(1,000) = 10,000$. It is reasonable to assume that the population of employees in manufacturing jobs is greater than 10,000.

$np = 1,000(0.19) = 190 > 10$, and $nq = 1,000(0.81) = 810 > 10 \rightarrow$ Normal approximation holds.

Step 3:

$$z = \frac{\hat{p} - p}{\sqrt{\frac{p(1-p)}{n}}} = \frac{0.23 - 0.19}{\sqrt{\frac{0.19(1-0.19)}{1,000}}} = 3.22.$$

Calculate the *p*-value: $p(z > 3.22) = 1 - 0.9994 = 0.0006$. Since this is a two-sided test, we need to double this value to account for $p(z < -3.22)$. Thus, our *p*-value is $2(0.0006) = 0.0012$.

Compare the *p*-value to the significance level. Is the *p*-value less than the alpha level?

Step 4: Since the *p*-value is less than the alpha level, 0.05, we reject the null hypothesis and conclude that there is sufficient evidence to question the claim that 19% of manufacturing employees are concerned with losing their jobs.

LARGE SAMPLE TEST FOR THE DIFFERENCE BETWEEN TWO PROPORTIONS

When testing for a difference between two proportions, the claim will center on the amount of difference between two population proportions (p_1 and p_2). Usually, the null hypothesis will state that the two population proportions are equal ($H_0: p_1 = p_2$), or in other words, $p_1 - p_2 = 0$. In the alternative hypothesis, you could test for a difference ($H_a: p_1 \neq p_2$), or higher or lower proportion for one population ($H_a: p_1 > p_2$ or $H_a: p_1 < p_2$). The normal approximation holds, provided that the $n_1p_1 > 10$, $n_1q_1 > 10$, $n_2p_2 > 10$, and $n_2q_2 > 10$. Additionally, the inference procedures are reasonably accurate if the population is at least 10 times larger than the sample, or $10n_1 < N$ and $10n_2 < N$. Since values for p_1 and p_2 are generally not expressed in the hypotheses, they can be approximated using the sample proportions $\hat{p}_1$ and $\hat{p}_2$.

The test statistic for the comparing the difference of two proportions is

$$z = \frac{(\hat{p}_1 - \hat{p}_2) - (p_1 - p_2)}{\sqrt{\frac{p_1 q_1}{n_1} + \frac{p_2 q_2}{n_2}}}$$

where $q_1 = 1 - p_1$ and $q_2 = 1 - p_2$.

Again, it is necessary to note that the population proportions are usually not expressed in the problem—only the difference between them. Therefore, you will have to estimate the population proportions (p_1 and p_2) using the sample proportions ($\hat{p}_1$ and $\hat{p}_2$).

Example:

A university decided to test the effectiveness of two teaching methods utilized by two different faculty members. Sixteen out of 25 students in Instructor A's class passed a standardized final exam. In comparison, 59 out of 72 students in Instructor B's class passed the same exam. Is Instructor A's success rate the same as Instructor B's? Use a 0.10 level of significance.

Solution:

Step 1:

H_0: The percentage of students who passed the final exam was the same for both professors.

H_a: The percentage of students who passed the final exam was different for the classes of the two professors.

$H_0: p_1 = p_2$, $H_a: p_1 \neq p_2$

$$\hat{p}_1 = \frac{16}{25} = 0.640, \ \hat{p}_2 = \frac{59}{72} = 0.820, \ \alpha = 0.10$$

Step 2: This is a two-tailed z-test for the difference between two sample proportions. There are two rejection regions, $\frac{\alpha}{2}$, in each tail. In this example, that is 0.05 in each tail. (See Figure 17.7.)

In order to perform this test, you must first ensure that the requirements for using the normal approximation have been met, and calculate the sample proportions $\hat{p}_1$ and $\hat{p}_2$.

$10n_1 = 10(25) = 250$ and $10n_2 = 10(72) = 720$. It is reasonable to assume that the population of the university is greater than 10 times either of these samples.

$\hat{q}_1 = 1 - 0.640 = 0.360,\ \hat{q}_2 = 1 - 0.820 = 0.180$

$n_1\hat{p}_1 = 16 > 5, n\hat{q}_1 = 9 > 5, n_2\hat{p}_2 = 59 > 5$, and $n_2\hat{q}_2 = 13 > 5 \rightarrow$ Normal approximation holds.

Figure 17.7

Two Rejection Regions, $\alpha = 0.05$

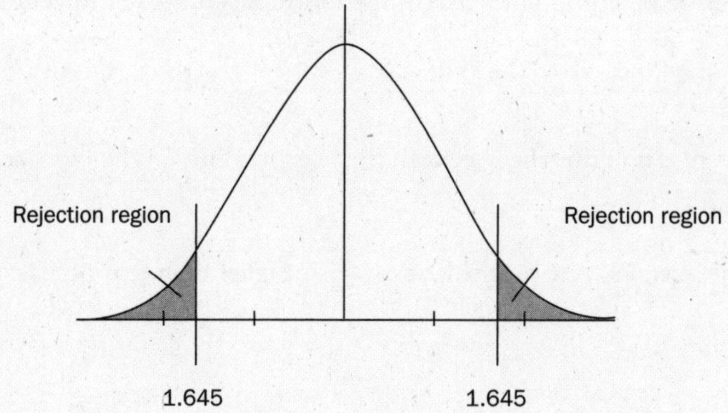

Step 3:

$$z = \frac{(\hat{p}_1 - \hat{p}_2) - (p_1 - p_2)}{\sqrt{\frac{p_1q_1}{n_1} + \frac{p_2q_2}{n_2}}} = \frac{0.640 - 0.82 - 0}{\sqrt{\frac{(0.64)(0.36)}{25} + \frac{(0.82)(0.18)}{72}}} = -1.70.$$

Calculate the p-value:

$$2P(z < -1.70) = 0.0649.$$

We multiply by two because this is a two-tailed test. Finding the probability of $z < -1.70$ would only give us the probability in the left tail.

Is the p-value $< \alpha$? Compare: $0.0649 < 0.10$

Step 4: Since the p-value is less than 0.10, we have evidence against the null hypothesis and can conclude that the proportions of success of the students in the two professors' classes are different.

Example:

The owner of an automotive garage wants to compare his rejection rates for state inspections with one of his competitors. He has had complaints from several customers that his inspections are too tough. Examining the rejection rates for a random sample of inspections for each establishment for the past year, he found that he rejected 28 out of 350 automobiles, and his competitor rejected 32 out of 500 automobiles. At the 10% significance level, is there a significant difference in the rejection rates between the two garages?

Solution:

Step 1:

p_1 = The proportion of automobiles the owner rejects on state inspections.

p_2 = The proportion of automobiles his competitor rejects on state inspections.

$\hat{p}_1 = \dfrac{28}{350} = 0.080 \quad \hat{p}_2 = \dfrac{32}{500} = 0.064$

H_0: $p_1 = p_2$ (There is no difference between the rejection rates of the two garages for state inspections.)

H_a: $p_1 > p_2$ (The rejection rate for the first garage is higher than that of his competitor for state inspections.)

α: = 0.10

Step 2: This is a one-tailed *z*-test for the difference between two proportions.

Checking the conditions/assumptions:

Both proportions were obtained from independent random samples. This was given.

$n_1 p_1 = 350(0.08) = 28 \qquad n_1(1 - p_1) = 350(0.92) = 322$

$n_2 p_2 = 500(0.064) = 32 \qquad n_2(1 - p_2) = 500(0.936) = 468$

Thus,

$n_1 p_1 > 10 \qquad n_1(1 - p_1) > 10$

$n_2 p_2 > 10 \qquad n_2(1 - p_2) > 10$

Step 3: Calculate the *p*-value.

$$z = \frac{(\hat{p}_1 - \hat{p}_2)}{\sqrt{\dfrac{\hat{p}_1(1-\hat{p}_1)}{n_1} + \dfrac{\hat{p}_2(1-\hat{p}_2)}{n_2}}} = \frac{0.08 - 0.064}{\sqrt{\dfrac{0.08(1-0.08)}{350} + \dfrac{0.064(1-0.064)}{500}}} = 0.88064597 \approx 0.8806.$$

$p(z > 0.8806) \approx 0.1894.$

Since 0.1894 > 0.10, the probability of getting a value as extreme as the one we observed is greater than the alpha level of 0.10.

Step 4:

Since the *p*-value is greater than 0.10, there is not sufficient evidence against the null hypothesis. The owner does not have enough evidence to support the customer claim that he rejects more automobiles than his competitor does.

TEST FOR A MEAN

There are two tests for claims about a mean: the *z*-test and the *t*-test. The *z*-test is used when $n \geq 30$ and σ is known (which is rarely the case). The *t*-test is used when the population is normally distributed, $n < 30$, and σ is unknown. In this situation, we substitute *s*, the sample standard deviation, for σ.

The *z*- and *t*-statistic for one-sample hypothesis tests about the mean are

$$z = \frac{\bar{x} - \mu}{\frac{\sigma}{\sqrt{n}}} \text{ and } t = \frac{\bar{x} - \mu}{\frac{s}{\sqrt{n}}} \text{ where } \mu \text{ is the hypothesized mean.}$$

In both cases, the samples must be from a simple random sample design, with $n \geq 30$, or the population approximately normally distributed.

Example:

A random sample of 50 adults identified by their primary care physicians as being overweight were fed a special high-protein diet for a month and had an average weight loss of 8.5 pounds. If the standard deviation of the sample was 3.7, what is the probability that the average weight change for the population of adults identified by their primary care physicians as being overweight would be less than 10 pounds? Use a 0.01 level of significance.

Solution:

Step 1:

H_0: The mean weight loss is 10 pounds for a month on the special high protein diet.

H_a: The mean weight loss is less than 10 pounds for a month on the special high protein diet.

$\mu = 10, H_a: \mu < 10$

$\alpha = 0.01$

Step 2: This is a left-tailed, one-sample *t*-test for the mean where σ is unknown. (See Figure 17.8.)

Since $n > 30$, and we are given that a random sample was used, we can use the *t*-test to validate this claim.

Step 3: $t = \dfrac{\bar{x} - \mu}{\frac{s}{\sqrt{n}}} = \dfrac{8.5 - 10}{\frac{3.7}{\sqrt{50}}} = -2.87$

Calculate the p-value: Given $t = -2.87$ with $df = 49$, we find our p-value $= 0.003$. Compare the p-value to α: $0.003 < 0.01$.

Step 4: Since the p-value is less than 0.01, we have evidence against the null hypothesis and conclude that the average weight loss for the population is less than 10 pounds.

Figure 17.8
Left-Tailed, One-Sample *t*-Test

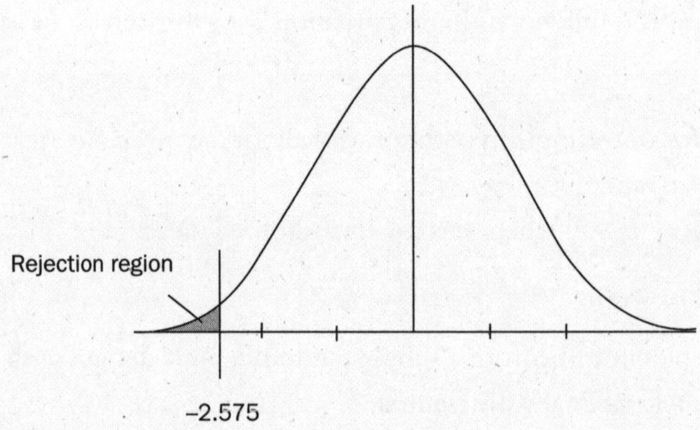

Example:

Previous classes have shown that a standardized final exam has a mean score of 68. A class of 20 students takes the test and has a mean score of 65 with a standard deviation of 13. Is this class typical of other students who have taken the test? Use an $\alpha = 0.05$ level of significance.

Solution:

Step 1:

H_0: The mean score on a standardized final exam is 68.

H_0: The mean score on a standardized final exam is not 68.

H_0: $\mu = 68$, H_a: $\mu \neq 68$

$\mu = 68$, $\alpha = 0.05$

Step 2: This is a two-tailed t-test for a population mean. (See Figure 17.9.)

Figure 17.9
Two-Tailed *t*-Test

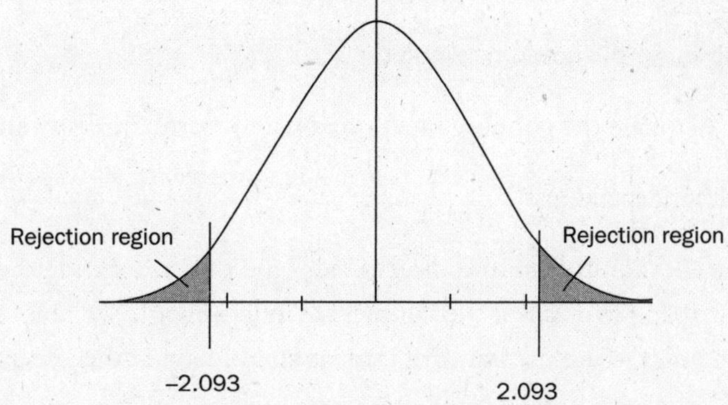

Since $n < 30$, σ is unknown, and we suspect test scores to be somewhat normally distributed, the *t*-test will be most appropriate for this problem.

Step 3: $t = \dfrac{\bar{x} - \mu}{\dfrac{s}{\sqrt{n}}} = \dfrac{65 - 68}{\dfrac{13}{\sqrt{20}}} = -1.03.$

Calculate the *p*-value. *p*-value for $t = -1.03$ with $df = 19$ is between 0.15 and 0.20, but closer to 0.15. Since this is a two-sided test, this probability would need to be doubled. Thus, $0.30 < p\text{-value} < 0.40$. The actual *p*-value is approximately 0.3150. Compare the *p*-value with α: $0.3150 > 0.05$

Step 4: The *p*-value is greater than 0.05, we fail to reject the null hypothesis; we have no reason, to believe the class was not typical of others who had taken the test. Note in the last example that we cannot draw the conclusion that the mean is still 68 based on this sample. The mean could indeed have decreased slightly, but our test cannot validate that claim.

TESTS FOR THE DIFFERENCE BETWEEN TWO MEANS

When testing the difference between two population means, the *z*-test is appropriate only when n_1 and n_2 are both greater than or equal to 30, or the population standard deviations (σ_1 and σ_2) are known. Otherwise, the *t*-test is used. When using the *t*-test, the degree of freedom is the smaller of $n_1 - 1$ and $n_2 - 1$.

The test statistic is:

$$z = \frac{(\bar{x}_1 - \bar{x}_2) - (\mu_1 - \mu_2)}{\sqrt{\dfrac{\sigma_1^2}{n_1} + \dfrac{\sigma_2^2}{n_2}}} \quad \text{and } t = \frac{(\bar{x}_1 - \bar{x}_2) - (\mu_1 - \mu_2)}{\sqrt{\dfrac{s_1^2}{n_1} + \dfrac{s_2^2}{n_2}}}.$$

The assumptions/conditions that need to be verified are the following:

(1) Are σ_1 and σ_2 known?

(2) Were the samples simple random samples?

(3) Is $n_1 + n_2 > 40$ or are the populations approximately normally distributed?

(4) Are the samples independent?*

* We can only use the addition rule to find the variance (and then the standard deviation) of the difference between two sample means if the samples are independent. We must therefore assume independence here in order to use the standard error as an estimate of the standard deviation.

If there is reason to believe that $\sigma_1 = \sigma_2$, you can perform a **pooled test**. A pooled test is used only when the standard deviations of the two populations are equal, or, if they are unknown, there is reason to believe that they are equal. In this case, $df = n_1 + n_2 - 2$.

Example:

A researcher testing sodium content of snack foods examines samples of two brands of potato chips that both claim to be low in sodium. A random sample of 25 lunch-size bags of chips from Super Chips indicates an average of 80 milligrams of sodium per bag, with a standard deviation of 5 milligrams. A second sample of 22 lunch-size bags of Real Potato Snacks shows an average of 77 milligrams of sodium per bag, with a standard deviation of 6 milligrams. Is the difference between the sodium levels of these two brands significant at the 5% level?

Solution:

Step 1:

$n_1 = 25 \quad \bar{x}_1 = 80 \quad s_1 = 5$

$n_2 = 22 \quad \bar{x}_2 = 77 \quad s_2 = 6$

$H_o: \mu_1 = \mu_2$ (There is no difference in the mean amount of sodium between the two brands of potato chips.)

$H_a: \mu_1 \neq \mu_2$ (There is a difference in the mean amount of sodium between the two brands of potato chips.)

$\alpha = 0.05 \quad df = \min\{n_1 - 1, n_2 - 1\} = \min\{25 - 1, 22 - 1\} = \min\{24, 21\} = 21$

Step 2: This is a two-sample t-test for the difference between two means.

Checking the assumptions/conditions:

Simple random sample and independent samples. It is given that both samples were random and independent.

Normal populations $n_1 + n_2 \geq 40 : n_1 + n_2 = 25 + 22 = 47 \geq 40$.

Step 3: Calculate the *t*-statistic.

$$t = \frac{(x_1 - x_2)}{\sqrt{\frac{s_1^2}{n_1} + \frac{s_2^2}{n_2}}} = \frac{80 - 77}{\sqrt{\frac{5^2}{25} + \frac{6^2}{22}}} = 1.847646 \approx 1.8476$$

Calculate the *p*-value

$P(t > 1.721) > P(t > 1.8476) > P(t > 2.080)$
$0.05 > P(t > 1.8476) > 0.025.$

This is a two-sided test, so:

$2(0.05) > 2P(t > 1.8476) > 2(0.025)$
$0.10 > 2P(t > 1.8476) > 0.05.$

Step 4: Since the *p*-value is not less than the alpha value of 0.05, there is not evidence against the null hypothesis. The researcher has no evidence to conclude that the sodium content in the two chips is significantly different.

MATCHED PAIRS

It is not uncommon for two samples to be related in some way, and thus not truly independent. For example, what if we want to compare two methods of doing a particular task on an assembly line, or measure the effectiveness of a test preparation program? Controlling for the difference between workers or students is difficult. However, being able to block out the effects of individuals is essential when determining if one method is really better than another method.

Example:

A consulting company claims that their method of assembly cuts the production time required to build a wooden chair. To test their claim, a manufacturing firm randomly selects 12 of their assembly workers and measures the time it takes for them to build a wooden chair using the firm's current method. The workers are then trained to use the new method, and the firm measures the time it takes for them to build a wooden chair under the new method. The time required for assembly is provided in the table below. Times are in minutes.

Worker Number	1	2	3	4	5	6	7	8	9	10	11	12
Current Method	55	77	89	57	36	64	69	49	56	55	55	57
New Method	48	65	81	49	38	69	58	43	57	48	54	59

Is there a significant improvement in the assembly time with the new method at a 5% significance level?

Solution:

To answer this, we can look at the difference in the performance of each of the 12 workers.

Worker Number	1	2	3	4	5	6	7	8	9	10	11	12
Current Method	55	77	89	57	36	64	69	49	56	55	55	57
New Method	48	65	81	49	38	69	58	43	57	48	54	59
Difference	7	12	8	8	–2	–5	11	6	–1	7	1	–2

When we are looking at the differences for a worker for the two methods, the problem is reduced to a one-sample t-test on the differences. The null hypothesis for matched pairs is that there is no difference between the two methods, or: $H_0: \mu_d = 0$

Step 1:

H_0: There is no difference in assembly time between the two methods.

H_a: It takes less time to assemble a chair with the new method.

$H_0: \mu_d = 0 \quad H_a: \mu_d > 0$

$\alpha = 0.05$

Step 2: This is a right-tailed, matched-pairs t-test for the mean where σ of the differences is unknown.

We are told that this was a random sample of workers.

We can assume that the population of differences is normally distributed.

Step 3: Calculate the mean of the differences.

$\bar{d} = 4.1666 \quad s_{\bar{d}} = 5.6702 \quad n = 12$

$$t = \frac{\bar{d} - d_0}{\frac{s_{\bar{d}}}{\sqrt{n}}} = \frac{4.1666 - 0}{\frac{5.6702}{\sqrt{12}}} = 2.5455 \quad df = n - 1 = 12 - 1 = 11$$

Using the table, we find that the *p*-value for $t = 2.5455$ with $df = 11$ is between 0.02 and 0.01. $0.01 \le p\text{-value} \le 0.02$. (See Figure 17.10.)

Figure 17.10

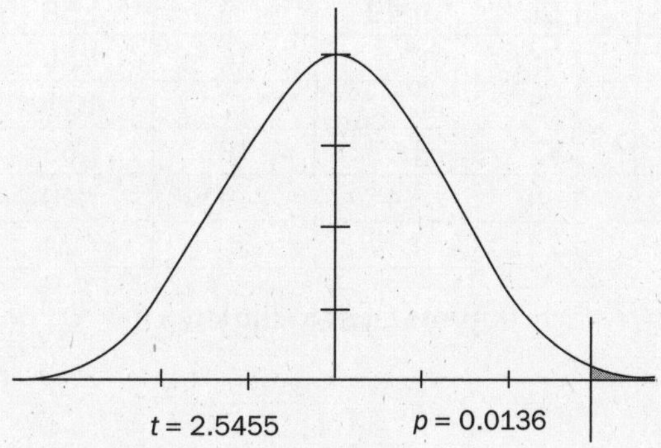

Step 4: Since our *p*-value < 0.05, we reject the null hypothesis and conclude that there is a difference in the two methods of assembling wooden chairs. Assembling a chair with the new method does take significantly less time.

Another type of matched pairs test is what is commonly referred to as **before and after**. Here a measurement would be taken, and then some treatment would be imposed, with the measurement taken again afterward. The two measurements would then be compared to determine if the treatment improved performance, improved a condition, or changed the subject with respect to the desired attribute. For example, the imposed treatment could be a training program, and the measurement would be a test for improvement in performance.

Why not treat these as two samples and use a two-sample *t*-test? Because the samples would not be independent, and thus would violate one of the basic conditions for a two-sample *t*-test. Since we are really interested in the difference, it makes sense to create a single data set based on the difference. The hypothesis becomes that there is no difference between the two procedures, tests, or programs, and thus the mean difference would be zero.

Example:

It has been stated that playing video games has a positive effect on hand-eye coordination. To test this, a physician tested the hand-eye coordination reaction time on a random sample of 20 high school students using a specifically designed test. (For the test, the lower the participant scores, the better his or her reaction time is.) She then had the students play several selected video games for a period of one month, and then tested their hand-eye coordination reaction time again. The results are provided in the following table.

Student	Before	After	Student	Before	After
1	1.57	1.43	11	1.48	1.42
2	1.48	1.52	12	1.75	1.87
3	1.55	1.61	13	1.31	1.34
4	1.54	1.37	14	1.22	0.98
5	1.37	1.49	15	1.58	1.51
6	1.07	0.95	16	1.85	1.59
7	1.35	1.32	17	1.11	0.91
8	1.77	1.68	18	1.65	1.71
9	1.44	1.44	19	1.40	1.38
10	1.27	1.17	20	1.24	1.12

At the 5% significance level, is there improvement in the student's reaction time after playing the video games?

Solution:

Step 1:

H_0: There is no difference in reaction time after using the video games.

H_a: Reaction times are shorter after using the video games.

$H_0: \mu_d = 0 \quad H_a: \mu_d < 0$

$\alpha = 0.05$

Step 2: This is a left-tailed, matched-pairs *t*-test for the mean where σ of the differences is unknown.

We are told that it is a random sample of students.

We can assume that the population of differences is normally distributed.

Step 3: Calculate the mean of the differences.

Student	Before	After	Difference	Student	Before	After	Difference
1	1.57	1.43	0.14	11	1.48	1.42	0.06
2	1.48	1.52	–0.04	12	1.75	1.87	–0.12
3	1.55	1.61	–0.06	13	1.31	1.34	–0.03
4	1.54	1.37	0.17	14	1.22	0.98	0.24
5	1.37	1.49	–0.12	15	1.58	1.51	0.07
6	1.07	0.95	0.12	16	1.85	1.59	0.26
7	1.35	1.32	0.03	17	1.11	0.91	0.20
8	1.77	1.68	0.09	18	1.65	1.71	–0.06
9	1.44	1.44	0.0	19	1.40	1.38	0.02
10	1.27	1.17	0.10	20	1.24	1.12	0.12

$\bar{d} = 0.0595 \quad s_{\bar{d}} = 0.1119 \quad n = 20$

$$t = \frac{\bar{d} - d_0}{\frac{s_{\bar{d}}}{\sqrt{n}}} = \frac{0.0595 - 0}{\frac{0.1119}{\sqrt{20}}} = 2.3379 \quad df = n - 1 = 20 - 1 = 19$$

Using the table, we find that the p-value for $t = 2.3779$ with $df = 19$.

$P(t > 2.205) > P(t > 2.3379) > P(t > 2.539)$

$0.02 > P(t > 2.3379) > 0.01$

Step 4: Since the p-value is between 0.01 and 0.02, it is less than the alpha level of 0.05. Thus, there is evidence against the null hypothesis; the researcher can conclude that there is a difference in the response times after using video games.

Hypothesis Testing and Confidence Intervals: One Sample

Inference Procedure	Assumptions	Test Statistic	Confidence Interval
Mean of population (σ) known	SRS large sample size ($n \geq 30$) or population approximately normally distributed	$z = \frac{\bar{x} - \mu}{\left(\frac{\sigma}{\sqrt{n}}\right)}$	$\bar{x} \pm z^*\left(\frac{\sigma}{\sqrt{n}}\right)$
Mean of population (σ) NOT known	SRS large sample size ($n \geq 30$) or population approximately normally distributed	$t = \frac{\bar{x} - \mu}{\left(\frac{s}{\sqrt{n}}\right)}$ $df = n - 1$	$\bar{x} \pm t^*\left(\frac{s}{\sqrt{n}}\right)$ $df = n - 1$
Proportion	SRS $10n \leq N$ $np \geq 10$ $n(1 - p) \geq 10$	$z = \frac{\hat{p} - p}{\sqrt{\frac{p(1-p)}{n}}}$	$\hat{p} \pm z^*\sqrt{\frac{\hat{p}(1-\hat{p})}{n}}$

Hypothesis Testing and Confidence Intervals: Two Samples

Inference Procedure	Assumptions	Test Statistic	Confidence Interval
Mean of populations σ_1 and σ_2 are NOT known Not pooled	SRS $\sigma_1 \neq \sigma_2$ $n_1 + n_2 > 40$ or population approximately normally distributed Samples are independent	$t = \left(\dfrac{(\bar{x}_1 - \bar{x}_2) - (\mu_1 - \mu_2)}{\sqrt{\dfrac{s_1^2}{n_1} + \dfrac{s_2^2}{n_2}}} \right)$ $df = \min\{n_1 - 1, n_2 - 1\}$	$(\bar{x}_1 - \bar{x}_2) \pm t^* \sqrt{\dfrac{s_1^2}{n_1} + \dfrac{s_2^2}{n_2}}$ $df = \min\{n_1 - 1, n_2 - 1\}$
Mean of population σ_1 and σ_2 are NOT known Pooled	SRS $\sigma_1 = \sigma_2$ $n_1 + n_2 > 40$ or population approximately normally distributed	$t = \dfrac{(\bar{x}_1 - \bar{x}_2) - (\mu_1 - \mu_2)}{\sqrt{s_p^2\left(\dfrac{1}{n_1} + \dfrac{1}{n_2}\right)}}$ $df = n_1 + n_2 - 2$	$(\bar{x}_1 - \bar{x}_2) \pm t^* \sqrt{s_p^2\left(\dfrac{1}{n_1} + \dfrac{1}{n_2}\right)}$ $df = \min\{n_1 - 1, n_2 - 1\}$
Proportion	SRS Independent samples $N \geq 10n_1$ and $N \geq 10n_2$ n_1p_1 and $n_2p_2 \geq 10$ $n_1(1 - p_1)$ and $n_2(1 - p_2) \geq 10$	$z = \dfrac{(\hat{p}_1 - \hat{p}_2) - (p_1 - p_2)}{\sqrt{\dfrac{\hat{p}_1(1 - \hat{p})}{n_1} + \dfrac{\hat{p}_2(1 - \hat{p}_2)}{n_2}}}$	$(\hat{p}_1 - \hat{p}_2) \pm z^* SE$

REVIEW QUESTIONS

1. A bottling company claims there are 2 liters of soda in a large bottle. The Bureau of Weights and Measures believes that the company is cheating the consumer by putting less than 2 liters in a bottle. The bureau decides to conduct an experiment to determine if the consumer is being cheated. Which of the following hypotheses would be appropriate?

   (A) H_0: $\mu = 2$, H_a: $\mu \neq 2$
   (B) H_0: $\mu = 2$, H_a: $\mu < 2$
   (C) H_0: $\mu = 2$, H_a: $\mu > 2$
   (D) H_0: $\mu \neq 2$, H_a: $\mu = 2$
   (E) H_0: $\mu < 2$, H_a: $\mu = 2$

2. A public relations expert is interested in comparing the views of Democrats and Republicans regarding a specific government regulation. Five hundred registered voters are randomly selected. Half of the voters are registered Democrat and half are registered Republican. Each voter is asked if he or she favors a particular regulation. Which of the following tests would be most appropriate for this situation?

   (A) One-sample *z*-test
   (B) Two-sample *z*-test
   (C) Paired *t*-test
   (D) Chi-squared goodness of fit test
   (E) One-sample *t*-test

3. The *z*-test may not be used when

   I. The sample is too small.
   II. The standard deviation of the population is unknown.
   III. The population is not normally distributed.
   IV. The sample is not normally distributed.

   (A) I only
   (B) II only
   (C) III only
   (D) II and IV
   (E) I and IV

4. A farmer is trying to determine if a new chicken feed will put weight on his chickens faster than his current brand. He randomly selected 10 chickens and fed them his standard brand. He took a second sample of 15 chickens and fed them the new brand. After a two-week period, he weighed the chickens. The chickens fed his regular brand had an average weight gain of 4.5 pounds with a standard deviation of 0.8 pounds. The chickens fed the new brand had an average weight gain of 5.5 pounds with a standard deviation of 0.9 pounds. If the farmer tests at the 98% significance level, what is the appropriate conclusion he can draw from this test?

   (A) There is less than a 2% chance that he would get data such as he got if the two brands were the same in producing weight gain.
   (B) There is less than a 2% chance that the two brands are the same.
   (C) There is less than a 2% chance that the new brand is better at producing a weight gain in the chickens.
   (D) There is less than a 2% chance that the farmer's normal brand is better at producing a weight gain in the chickens.
   (E) There is less than a 1% chance there is a difference between the two brands.

5. A researcher conducted an experiment regarding the effectiveness of a new drug. Following the statistical analysis, the results were reported with a p-value of 0.12. Based on this p-value and a significance level of 0.05, which of the following conclusions should the researcher reach?

   (A) Reject the null hypothesis, since p-value of 0.12 is greater than the significance level of 0.05.

   (B) Reject the null hypothesis, since $1 - p$-value is 0.88, which is greater than the significance level of 0.05.

   (C) Fail to reject the null hypothesis, since there is a 12% chance that you could obtain these results when H_0 is true, which is higher than the significance level of 0.05.

   (D) Fail to reject the null hypothesis, since there is an 88% chance that you could obtain these results when H_0 is true, which is higher than the significance level of 0.05.

   (E) Accept the null hypothesis, since the p-value is too large.

6. The analysis of a sample of 250 shoppers at a mall in a large metropolitan area produced a 99% confidence interval that the mean amount spent that day was ($124, $154). Suppose you wish to test the null hypothesis that H_0: $\mu = \$160$ at the $\alpha = 0.01$ level of significance. Can you use the data provided to draw a conclusion?

   (A) Yes; it can be concluded that the mean amount spent is significantly different from $160, since this value is not in the 99% confidence interval.

   (B) Yes; it can be concluded that the mean amount spent is not significantly different from $160, since this value is not in the 99% confidence interval.

   (C) No; the distribution of the population must be known before a conclusion can be drawn.

   (D) No; the data is needed to properly conduct a hypothesis test.

   (E) No; hypotheses cannot be tested based on a confidence interval.

7. The probability of finding a true difference in a hypothesis test can be increased when which of the following is true?

   (A) n is increased and α is increased.

   (B) n is increased and α is decreased.

   (C) n is decreased and α is increased.

   (D) n is decreased and α is decreased.

   (E) None of the above

8. Two students perform a hypothesis test on the same data. One performs a two-tailed test at the $\alpha = 0.05$ level of significance and fails to reject H_0. The other student calculates the same test statistic, but performs a left-tailed test and concludes that the null hypothesis should be rejected. Which of the following values could have been the z-test statistic?

(A) −1.950

(B) −1.610

(C) 1.610

(D) 1.950

(E) Not enough information is given to determine a possible value for the test statistic.

9. A student wished to compare the mean ages of the students in her morning economics class with those in her evening accounting class. There were 32 students in her economics class and 25 students in her accounting class. She asked all students in both classes their age, and calculated the mean age for each class. She then conducted a hypothesis test to determine if the mean ages of the two classes differed based on the t-statistic. This procedure is inappropriate because

(A) the size of the two classes is not equal.

(B) the z-statistic is the appropriate statistic when comparing two population means.

(C) means of two different types of classes cannot be compared.

(D) the ages of the students are probably skewed and therefore the t-test would not be appropriate.

(E) the entire class was polled, so statistical analysis tests would not be required.

10. A certified reference material of steel contains 2.31% of Ni. A new method is tested by analyzing this reference material six times. The mean value of the six measurements is 2.30% and the standard deviation is 0.053%. Which of the following hypotheses would be appropriate?

(A) H_0: $\bar{x} = 2.30$, H_a: $\bar{x} \neq 2.30$

(B) H_0: $\mu = 2.31$, H_a: $\mu < 2.31$

(C) H_0: $\mu = 2.30$, H_a: $\mu > 2.30$

(D) H_0: $\mu = 2.31$, H_a: $\mu \neq 2.31$

(E) H_0: $\mu \neq 2.31$, H_a: $\mu = 2.31$

11. In performing a two-tailed hypothesis test using a sample of size 6, the calculated t-value is 0.46. The t-critical value at alpha = 0.05 and 5 degrees of freedom is 2.571. What does this imply?

(A) Accept H_0

(B) Reject H_0

(C) Fail to accept H_0

(D) Fail to reject H_0

(E) Status of H_0 cannot be determined from the information given.

12. A piece of glass from the scene of a crime is compared with a piece found on the clothing of a suspect. The refractive indices of both pieces are measured five times and the mean and standard deviation for each sample are calculated. A hypothesis test is conducted to determine if there is a difference in the samples at the $\alpha = 0.05$ level. The calculated t-value is 0.073. The t-critical value at alpha = 0.05 and 8 $(5 + 5 - 2)$ degrees of freedom is 2.306. Which of the following statements is true?

    (A) The samples are identical.
    (B) We cannot detect a difference between the two samples.
    (C) This is a one-tailed test.
    (D) There is a difference between the two samples.
    (E) H_0 can be accepted.

13. A study of local elementary school students conducted by the state indicated that students in grades 1 through 3 watched an average of 3.7 hours of television daily, with a standard deviation of 0.8 hours. The Parent-Teacher Organization doubted the results, feeling the average was much lower. If they wanted to challenge the study, the null and alternative hypotheses for their test should be

    (A) $H_0: \bar{x} = 3.7, \quad H_a: \bar{x} < 3.$
    (B) $H_0: \bar{x} = \mu, \quad H_a: \bar{x} < \mu.$
    (C) $H_0: \mu \neq 3.7, \quad H_a: \mu < 3.7.$
    (D) $H_0: \mu = 3.7, \quad H_a: \mu < 3.7.$
    (E) $H_0: \mu < 3.7, \quad H_a: \mu \geq 3.7.$

14. A pharmaceutical company reported a p-value of 0.13 on one of its tests for the effectiveness of a new drug. A physician reading this report could conclude

    (A) the test was statistically significant because a p-value of 0.13 is greater than a significance level of 0.05.
    (B) the test was statistically significant because $p = 1 - 0.13 = 0.87$ and this is greater than a significance level of 0.05.
    (C) the test was not statistically significant because it was a two-sided test and the p-value is really 0.26.
    (D) the test was statistically significant because if the null hypothesis were true, one could expect to get a test statistic at least as extreme as that observed 13% of the time.
    (E) the test was not statistically significant because if the null hypothesis were true, one could expect to get a test statistic at least as extreme as that observed 13% of the time.

15. Which of the following is a criterion for choosing a t-test rather than a z-test when making an inference about the mean of a population?

    (A) The standard deviation of the population is unknown.
    (B) The population is not normally distributed.
    (C) The sample may not have been a simple random sample.
    (D) The sample is not normally distributed.
    (E) The sample size is less than 50.

16. In a test of the null hypothesis H_o: $\mu = 107$ against the alternative hypothesis H_a: $\mu > 107$, a sample from a normal population produces a mean of 113.4. The p-value is 0.0015. Based on these statistics, which of the following conclusion could be drawn?

(A) There is reason to conclude that the alternative hypothesis is true.

(B) Rejecting the alternative hypothesis is an error 1.5% of the time.

(C) The mean is above 107 1.5% of the time.

(D) The mean is below 107 98.5% of the time.

(E) There is reason to be 98.5% certain that the mean is 107.

17. Ace Battery Company produces batteries for automobiles. The company claims that its top of the line battery has an average life expectancy of 65 months. After several complaints about the product, a local consumer reporter tested 18 of the batteries to check this claim. The reporter found that the mean life of these 18 batteries was 63.6 months, with a standard deviation of 2.4 months. What is the approximate p-value for the appropriate test of significance?

(A) 0.00667

(B) 0.99333

(C) 0.01208

(D) 0.02416

(E) 0.98792

18. If a null hypothesis is rejected when it is actually true, then

(A) a Type II error occurs.

(B) a Type I error occurs.

(C) a β error occurs.

(D) a random error occurs.

(E) a power error occurs.

19. The union claims 4 out of 5 of its members do not support a proposed dress code. Management doubts this claim, and randomly samples 50 employees; management finds that 36 of the 50 employees do not support the dress code. Which of the following is an appropriate test outcome?

(A) $z = -1.768$ $p = 0.039$

(B) $t = -1.768$ $p = 0.039$

(C) $z = -1.768$ $p = 0.077$

(D) $z = -1.768$ $p = 0.961$

(E) $t = -1.768$ $p = 0.077$

Use the following information to answer questions 20 and 21.

A doctor wants to test the effect of a restrictive diet on systolic blood pressure. A sample of 8 patients currently being treated for high blood pressure is randomly selected, and all are placed on the diet for one month. The following table gives the patients' blood pressure prior to being put on the diet and after a month on the diet.

Before	187	210	180	224	195	220	231	199
After	183	193	186	233	186	223	220	183

20. The appropriate test for this situation is

(A) a one-sample z-test.

(B) a two-sample z-test.

(C) a matched pairs t-test.

(D) a chi-square goodness of fit test.

(E) a two-sample t-test.

21. The appropriate null hypothesis for this situation is

(A) H_0: $\mu_d = 4.875$.

(B) H_0: $\mu = 205.75$.

(C) H_0: $\mu = 203.313$.

(D) H_0: $\mu_d \neq 0$.

(E) H_0: $\mu_d = 0$.

22. In testing a claim at the 5% significance level about the percentage of the population that supports a ban on using cell phones while driving, the alternative hypothesis is $H_a: p \neq 0.6$. What is the appropriate decision rule?

    (A) p-value < 0.025
    (B) p-value < 0.10
    (C) 2(p-value) < 0.025
    (D) p-value < 0.05
    (E) 2(p-value) < 0.10

23. A student claimed that it took an average of 50 minutes for each student in the class to complete his or her homework assignment. The teacher took issue with this claim, stating that 50 minutes simply did not sound correct. A sample of 40 students was taken and the average amount of time to complete the nightly homework assignment was 47 minutes with a standard deviation of 7 minutes. What is the p-value for this test?

    (A) 0.0049
    (B) 0.0067
    (C) 0.0033
    (D) 0.9950
    (E) 0.0099

24. The power of a significance test against a particular alternative is 91%. Which of the following is true?

    (A) The probability of a Type I error is 91%.
    (B) The probability of a Type II error is 91%.
    (C) The probability of a Type I error is 9%
    (D) The probability of a Type II error is 9%.
    (E) The probability of an alpha error is 9%.

FREE-RESPONSE QUESTIONS

1. School administrators want to know if gender affects the rate of high school students who smoke. They randomly survey 278 students in the local high schools and ask them if they smoke. Of the 125 boys interviewed, 47 are smokers. One hundred fifty-three girls are interviewed, and 52 are smokers. What conclusion can you make?

2. A little league baseball coach wants to compare his team to the national average in scoring runs. Nationally, the average number of runs scored by a little league team in a game is 5.7. He chooses five games at random; in those games, his team scored 5, 9, 4, 11, and 8 runs. How does his team compare to the national average?

3. The Student Council Association (SCA) of a large suburban high school claims that 75% of the student body supports a proposal to reinstate various senior privileges. The principal believes the true percentage is lower, and runs a hypothesis test at the 10% significance level. What is the conclusion if 108 of 150 randomly selected students say that they would support reinstating senior privileges?

4. To test the effectiveness of two medicated ointments for the treatment of poison ivy, a dermatologist randomly selects 12 of her patients suffering from a severe case of poison ivy. The dermatologist chooses two areas of approximately the same size and severity of outbreak on each of the 12 patients, and treats one area with medicated ointment A and the other with medicated ointment B. The number of hours it took the outbreak to subside was recorded for each medication on each patient. The data is provided in the table below.

Patient	1	2	3	4	5	6	7	8	9	10	11	12
Med A	46	50	46	51	43	45	47	48	46	48	37	39
Med B	43	49	48	47	40	40	47	44	41	45	37	43

At the 1% significance level, is there sufficient evidence to conclude that a difference exists in the mean time required to alleviate the poison ivy rash for the two medicated ointments?

ANSWERS AND EXPLANATIONS

1. B

Remember that the null hypothesis *always* contains an equal sign. Therefore, choices (D) and (E) cannot be correct. Since the claim is being made that the consumer is being cheated because there is not enough soda in the bottle, we are dealing with a "less than" situation. Therefore, the alternate hypothesis should contain a < sign, making choice (B) the correct choice.

2. B

Since we are interested in comparing two population proportions, we would need to perform a two-sample test. The only two-sample test we have for comparing proportions is the two-sample *z*-test.

3. B

The *z*-test requires that the data come from a simple random sample; that there is a large sample ($n \geq 30$); or that the population is approximately normally distributed and σ known. Statement I can be overcome if the population is normally distributed. Statement III can be overcome if the sample size is greater than 30. Statement IV is not a necessary condition for use of the *z*-test. Statement II is the only one of the choices that would be at issue in using the *z*-test. The population standard deviation must be known to use the *z*-test, or else you would have to use a *t*-test.

4. A

μ_1 = the mean weight gain with the current brand used by the farmer

μ_2 = the mean weight gain with the new brand

$H_0: \mu_1 = \mu_2$
$H_a: \mu_1 < \mu_2$

This is a two-sample *t*-test with

$df = \min\{n_1 - 1, n_2 - 1\} = 9$

$$t = \frac{(\bar{x}_1 - \bar{x}_2) - (\mu_1 - \mu_2)}{\sqrt{\frac{s_1^2}{n_1} + \frac{s_2^2}{n_2}}} = \frac{4.5 - 5.5}{\sqrt{\frac{0.8^2}{10} + \frac{0.9^2}{15}}}$$
$$= -2.9111.$$

From the *t*-table, the *p*-value for $t = -2.9111$ is between 0.01 and 0.02. Therefore, there is sufficient evidence to reject the null hypothesis and conclude that the new brand does result in a higher weight gain. Specifically, there is less than a 2% chance based on the *t*-test that these results would have been collected if the true weight gain for the two brands were the same. Thus, A is the correct choice.

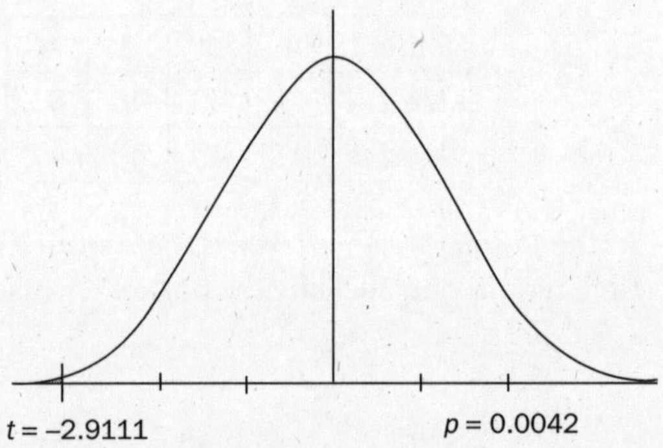

Remember the area defined under the curve represents the probability of getting a result like we got, $\mu_1 = \mu_2$.

5. C

Keep in mind that the *p*-value gives the actual probability of making an error in rejecting the null hypothesis. The value 1 – *p*-value is meaningless in statistical analysis. Also, remember that you *never* accept the null hypothesis. Since the *p*-value, which describes the probability of obtaining your sample results when H_0 is true, is higher than the significance level, you should fail to reject H_0.

6. A

A confidence interval can be used to test a hypothesis provided the confidence level is $1 - \alpha$. There is a direct relationship between hypothesis tests and confidence intervals. If the hypothesized value falls outside the $1 - \alpha$ confidence interval, then the null hypothesis would be rejected. Otherwise, the conclusion is to fail to reject H_0.

7. B

Recall that power is $1 - \beta$. This is defined as the probability of finding a difference if one does indeed exist. As the *P*(Type II error) decreases, the power of the test increases. To decrease the *P*(Type II error), we must either increase *n*, decrease α, or do both.

8. A

When a test rejects for a one-tailed test but not for a two-tailed test, the test statistic must fall between the critical value(s) for these two different tests. A quick check for the bottom line of the *t*-distribution table will give us those two critical values. For the one-tailed test, all the α probability is in one tail. With 5% probability in the tail, the *z*-critical value is 1.645. For the two-tailed test, the α probability must be divided evenly between each tail, leaving 2.5% probability in the tail. The *z*-critical value for this situation is 1.96. The test statistic in this problem must be between 1.645 and 1.96. Since the one-tailed test is left-tailed, the test statistic must be negative. Therefore, the answer would be −1.95.

9. E

Statistical analyses are only used when trying to infer information about a population based on a sample. This student was not interested in using her two classes as samples to test the difference between morning and evening students, but rather was only interested in the mean age of her two classes. Since she interviewed all the students in both classes, she conducted a census. With the results of a census, a statement can be made without the need of inferential statistics.

Besides the fact that the census negates the need for hypothesis tests and confidence interval estimates, the other statements are obviously false. A hypothesis test to compare two means does not require equal sample sizes. The *z*-statistic would not be the appropriate test when either of the samples is < 30. Obviously, comparing two means would involve two different groups of people, in this case the two classes. Finally, ages tend to be normally distributed. In light of the fact that a census was performed on each group and the other statements are false, the correct answer is choice (E).

10. D

Since we already *know* that the sample mean value is (2.31), what is interesting is the *true* value which we are claiming to be 2.31. Therefore, choices (A) and (C) are incorrect, since they make a statement about the sample mean, not the hypothesized mean. We are not interested in a one-tailed difference as shown in (B); if the mean had been larger than the certified value, the analysis method would still be inappropriate. Choice (E) cannot be correct since it includes a statement of inequality in the null hypothesis. Therefore, the correct answer is (D).

11. D

The only valid answers are (B) and (D) since we *never* accept or fail to accept the null hypothesis. The critical region for a two-tailed test is less than −2.571 and greater than 2.571. Since the calculated *t*-value falls between these two values, it is in the fail to reject region. Thus, the correct answer is (D).

12. B

Since the test is being conducted to determine if there is a difference, this is a two-tailed test, eliminating choice (C). The conclusion would be to fail to reject the null hypothesis—which is never "accepted"—eliminating choice (E) and choice (D). Between choice (A) and (B), the latter is correct since we cannot prove the samples are identical—we can only state that there is no detected difference between them.

13. D

The null hypothesis and the alternative hypothesis are always stated with respect to the population parameter. Thus, choices (A) and (B) are incorrect. Further, the null hypothesis proposes a value for the population parameter. Choice (D) proposes a value of 3.7. Then, the alternative hypothesis states what we really feel is true: that it is less than, greater than, or simply not equal to.

14. E

The *p*-value gives the probability that a result as extreme as the one we observed would occur if the true mean was the value being tested in the null hypothesis. In this case, that means that if the null hypothesis were true, we would expect to see a value like the one we observed about 13% of the time. A difference is statistically significant if the *p*-value obtained from the hypothesis test is less than a specified alpha value. Generally, the alpha value is set at 0.05 or 0.10. Choice (A) is incorrect because the test result is significant if the *p*-value is *less than* alpha, not greater than alpha. Choice (D) is incorrect because it states the test is significant if the *p*-value is greater than the alpha value. Choice (B) is incorrect because it finds the complement of the *p*-value. Choice (C) erroneously states that this was a two-sided test, and that the *p*-value should be doubled.

15. A

For both the *z*-test and the *t*-test, the conditions that must be met are that the data come from a simple random sample; that the sample is large ($n \geq 30$); or that the population is normally distributed. The difference between the two tests is that the *z*-test is used when the population standard deviation is known, and the *t*-test is used when the population standard deviation is unknown.

16. A

Recall that the *p*-value indicates the probability that you would observe a value as extreme as the one you observed. If the *p*-value is less than a set alpha value, the error you are willing to live with, then there is evidence against the null hypothesis. Choice (A) states that there is reason to conclude that $\mu > 107$. This is correct since the *p*-value is less than 0.05, a common alpha level. Choices (B) through (E) provide incorrect interpretations of the *p*-value.

17. C

This would be a one-sample *t*-test.

$H_0: \mu = 65 \quad H_a: \mu < 65$

The reporter is testing the claim that the average life of the battery is 65 months; based on complaints, he believes that the average is actually less than 65 months. The population standard deviation is not known, so we must use the *t*-test and not the *z*-test:

$$t = \frac{\bar{x} - \mu}{\frac{s}{\sqrt{n}}} = \frac{63.6 - 65}{\frac{2.4}{\sqrt{18}}} = -2.47487$$

$$df = n - 1 = 17.$$

From the table, $0.01 < p\text{-}value < 0.02$. Using the calculator, we find the *p*-value to be approximately 0.012078.

18. B

A Type I error occurs when a true null hypothesis is rejected. This is also called an alpha error. A Type II error, or beta error, occurs when a false null hypothesis is not rejected. Power is not an error, but the probability of not making a Type II error.

19. A

This is a one-sample z-test for a proportion:

$$H_0: p = \frac{4}{5} = 0.8 \quad H_a: p < 0.8$$

$$z = \frac{\hat{p} - p}{\sqrt{\frac{p(1-p)}{n}}} = \frac{0.7 - 0.8}{\sqrt{\frac{0.8(1-0.0)}{50}}} = -1.768.$$

The p-value is $P(z < -1.768) \approx 0.039$.

20. C

Why not treat these as two samples and use a two-sample t-test? The samples would not be independent, and thus would violate one of the basic conditions for a two-sample t-test. Since we are really interested in the difference, it makes sense to create a single data set based on the difference.

21. E

The null hypothesis for a matched pairs t-test states that there is no difference between the two procedures, tests, or programs, and thus the mean difference would be zero.

22. A

This is a two-tailed, one-sample z-test. At the 5% significance level, there would be $\frac{1}{2}\alpha$ in each tail of the test. Thus, there would be 0.025 in the rejection region at each tail.

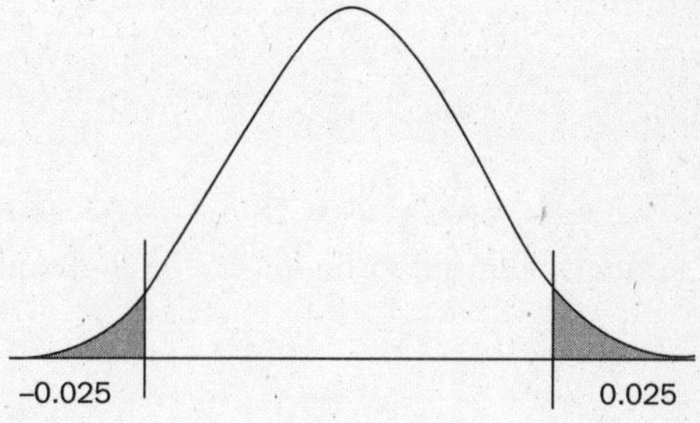

23. E

This is a two-sided test:

$H_0 : \mu = 50 \quad H_a : \mu \neq 50$

$$t = \frac{\bar{x} - \mu}{\frac{s}{\sqrt{n}}} = \frac{47 - 50}{\frac{7}{\sqrt{40}}} = -2.7105$$

$df = n - 1 = 49$

Using the table, the p-value is between 0.0025 and 0.005. This would represent only one tail, so would need to be doubled. Thus, $2(0.0025) < p\text{-value} < 2(0.005)$, $0.005 < p\text{-value} < 0.01$.

Using the calculator, we find that for a two-sided t-test, the p-value would be 0.0099.

24. D

Power is defined as $(1 - \beta)$, where β is the probability of a Type II error. Thus, $91\% = 1 - \beta$ and $\beta = 9\%$.

Free-Response Answers

1. Perform a test of significance to determine if there is a difference between the proportion of boys and girls who smoke. Use a 10% level of significance.

   Let population 1 represent the boys and population 2 represent the girls.

   $$n_1 = 125,\ \hat{p}_1 = \frac{47}{125} = 0.376,\ \hat{q}_1 = 1 - 0.376 = 0.624.$$

   $$n_2 = 153,\ \hat{p}_2 = \frac{52}{153} = 0.340,\ \hat{q}_2 = 1 - 0.340 = 0.660.$$

   Since $n_1\hat{p}_1 = 47$, $n_1\hat{q}_1 = 79$, $n_2\hat{p}_2 = 52$, and $n_2\hat{q}_2 = 101$, all of which are > 10, the normal approximation for the binomial distribution can be used. Appropriate test statistic is z.

   $H_0: p_1 = p_2$

   $H_a: p_1 \neq p_2$

   $\alpha = 0.10$

   Two-sample z-tests for proportions

   $$z = \frac{(\hat{p}_1 - \hat{p}_2) - (p_1 - p_2)}{\sqrt{\frac{p_1q_1}{n_1} + \frac{p_2q_2}{n_2}}} = \frac{0.376 - 0.340}{\sqrt{\frac{0.376 \times 0.624}{125} + \frac{0.340 \times 0.660}{153}}} = \frac{0.036}{0.05782} = 0.623.$$

   p-value for a two-tailed test $= 2 \times P(z > 0.623) = 2 \times 0.2676 = 0.5352.$

   Fail to reject H_0, since the z-statistic falls outside the rejection regions.

   The difference between the proportion of girl and boy smokers could be observed by chance when the proportions are equal. Therefore, the test indicates that there is no significant difference between the smoking rates of boys and girls.

2. $H_0: \mu = 5.7$

   $H_a: \mu \neq 5.7$

   $$\bar{x} = \frac{5 + 9 + 4 + 11 + 8}{5} = \frac{37}{5} = 7.4$$

   $$s = \sqrt{\frac{(5 - 7.4)^2 + (9 - 7.4)^2 + (4 - 7.4)^2 + (11 - 7.4)^2 + (8 - 7.4)^2}{4}}$$

   $$= \sqrt{\frac{33.2}{4}} = 2.88$$

Assume the game scores are somewhat normally distributed. Since $n < 30$, the t-distribution is used to test the mean.

One-sample *t*-test for the mean:

$$t = \frac{\bar{x} - \mu}{\frac{s}{\sqrt{n}}} = \frac{7.4 - 5.7}{\frac{2.88}{\sqrt{5}}} = \frac{1.7}{1.29} = 1.32$$

degrees of freedom = 5 – 1 = 4.

Assume $\alpha = 0.05$ since it is not specified in the problem.

p-value for a two-tailed test = $2P(t > 1.32)$ $0.10 < P(t > 1.32) < 0.15$

$0.20 < 2P(t > 1.32) < 0.30$

p-value > 0.05.

Since the p-value is greater than 0.05, we fail to reject the null hypothesis. Therefore, it appears that the team's scoring is no different from that of the national average.

3. Check conditions/assumptions:

Data are from an SRS. This is given.

$np \geq 10$ $150(0.75) = 112.5 \geq 10$
$n(1 - p) \geq 10$ $150(1 - 0.75) = 37.5 \geq 10$

H_0: 75% of the student body support reinstating senior privileges.
H_a: The % of the student body who support reinstating senior privileges is less than 75%.
H_0: $p = 0.75$
H_a: $p < 0.75$
$\alpha = 0.10$

$$\hat{p} = \frac{108}{150} = 0.72$$

This is a one-sample *z*-test.

$$z = \frac{\hat{p} - p}{\sqrt{\frac{p(1-p)}{n}}} = \frac{0.72 - 0.75}{\sqrt{\frac{0.75(1-0.75)}{150}}} = -0.8485.$$

$P(z \leq -0.8485) \approx 0.1980.$

Since $P(z \leq 0.1980) > \alpha$, there is not sufficient evidence to reject the null hypothesis at the 10% significance level. Thus, there is not sufficient evidence to question the SCA president's claim that 75% of the student body supports reinstating senior privileges.

4. This problem is a matched pairs or paired *t*-test. It tests the difference in the means of the two medications, and is a two-sided test.

H_0: There is no difference between the two ointments in the time it takes to improve the rash from poison ivy.

H_a: There is a difference between the two ointments in the time it takes to improve the rash from poison ivy.

$H_0: d = 0$

$H_a: d \neq 0$

$\alpha = 0.01$

Check conditions/assumptions:

σ is unknown.

Normal population, or $n \geq 30$. It is reasonable to assume that the population of differences would be approximately normally distributed. Since n is small, we should also look at a graphic display of the sampled differences.

The histogram and boxplot of the sampled differences indicate a slight skew, but nothing to prevent the use of the *t*-test for this data.

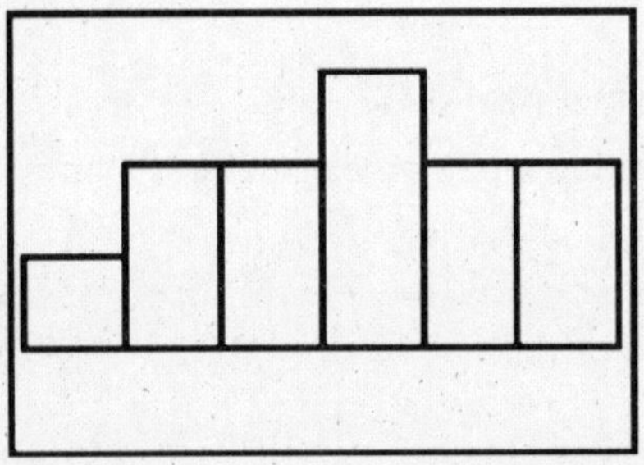

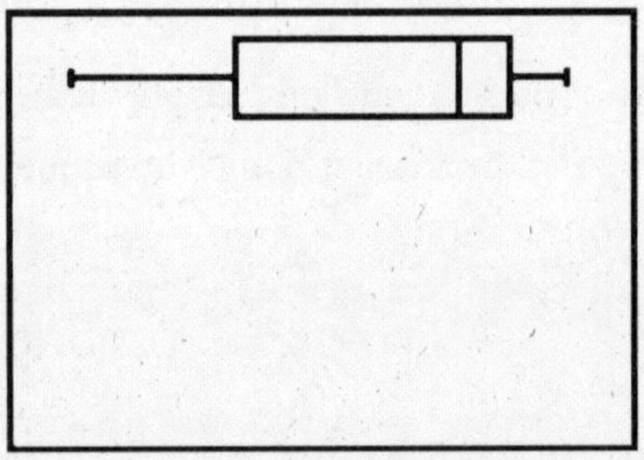

Calculate the differences by patient between the number of hours it took for improvement using the two ointments.

Patient	1	2	3	4	5	6	7	8	9	10	11	12
Med A	46	50	46	51	43	45	47	48	46	48	37	39
Med B	43	49	48	47	40	40	47	44	41	45	37	43
Difference	3	1	–2	4	3	5	0	4	5	3	0	–4

Calculate the mean difference and standard deviation of the difference: $\bar{d} = 1.833$, $s_d = 2.8551$.

$$t = \frac{\bar{d} - d}{\frac{s_d}{\sqrt{n}}} = \frac{1.833 - 0}{\frac{2.8551}{\sqrt{12}}} = 2.2244.$$

Using the t-table, we find that

$df = (n - 1) = 12 - 1 = 11,$

$0.025 > p\text{-value} > 0.02.$

Since the p-value is greater than α, there is not sufficient evidence to reject the null hypothesis at the 1% significance level. Thus, there is not sufficient evidence to say there is a difference in the two ointments in the time it takes the patients to see improvement in their poison ivy.

CHAPTER 18: CHI-SQUARE AND INFERENCES FOR THE SLOPE OF A REGRESSION LINE

IF YOU LEARN ONLY THREE THINGS IN THIS CHAPTER . . .

1. Chi-square analysis is based on the following calculation.

$$\chi^2 = \sum_{i=1}^{n} \frac{(\text{observed value} - \text{expected value})^2}{\text{expected value}} = \sum_{i=1}^{n} \frac{(O-E)^2}{E}.$$

2. There are three basic types of chi-square tests: (1) test for goodness of fit, (2) test for homogeneity, and (3) test of association/independence.
3. The least-squares regression line, $\hat{y} = a + bx$, is the line that best fits the data based on the criteria that the sum of the squares of the residuals was minimized, or the line of averages.

CHI-SQUARE TEST FOR GOODNESS OF FIT, HOMOGENEITY OF PROPORTIONS, AND INDEPENDENCE

The statistical procedures we have thus far reviewed are appropriate at the interval or ratio level of measurement. Sometimes, we may be interested in the relationship between two nominal and/or ordinal variables. In this case, we would utilize the **chi-square** test.

Chi-square techniques help answer questions like the following:

- How do we test for differences among more than two proportions? For example, the racial and ethnic mix of the student population at various college campuses?
- How do we establish the existence of a relationship between two categorical variables?
- How do we judge if the observed data are consistent with the expected patterns?

Chi-square analysis is based on the following calculation.

$$\chi^2 = \sum_{i=1}^{n} \frac{(\text{observed value} - \text{expected value})^2}{\text{expected value}} = \sum_{i=1}^{n} \frac{(O - E)^2}{E}.$$

The p-value is obtained from the χ^2 table and compared to the desired level of significance. Note that the chi-square test is always a one-sided, or one-tail, test. The chi-square distribution is appropriate to use when all individual expected counts are at least 1, and no more than $\frac{1}{5}$, or 20%, of the expected counts are less than 5.

There are three basic types of chi-square tests. They are:

AP EXPERT TIP

Most calculators can quickly perform a chi-squared test (see chapter 19), but it is important to understand all the steps and the formulas in the test.

1. χ^2 **Test for Goodness of Fit:** This test compares the observed sample distribution with the population distribution. Generally, we are testing how well the observations "fit" what we expect. $df = n–1$.
2. χ^2 **Test for Homogeneity:** This test compares more than two groups. This is also referred to as a two-way table, or a contingency table, as the data are in a table or matrix. $df = (r - 1)(c - 1)$ where r = number of rows and c = number of columns.
3. χ^2 **Test of Association/Independence:** This type tests the null hypothesis that there is no relationship between two categorical variables from a simple

random sample with each individual classified according to both of the categorical variables. $df = (r - 1)(c - 1)$ where r = number of rows and c = number of columns.

Recall that the χ^2 distribution is a family of distributions governed by one parameter, the degrees of freedom. At fewer degrees of freedom, the distribution is skewed strongly to the right. As the degrees of freedom increase, the distribution gets closer to a symmetric shape, but never becomes completely symmetric; the χ^2 distribution always has a longer tail to the right. As with all probability density functions, the total area under the curve is equal to 1. Figure 18.1 shows four χ^2 distributions. These curves have 1, 5, 10, and 20 degrees of freedom. Notice how the skew dissipates as the degrees of freedom increase.

Figure 18.1

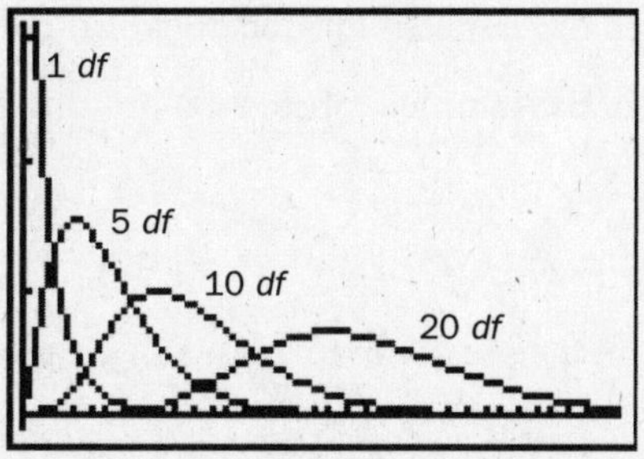

CHI-SQUARE TEST FOR GOODNESS OF FIT

Example:

In one experiment, a scientist observed certain genetic alterations in offspring of lobsters in the Gulf of Maine. She found that 315 had alteration A, 108 had alteration B, 101 had alteration C, and 32 had alteration D. According to her theory, the expected frequencies should follow the ratio 9:3:3:1. Does the sample data lend confirmation to her theory?

This is a χ^2 test for goodness of fit. Of interest is whether the observed genetic alterations found in the lobsters match the theorized distribution of genetic alterations.

Solution:

First, we need to calculate the expected values. Find the sum of the observations.

$$315 + 108 + 101 + 32 = 556.$$

Applying the theorized ratio of 9:3:3:1 to the total, we get the expected values as provided in the table below.

Alteration	Experiment	Theoretical
A	315	312.75
B	108	104.25
C	101	104.25
D	32	34.75

Now we will follow the same steps that we did in our other tests of significance.

Step 1:

H_0: The genetic alterations in the sampled lobsters mirror those of the theorized ratio 9:3:3:1.

Ha: The genetic alterations in the sampled lobsters do not mirror those of the theorized ratio 9:3:3:1.

Step 2: χ^2 test for goodness of fit with $df = n - 1 = 4 - 1 = 3$.

All individual expected counts are greater than 1, and all are greater than 5 as well.

Step 3:

$$\chi^2 = \sum_{i=1}^{n} \frac{(\text{observed value} - \text{expected value})^2}{\text{expected value}}.$$

$$\chi^2 = \frac{(315 - 312.75)^2}{312.75} + \frac{(108 - 104.25)^2}{104.25} + \frac{(101 - 104.25)^2}{104.25} + \frac{(32 - 34.75)^2}{34.75} = 0.470.$$

The chi-square table gives the probability (p-value) of $\chi^2 \geq 0.470$ with $df = 3$ between 0.9 and 0.95. That is $0.9 < P(\chi^2 \geq 0.470) < 0.95$.

p-value > 0.05.

Step 4: The large p-value gives no good evidence to conclude that there is a difference in the distribution of the genetic alterations in the experiment from the theorized distribution.

Example:

The State University at Center City claims that their student body is accepted in the same proportion as the population of the four key areas of the state. The university claims that their acceptance rate adheres to the following pattern, consistent with where the population of the state lives: 45% live in the northern section; 25% in the central section; 10% in the southwestern section; 15% in the southeastern section; and 5% are from outside the state. A group of students question if State University really follows their stated policy. A researcher takes a random sample of 1,000 incoming freshmen and finds the following

OBSERVED	*Northern*	*Central*	*Southwest*	*Southeast*	*Out of State*
Number of Students	487	218	89	147	59

Do these data provide evidence at the 5% significance level that the students are correct in doubting the university's claim?

This is a χ^2 test for goodness of fit. The question of interest here is if the observed distribution of students matches the overall geographical distribution of the state's population.

Solution:

First, we need to calculate the expected values. There are 1,000 students in the sample, so we would expect the numbers to be as follows:

EXPECTED	*Northern*	*Central*	*Southwest*	*Southeast*	*Out of State*
Number of Students	1,000(0.45)=450	1,000(0.25)=250	1,000(0.10)=100	1,000(0.15)=150	1,000(0.05)=50

Step 1:

H_0: The distribution of students at State University at Center City is consistent with the distribution of the population of the state by geographical region.

H_a: The distribution of students at the State University at Center City is not consistent with the distribution of the population of the state by geographical region.

Step 2:

This is a χ^2 test for goodness of fit with $df = n - 1 = 5 - 1 = 4$.

All individual expected counts are greater than 1, and all are greater than 5 as well.

Step 3:

$$\chi^2 = \sum_{i=1}^{n} \frac{(\text{observed value} - \text{expected value})^2}{\text{expected value}}.$$

$$\chi^2 = \frac{(487-450)^2}{450} + \frac{(218-250)^2}{250} + \frac{(89-100)^2}{100} + \frac{(147-150)^2}{150} + \frac{(59-50)^2}{50} = 10.0282.$$

The chi-square table gives the probability (p-value) of $\chi^2 \geq 10.0282$ with $df = 4$ between 0.025 and 0.05. That is, $0.25 \leq p(\chi^2 \geq 10.0282) \leq 0.05$.

Step 4:

Since the p-value is less than the significance level of 0.05, there is sufficient evidence to reject the claim of the university that their acceptance of students matches that of the geographic distribution of the state, and conclude that the students are correct.

CHI-SQUARE TEST FOR HOMOGENEITY OF PROPORTIONS

When there are only two groups, the z-test for the difference between two proportions does the job. But what if we have more than two groups? The chi-square test for homogeneity provides the means for comparing more than two groups. The term homogeneity itself means that things are the same. Thus, the null hypothesis for the chi-square test for homogeneity is that the distribution does not change from one group to the other. The example below tests whether or not there is a difference between boys and girls in their preference for three toys.

Example:

The following table provides the responses of a group of 100 children shown three different toys and asked which one they liked the best. Based on the data, is there evidence that one of the three shows a difference in the preference between the boys and girls at a 5% significance level?

	Toy A	Toy B	Toy C	Total
Boys	25	24	11	60
Girls	9	25	6	40
Total	34	49	17	100

Solution:

To find the test statistic, you must first know the expected frequency for each cell in the contingency table. The expected frequencies are found using the following formula.

$$\textit{Expected Frequency} = \frac{(\textit{row total})(\textit{column total})}{(\textit{grand total})}.$$

In our example, we would find the expected value for the first cell (boys who chose toy A) using the following formula.

$$\textit{Expected Frequency} = \frac{(60)(34)}{(100)} = 20.4.$$

The rest of the expected frequencies are shown in the table below.

	Toy A	Toy B	Toy C	Total
Boys	20.4	29.4	10.2	60
Girls	13.6	19.6	6.8	40
Total	34	49	17	100

Step 1:

H_0: There is no difference in the preference for toys between boys and girls.

H_a: There is a difference in the preference for toys between boys and girls.

$\alpha = 0.05$

Step 2: χ^2 test for homogeneity with $df = (r-1)(c-1) = (2-1)(3-1) = 2$.

All expected cell counts are greater than 1, and there are none less than 5.

Step 3: $\chi^2 = \sum \frac{(O-E)^2}{E}$.

$$\chi^2 = \frac{(25-20.4)^2}{20.4} + \frac{(24-29.4)^2}{29.4} + \frac{(11-10.2)^2}{10.2} + \frac{(9-13.6)^2}{13.6} + \frac{(25-19.6)^2}{19.6} + \frac{(6-6.8)^2}{6.8} = 5.23.$$

p-value $= P(\chi^2 \geq 5.23)$, $0.05 < P(\chi^2 \geq 5.23) < 0.10$.

p-value > 0.05.

Step 4: Since the p-value is greater than 0.05, we do not have evidence against the null hypothesis; therefore, boys and girls do not differ in their choice of toys.

CHI-SQUARE TEST FOR INDEPENDENCE

The chi-square test for independence is performed when the independence between two variables is in question. The tables that lay out the data are called contingency tables. The term contingency table is derived from the concept that what is being tested is whether the distribution of one variable is contingent upon another variable.

Example:

In a study of exercise habits in men working in the health care profession in Chicago, researchers classified the 356 sampled employees according to the level of education they completed and their exercise habits. The researchers want to ascertain if there is an association between the level of education completed and exercise habits at the 5% significance level. The data from the study are provided in the table below.

	Observed Counts of Education Level Completed			
Exercise Habits	College	Some College	HS	Total
Regularly	51	22	43	116
Occasionally	92	21	28	141
Never	68	9	22	99
Total	211	52	93	356

Solution:

Step 1:

H_0: There is no association between educational level completed and exercise habits for men working in the health care profession in Chicago.

H_a: There is an association between educational level completed and exercise habits for men working in the health care profession in Chicago.

$\alpha = 0.05$

Again, we will calculate our expected counts using the following formula.

$$\textit{Expected Frequency} = \frac{(\textit{row total})(\textit{column total})}{(\textit{grand total})}.$$

	Expected Counts of Education Level Completed			
Exercise Habits	College	Some College	HS	Total
Regularly	68.75	16.94	30.30	115.99
Occasionally	83.57	20.60	36.83	141.00
Never	58.68	14.46	25.86	99.00
Total	211.0	52.0	92.99	355.99

Step 2: χ^2 test of association/independence with $df = (r-1)(c-1) = (3-1)(3-1) = 4$.

All expected cell counts are greater than 1, and there are none less than 5.

Step 3: $\chi^2 = \Sigma \frac{(O-E)^2}{E} = 18.5097.$

$p\text{-value} = P(\chi^2 \geq 18.5097),\ 0.0005 < P(\chi^2 \geq 18.5097) < 0.001.$

$p\text{-value} < 0.05.$

Step 4: Since the *p*-value is less than 0.05, there is evidence to reject the null hypothesis; in other words, there is evidence of an association between level of education and exercise habits among men working in the health care profession in Chicago.

CHI-SQUARE TEST FOR INDEPENDENCE VS. CHI-SQUARE TEST FOR HOMOGENEITY

How do you tell which test is the right test? The key to keeping these two tests straight is to focus on the population being tested. In the test for independence, there is one population and you are examining the association or dependence between two different categorical variables about that population. This could involve looking at the adult population in the United States and determining if a particular health condition is associated with gender. Here both gender and the health condition (you have it, or you do not have it) are categorical variables.

The test for homogeneity, on the other hand, compares multiple populations on the same categorical variable. This could involve comparing the number of adults with a certain disease at different points in time—say, at ten-year intervals. For example, the adult populations in 1970, 1980, 1990, and 2000 would be four different populations, and you would compare the proportion of each of these populations that had a disease.

Example:

A political candidate wants to know how a certain issue falls along party affiliation. He commissions a poll to examine the opinions of Democrats, Republicans, and Independents regarding the issue of life appointments for Supreme Court Justices. Which test would he employ?

Solution:

This would be a χ^2 test for independence. We have one population, voting adults, and we are looking at the two categorical variables: political party affiliation and their opinion on the life appointments for Supreme Court Justices.

Example:

An economist is studying the distribution of income. He wants to see if the income distribution in the states of New York, California, Texas, and Florida are comparable. Which test would he employ?

Solution:

This would be a χ^2 test for homogeneity. Here we have four different populations; these are the populations of New York, California, Texas, and Florida. The economist then compares the distribution of income for these four populations. Thus, we have one categorical variable—distribution of income—and four populations.

Example:

Students are comparing the grade distributions of three professors who teach the same course. Which test would they employ?

Solution:

This would be a χ^2 test for homogeneity. Here, the categorical variable is the grade distribution. The three professors are three different populations. The test is to determine if their grade distributions were homogeneous. This setup actually mirrors that of the previous question about the states and income distribution.

Example:

A physician wants to test the effectiveness of two drugs, drug A and drug B, on patients with a particular medical condition. Which test would he employ?

Solution:

This would be a χ^2 test for independence. There is only one population mentioned: those patients who have a particular disease. The test is to determine if the improvement rate among the patients was better with drug A or drug B. The two categorical variables are the drug taken (drug A or drug B), and whether the patient improved or not.

TEST FOR THE SLOPE OF THE LEAST-SQUARES REGRESSION LINE

The least-squares regression line, $\hat{y} = a + bx$, is the line that best fits the data based on the criteria that the sum of the squares of the residuals was minimized. It is the line of averages. The slope, b, and the intercept, a, of the regression line are statistics. They are calculated from sample data. If we drew another sample, the values of *b* and *a* would be different.

The true regression line based on the population, $\mu_y = \alpha + \beta x$, is estimated by $\hat{y} = a + bx$. The question is, "How good are the estimates *b* and *a*?" We have already examined calculating a confidence interval for *b*, and are now going to examine a hypothesis test for the slope.

If there is a relationship between *x* and *y*, we would expect that the slope of the regression equation would not be zero. Therefore, the hypothesis test for the slope of a regression line tests just that. The null hypothesis states that there is no linear relationship between *x* and *y*, or H_0: $H_o: \beta = 0$. Note that we are testing β and not *b*. Hypothesis tests are testing the population parameters.

The alternative hypothesis could be $H_a: \beta(< \text{ or } \neq \text{ or } >)0$.

The hypothesis test to test the null hypothesis is a *t*-test with $df = n - 2$.

Recall that the *t*-value was $t = \dfrac{\overline{x} - \mu}{\frac{s}{\sqrt{n}}}$.

In this case, $t = \dfrac{b - \beta}{SE_b}$ with $df = n - 2$, $SE_b = \dfrac{s}{\sqrt{\sum (x - \bar{x})^2}}$ and

$$s = \sqrt{\frac{1}{n-2}\sum (\text{residuals})^2} = \sqrt{\frac{1}{n-2}\sum (y - \hat{y})^2}.$$

These values are often cumbersome to compute, and are generally provided on the computer output for regression models.

Prior to running a hypothesis test on the slope we need to verify the following:

4. The scatterplot indicates a reasonable linear relationship.
5. The set of observations represents the population, and was randomly selected.
6. The residual plot does not show any curved pattern and is reasonably scattered. It shows little skewness and no extreme outliers.
7. The errors around the regression line at each value of x follow a normal distribution.

Example:

The Fish and Wildlife Agency is interested in being able to estimate the weight of bears based on their lengths. Data was collected from a random sample of 143 bears and a least-squares regression line estimated. The output from this model is provided below.

Regression Analysis: Weight versus Length

The regression equation is: Weight = –422 + 10.1 Length

Predictor	Coef	SE Coef	T	P
Constant	–422.49	31.19	–13.55	0.000
Length	10.1487	0.5031	20.17	0.000

S = 56.07 R-Sq = 74.3% R-Sq(adj) = 74.1%

Analysis of Variance

Source	DF	SS	MS	F	P
Regression	1	1279275	1279275	406.90	0.000
Residual Error	141	443295	3144		
Total	142	1722569			

Based on this information, is there a significant relationship between the length of a bear and its weight? Use $\alpha = 0.01$.

Solution:

Step 1:

H_0: There is no linear relationship between the length of a bear and its weight.

H_a: There is a linear relationship between the length of a bear and its weight.

$H_0: \beta = 0$

$H_a: \beta \neq 0$

$n = 143,\ \alpha = 0.01$

Step 2:

t-test for the slope of a regression line with $df = n - 2 = 143 - 2 = 141$.

The scatterplot is fairly linear as shown in Figure 18.2.

Figure 18.2

Weight and Lengths of Bears

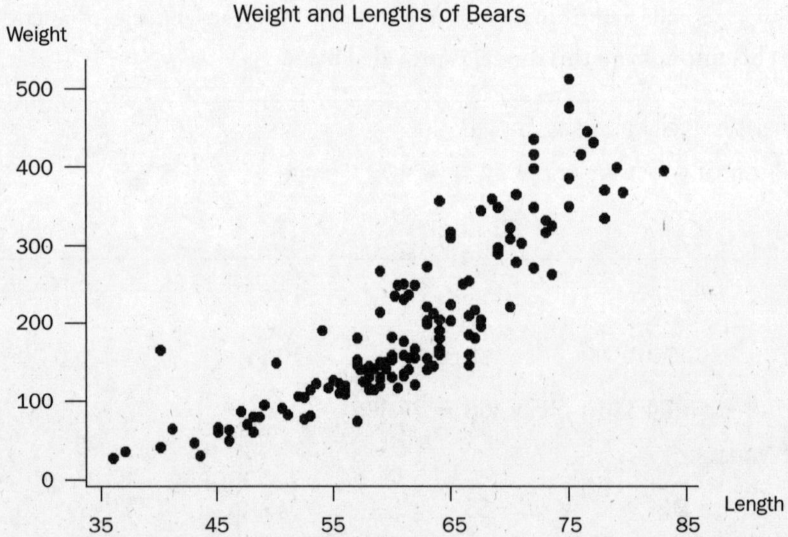

The sample was stated as a random sample of bears.

The residual plot shows no patterns, as shown in Figure 18.3.

Figure 18.3
Residuals Versus the Fitted Values
(response is weight)

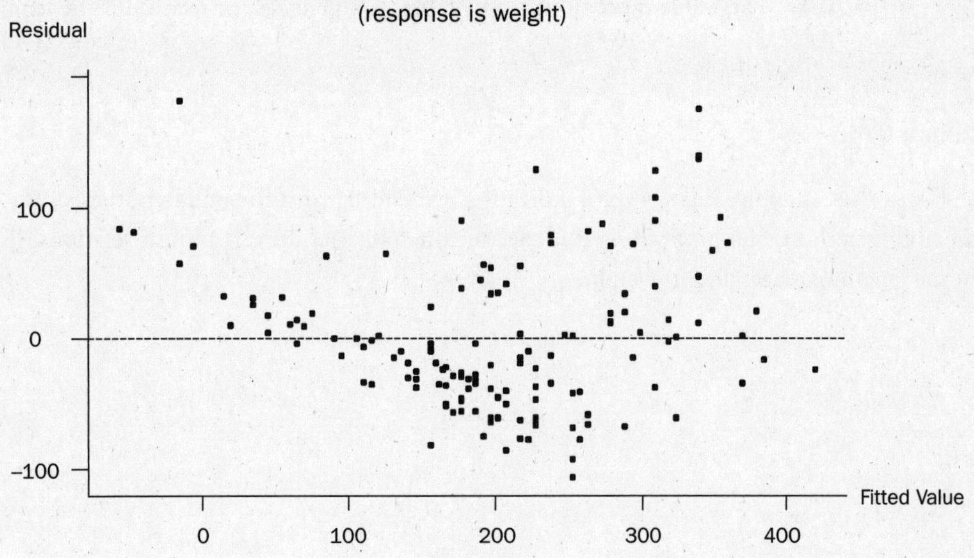

A normal plot of the residuals is fairly linear, indicating normal distribution, as shown in Figure 18.4.

Figure 18.4
Residuals Versus the Fitted Values
(response is weight)

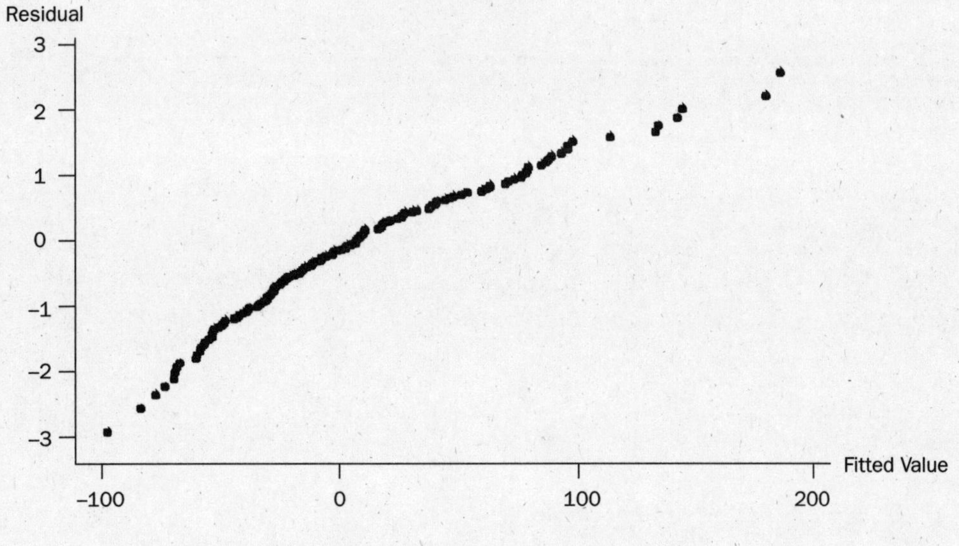

Step 3:

$t = \dfrac{b - \beta}{SE_b} = 20.17$. This value is provided in the table. It is given as the *t*-ratio for the length.

$p\text{-value} = 2P(t > 20.17) \approx 0.$

$p\text{-value} < 0.01.$

Step 4: The *p*-value is approximately zero, indicating evidence against the null hypothesis. We reject the null hypothesis and have strong evidence to conclude that there is a linear relationship between the length of a bear and its weight.

REVIEW QUESTIONS

1. A safety expert claims that twice as many accidents happen on Monday and Friday as happen on the other days of the work week. She randomly selects 98 accident reports and notes the day of the week in which the incident occurred. She finds the following distribution of accidents: Monday—29, Tuesday—15, Wednesday—17, Thursday—13, Friday—24. Which of the following statements is true regarding the χ^2 statistic?

   (A) $\chi^2 < 1$
   (B) $1 < \chi^2 < 10$
   (C) $10 < \chi^2 < 20$
   (D) $20 < \chi^2 < 30$
   (E) $\chi^2 > 30$

2. Using the information in the previous problem, describe the rejection region for the χ^2 test. Use a $\alpha = 0.05$ level of significance.

   (A) To the right of 9.49
   (B) To the right of 14.07
   (C) To the right of 14.45
   (D) To the right of 16.01
   (E) To the right of 12.56

3. The random variable χ^2 has values in what range?

   (A) All real numbers
   (B) −1 to 1
   (C) 0 to 1
   (D) Nonnegative values only
   (E) It depends on the degrees of freedom.

4. The χ^2 distribution is

   (A) symmetric.
   (B) skewed left.
   (C) bimodal.
   (D) unimodal.
   (E) skewed right.

5. You want to check the fairness of a regular six-sided die. You toss the die 100 times and record the results. What would be your expected value in each cell for a χ^2 test for goodness of fit for this data?

   (A) $\frac{1}{6}$
   (B) 0.1667
   (C) 16.667
   (D) 50
   (E) There is not sufficient information to calculate the expected value.

6. A χ^2 test of significance results in a test statistic of $\chi^2 = 32.89$ with $df = 20$. If the test compared the attitudes of proposed employment benefits for three occupational groups at a large metropolitan airport, which of the following conclusions would be valid?

   (A) There is evidence at the 5% significance level that the attitudes of the three groups are different.
   (B) Because there are three groups, the χ^2 test is not the appropriate test.
   (C) There is evidence at the 2.5% significance level that the attitudes of the three groups are different.
   (D) There is not sufficient evidence at the 5% significance level that the attitudes of the three groups are different.
   (E) The information given is not sufficient to draw a conclusion.

7. When performing a hypothesis test for the slope of a linear regression, which is the appropriate test to use?

(A) One-sample *z*-test
(B) One-sample *t*-test
(C) Two-sample *z*-test
(D) Two-sample *t*-test
(E) Chi-square test for goodness of fit

8. The following computer output is for a least-squares regression. Which of the following would be the appropriate null and alternative hypotheses for a test of significance about the slope of the regression line?

Predictor	Coef	SE Coef	T	P
Constant	–257.19	14.29	–18.00	0.000
Chest.G	12.5749	0.3838	32.76	0.000

(A) H_0: $\beta = 0$ H_a: $\beta \neq 0$
(B) H_0: $\beta = 0.3838$ H_a: $\beta \neq 0.3838$
(C) H_0: $\beta = 12.5749$ H_a: $\beta \neq 12.5749$
(D) H_0: $\beta = 32.76$ H_a: $\beta \neq 32.76$
(E) H_0: $\beta = 0.3838$ H_a: $\beta > 0.3838$

Free-Response Questions

1. The State University claims that the ethnic breakdown of its population mirrors that of the state as a whole. The breakdown of the state's population is as follows.

Ethnicity	Percent
White	65.8
Black	18.2
Hispanic	6.7
Asian/Pacific Islander	5.8
Native American	1.8
Not U.S. Citizen	1.7

Several minority groups question that claim. A random sample of 300 students at the university provides the following information.

Ethnicity	Sample
White	216
Black	46
Hispanic	9
Asian/Pacific Islander	14
Native American	3
Not U.S. Citizen	12

Conduct an appropriate test of significance at the 5% level to determine if the ethnic breakdown of the student body matches that of the state.

2. A simple random sample of high school students was asked if they watched the television show *Amateur Pop Singers.* The sample provided the following data.

	Yes	No
Female	91	56
Male	47	23

At the 5% significance level, what conclusion can be drawn if you are testing the null hypothesis that there is a relationship between watching *Amateur Pop Singers* and gender among high school students?

3. A farmer noted that his apple crop had been infested with worms. He took a random sample of trees, and calculated both the number of pieces of fruit per tree and the percent per tree that were infested. The following provides the output from a regression model that he ran. Note that the crop yield is in hundreds of pieces of fruit. Also provided are the scatterplot of the data and a residual plot for the regression.

Predictor	Coef	SE Coef	T	P
Constant	64.247	3.603	17.83	0.000
Crop Yield	−1.0130	0.1722	−5.88	0.000
r	−0.8809	r^2	0.766	

Scatterplot with LSRL

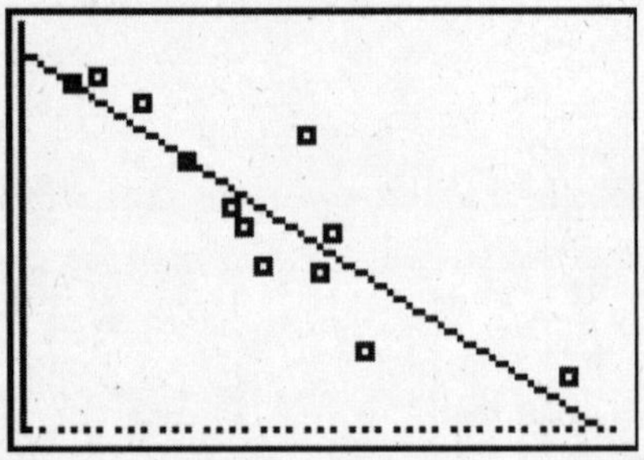

Residual Plot

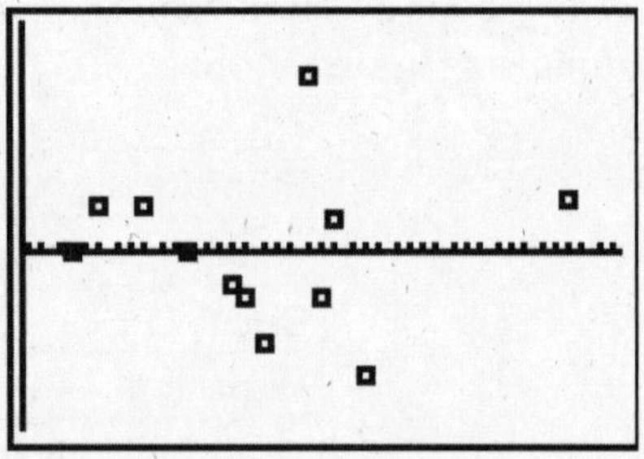

(a) What is the equation of the least-squares regression line (LSRL) that describes the relationship between the percent of wormy fruit and the crop yield? Define any variables used in this equation.

(b) What is the value of the correlation coefficient for percent of wormy fruit and crop yield? Interpret this correlation.

(c) Suppose you want to describe the relationship between the percent of wormy fruit and the crop yield. Does this model do a good job? Why or why not? Describe any improvements you would make to the model.

ANSWERS AND EXPLANATIONS

1. B

In order to compute the χ^2 statistic, you must first calculate the expected frequencies. Since twice as many accidents are expected on Monday and Friday, you must double the number for those days. By dividing the total number (98) of incidents by seven (five days plus two extra division to double weight Monday and Friday), you find that there should be 14 incidents per day with 28 on Monday and 28 on Friday. By using the formula for calculating the χ^2 statistic, you obtain 1.39. Therefore, the χ^2 is between 1 and 10, which is described in choice (B).

$$\chi^2 = \frac{(29-28)^2}{28} + \frac{(15-14)^2}{14} + \frac{(17-14)^2}{14} + \frac{(13-14)^2}{14} + \frac{(24-28)^2}{28} = 1.39$$

2. A

Since the χ^2 test is always right-tailed, all the α probability goes into the tail. Therefore, you would be looking up the tail probability (p) of 0.05.

3. D

The entire curve of the χ^2 distribution lies to the right of the y-axis, and thus assumes only nonnegative values.

4. E

The χ^2 distribution is strongly skewed to the right for small degrees of freedom, and becomes closer to symmetric (bell-shaped) as the degrees of freedom get larger. However, it never becomes truly symmetric, and always remains right-skewed.

5. C

The expected value would be the total number of tosses (100) times the expected distribution. Since you are testing that the die is fair, you would expect each face to come up $\frac{1}{6}$ of the time. Thus $(100)\left(\frac{1}{6}\right) = 16.667$.

6. A

With $df = 20$

$$P(\chi^2 = 31.41) > P(\chi^2 = 32.89) > P(\chi^2 = 34.17)$$
$$0.05 > P(\chi^2 = 32.89) > 0.025$$

Thus, the p-value is smaller than 0.05; we would reject the null hypothesis, which states that there is no difference in the attitudes of the employment groups regarding the proposed changes to the employee benefits.

7. B

The correct test for the hypothesis of the slope of a regression line is the one-sample t-test. Since you are dealing with sample data, you do not have the population values needed to use a z-test. Although you have two variables—one dependent and one independent—you only have one sample.

8. A

The t-test for the slope of a regression line is testing whether or not there is a relationship between x and y. If a relationship exists, the slope will not be zero. Therefore, the null hypothesis is that the slope is zero; the alternative hypothesis is that the slope is *not* zero.

Free-Response Answers

1. **Step 1:**

H_0: The ethnic breakdown of the student body at the State University matches the state's ethnic breakdown as a whole.

H_a: The ethnic breakdown of the student body at the State University does not match the state's ethnic breakdown as a whole.

Ethnicity	Expected Counts
White	300(0.658) = 197.4
Black	300(0.182) = 54.6
Hispanic	300(0.067) = 20.1
Asian/Pacific Islander	300(0.058) = 17.4
Native American	300(0.018) = 5.4
Not U.S. Citizen	300(0.017) = 5.1

Step 2: This is a χ^2 test for goodness of fit with $df = n - 1 = 6 - 1 = 5$.

All individual expected counts are greater than 1, and all are greater than 5 as well.

Step 3:

$$\chi^2 = \sum \frac{(\text{observed} - \text{expected})^2}{\text{expected}}$$

$$\chi^2 = \frac{(216 - 197.4)^2}{197.4} + \frac{(46 - 54.6)^2}{54.6} + \frac{(9 - 20.1)^2}{20.1} + \frac{(14 - 17.4)^2}{17.4} + \frac{(3 - 5.4)^2}{5.4} + \frac{(12 - 5.1)^2}{5.1} = 20.30.$$

$P(\chi^2 > 18.29) > P(\chi^2 > 20.30 > P(\chi^2 > 20.51)$

$0.0025 > P(\chi^2 > 20.51) > 0.001$

p-value < 0.05.

Step 4: Since the p-value is less than 0.05, there is sufficient evidence to reject the null hypothesis and conclude that the university's claim is incorrect. The ethnic breakdown of the campus does not match that of the state's population as a whole.

2. This is a χ^2 test for independence. We have one population (high school students), and we are questioning the association between two different categorical variables (watching *Amateur Pop Singers* and gender).

Step 1:

H_0: Watching *Amateur Pop Singers* and gender are independent.

H_a: Watching *Amateur Pop Singers* and gender are not independent.

Calculate expected values:

To calculate the expected counts, we need to find the sum and row totals from our observed counts. The expected counts will be $\frac{\textit{column total}}{\textit{grand total}}(\textit{row total})$.

	Yes	No	Total
Female	91	56	147
Male	47	23	70
Total	138	79	217

Expected Counts	Yes	No
Female	$\frac{138}{217}(147) = 93.48$	$\frac{79}{217}(147) = 53.52$
Male	$\frac{138}{217}(70) = 44.52$	$\frac{79}{217}(70) = 25.48$

Step 2: χ^2 test of independence with $df = (r-1)(c-1) = (2-1)(2-1) = 1$.

All expected cell counts are greater than 1, and there are none less than 5.

Conditions/assumptions are met.

Step 3:

$$\chi^2 = \sum \frac{(\text{observed} - \text{expected})^2}{\text{expected}} = \frac{(91-93.48)^2}{93.48} + \frac{(56-53.52)^2}{53.52} + \frac{(47-44.52)^2}{44.52} + \frac{(23-25.48)^2}{25.48} = 0.56022.$$

p-value $= P(\chi^2 > 0.56022) > 0.25$.

p-value > 0.05.

Step 4: Since the p-value is greater than 0.05, there is not sufficient evidence to reject the null hypothesis; in other words, there is not sufficient evidence that watching *Amateur Pop Singers* and gender are independent.

3. (a) The equation of the least-squares regression line is

% of wormy fruit = 64.247 – 1.013(crop yield), where crop yield is in hundreds.

(b) The correlation coefficient is $r = \sqrt{0.766} = -0.8809$.

There is a moderately strong negative linear relationship between the percentage of wormy fruit and crop yield.

(c) The model appears to be reasonable. First, the scatterplot shows the negative linear relationship that is supported by the correlation coefficient of –0.8809. The regression line runs nicely through the data. The residual plot shows a reasonable amount of scatter. There does appear to be an outlier in the crop yield data. Removing this may improve the model.

CHAPTER 19: USING THE GRAPHING CALCULATOR

The graphing calculator is an indispensable tool for any statistics student. Because you've already completed a course in AP Statistics, it's safe to assume you are at least familiar with the graphing calculator and its functions. This section is intended as a review of the primary functions of the graphing calculator because students taking the AP Statistics exam commonly use it. The functions described here are shown on a TI-83 plus graphing calculator; however, most major brands of graphing calculators feature similar functions. If you need information specific to your brand and model of calculator, please refer to your instruction manual. If you cannot find your instruction manual, check the Internet—many manufacturers provide electronic versions of manuals and user's guides on their Websites.

CALCULATOR USAGE ON THE AP STATISTICS EXAM

As stated in chapter 1, each student taking the AP Statistics exam is expected to bring a calculator capable of performing statistical functions. The more familiar you are with your calculator's statistics-related functions, the more quickly you will be able to work each problem, and, as with any standardized examination, time is often a critical determining factor in your ability to fully address every question on the exam. Calculation Questions tend to be time-consuming; knowing how to use your calculator could mean the difference between giving careful attention to each question and "last-minute guessing" as the deadline approaches.

AP EXPERT TIP

Calculator commands that are written on the exam are generally ignored by the AP exam graders. Be certain to show work that demonstrates that you understand the calculator commands that you are using.

Unlike some other AP subject exams, you are free to use your calculator on every part of the AP Statistics exam. It is important to remember, however, that the AP Statistics exam is not only about getting the answers right. For the free-response portion, getting a correct answer does not necessarily mean you will receive full credit for that question. You must show all relevant work at every step, as well as demonstrate an understanding of the results, to receive the most credit. When solving a problem with a powerful calculator, it is sometimes easy to forget to explain important steps as you go. It is also tempting to let the numbers "speak for themselves." Just remember: The AP exam graders are already well aware of what your calculator can do. They need to see what YOU can do—through your interpretations, explanations, and experiment designs. Your calculator is powerful, but it is a tool—not a crutch.

CALCULATOR TIPS FOR EXPLORING UNIVARIATE DATA

Summary Statistics

For univariate data, your calculator will provide summary data, boxplots, and histograms. This chapter contains many examples to help explain some of the calculator functions that may prove invaluable when taking the AP Statistics exam. The following data represents math SAT scores for a class of AP Statistics students.

570	730	490	510	740	590	560
570	540	670	500	550	590	540
640	650	720	510	600	790	560
610	740	530	500	660	580	670

How can we use a graphing calculator to process this data? First, enter the data into a list. To enter data, press STAT, then EDIT. This will bring up the List screen, where you can enter your data.

Calculator Screen

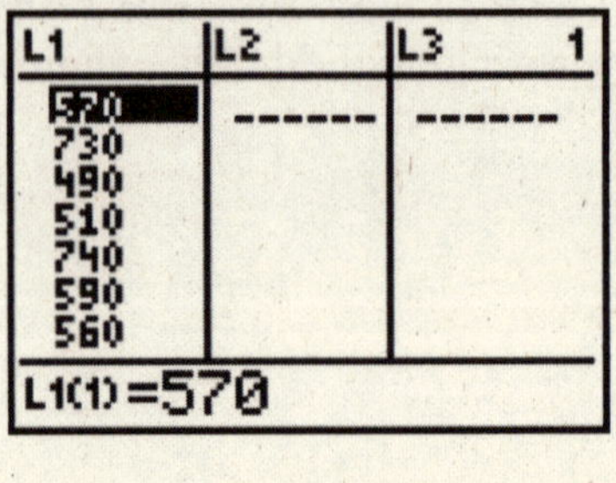

To get the summary data, press **STAT**, then press **CALC** and select **1 – Var Stats**. This will paste **1 – Var Stats** to the home screen. You now need to state where your data is stored. In this example, the data is stored in L_1. You can omit L_1 in this case because it is the default.

Calculator Screen

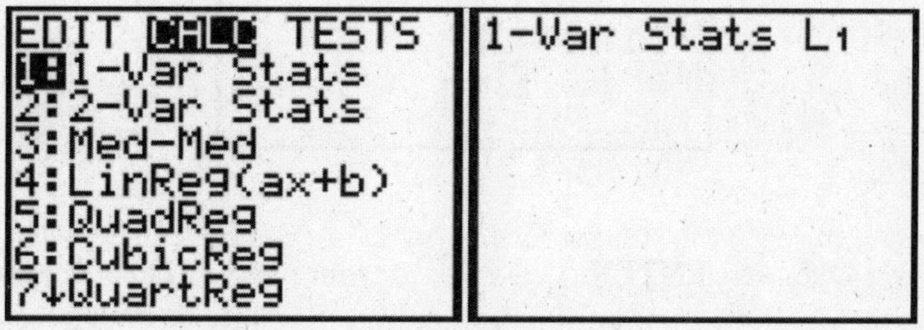

Press **ENTER** and the summary statistics are provided. Note that they are on two screens, and you can scroll between them.

Calculator Screen

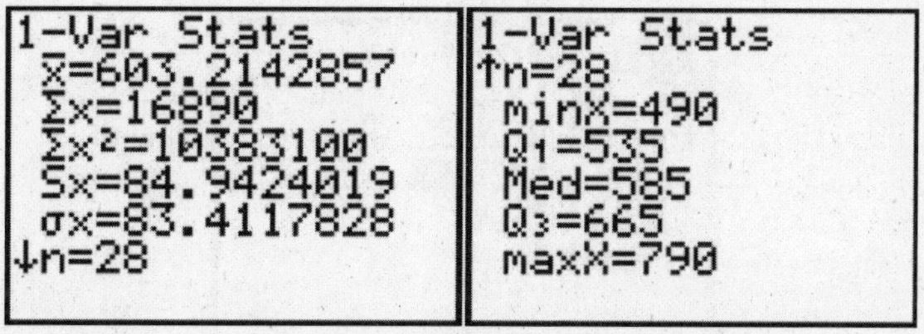

Creating a Boxplot

Boxplots are found under **STAT PLOT**. Press **2nd STAT PLOT** to reach the following screen.

Calculator Screen

Notice in this case that all plots are currently off. Press **ENTER**.

Calculator Screen

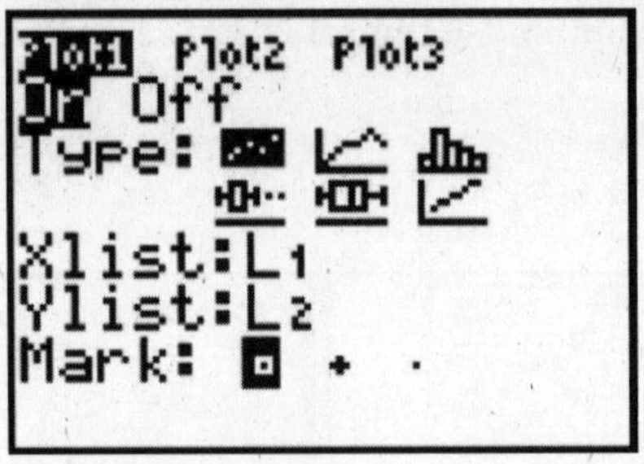

With the cursor over **ON**, press **ENTER**. This will turn your plot on. There are two choices for boxplots. The first places outliers as dots or points beyond the whiskers. The second extends the whiskers to the absolute maximum and absolute minimum values. Use the arrows to move to the boxplot you want, and press **ENTER** to select that choice.

Calculator Screen

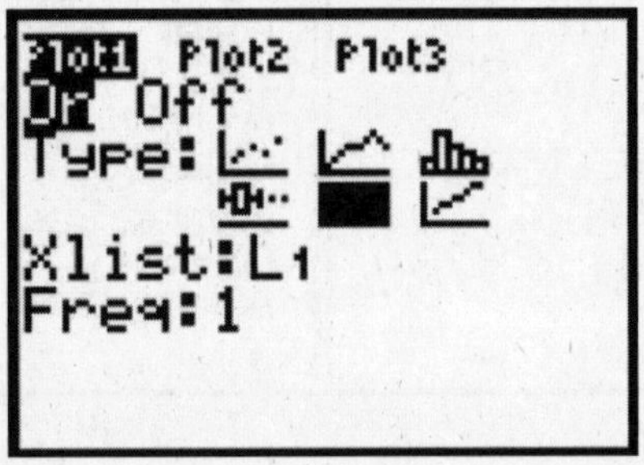

The **Xlist** is the location of the data you want to use to create the boxplot. The **Freq** is the frequency, or number of times you want the data counted. Except in rare instances, this will always be 1. Then press **ZOOM 9**. The **ZOOM 9** function will automatically adjust your window to your data, and your plot will be graphed. By pressing **TRACE** and then using your left and right arrows, you can read the various data values from the plot.

Calculator Screen

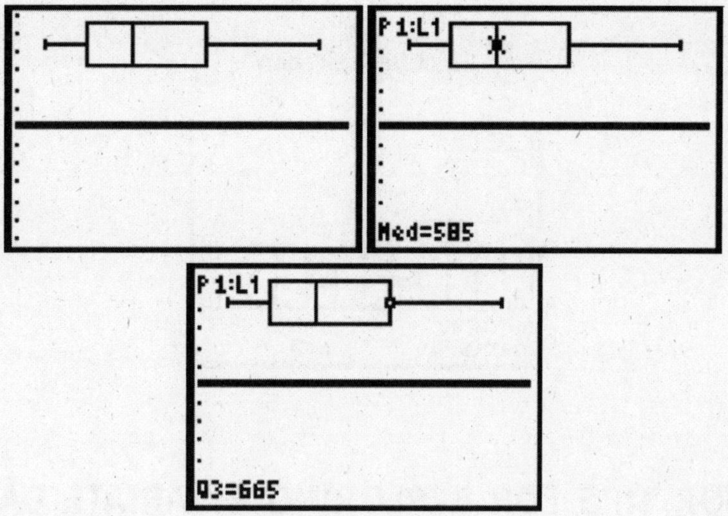

CREATING A HISTOGRAM

The histogram is another type of plot useful in AP Statistics. It can be found as one of the options under **Plot Type**. Use the arrow keys to highlight the histogram, and press **ENTER** to select it. Pressing **ZOOM 9** will automatically set the window to match the data; pressing **TRACE** and using your left/right arrows will give you both the minimum and maximum of the bar, and the number of cases that appear in that interval. You can change the size of the intervals by changing the x-scale in your **WINDOW**.

Calculator Screen

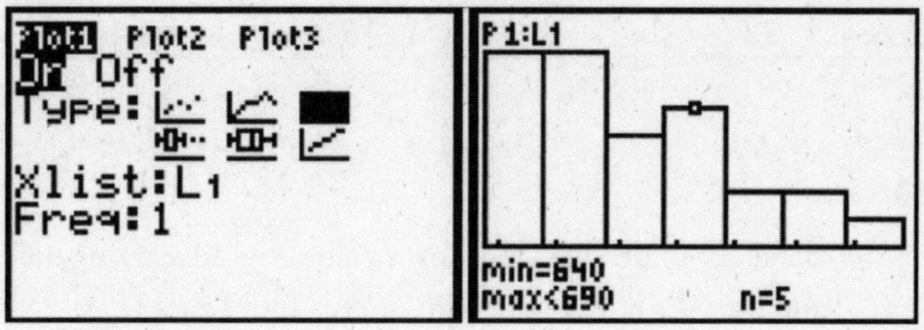

The plot above has an *x*-scale of 50, so the width of the interval is 50. Changing that to 30 gives us the following.

Calculator Screen

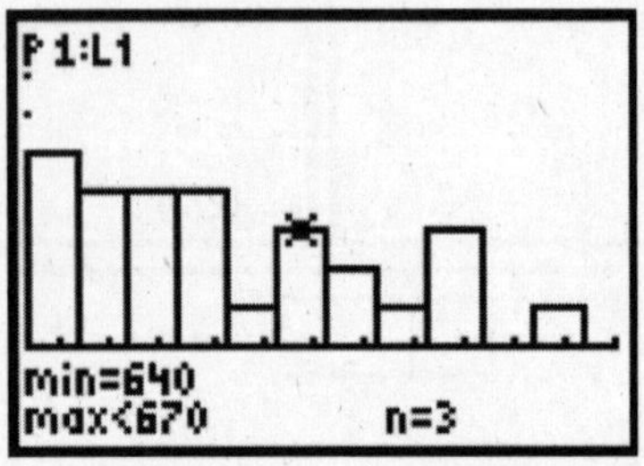

CALCULATOR TIPS FOR EXPLORING BIVARIATE DATA

With bivariate data, you are examining the relationships between two variables, generally a dependent variable and an independent variable. The following data represents the dollars spent on advertising and the dollars realized in sales over an 18-month period for a retail store, with all amounts expressed in hundreds of dollars.

Advertising $	Sales $	Advertising $	Sales $	Advertising $	Sales $
15	12	21	24.5	21	17.3
16	20.5	22	21.3	29	25.3
18	21	28	23.5	62	25
27	15.5	36	28	65	36.5
21	15.3	40	24	46	36.5
49	23.5	3	15.5	44	29.6

SUMMARY DATA

Enter the data into the calculator in two lists. As shown below, the advertising dollar values are entered into List 1 (L_1) and the sales dollar values are entered into List 2 (L_2).

Calculator Screen

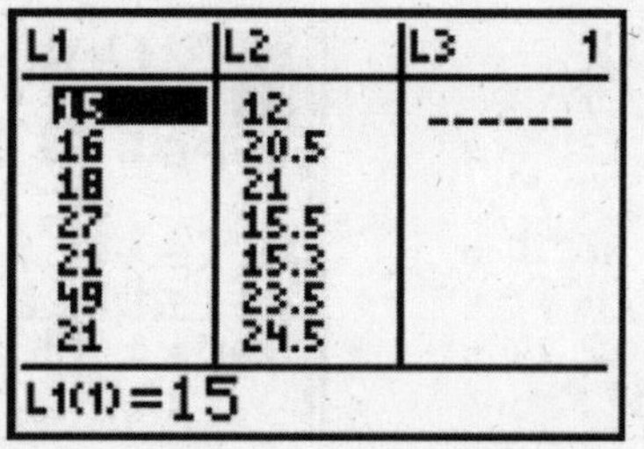

To get the summary data, press **STAT**; then press **CALC** and select **2 – Var Stats**. This will paste **2 – Var Stats** to the home screen. You now need to state where your data is stored. In this case, the data is stored in both L_1 and L_2. You can omit L_1 and L_2 since they are the default lists. Press **ENTER** and the summary statistics are provided. Note that they appear on several screens, and you can scroll between them. The data is labeled as *x* and *y*.

Calculator Screen

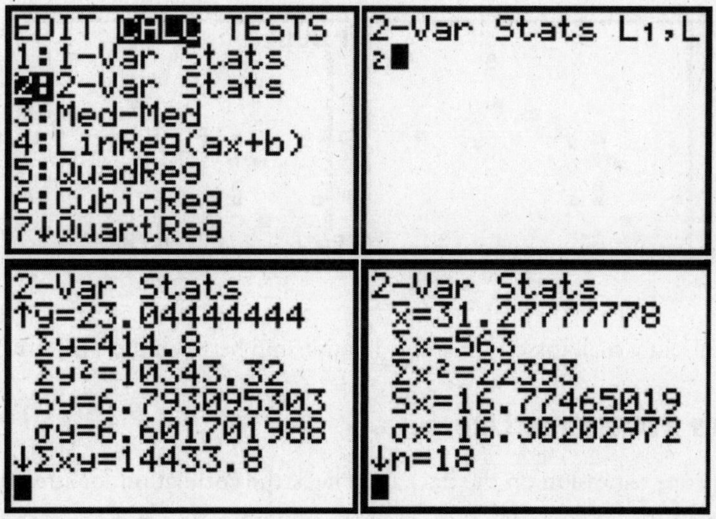

SCATTERPLOTS

Comparing the data by way of a plot is always useful. Press **2nd STAT PLOT** and you get the following screen. Press **ENTER** and the second screen appears.

Calculator Screen

The first option under **Type** is a scatterplot. Below that, you must specify where your *x* data and *y* data are located (which, in this case, means L_1 and L_2, respectively). **Mark** gives you the option of which shape you want the calculator to use to indicate the points. Pressing **ZOOM 9** will reset your window to match the data and plot the data. Pressing **TRACE** will allow you to move using your arrow keys from one point to another on the plot.

Calculator Screen

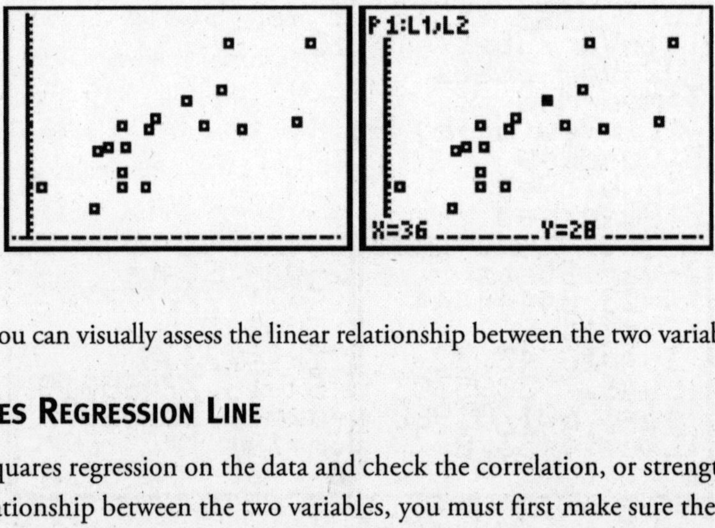

From the plot, you can visually assess the linear relationship between the two variables.

Least-Squares Regression Line

To run a least-squares regression on the data and check the correlation, or strength and direction of the linear relationship between the two variables, you must first make sure the **Diagnostics** are activated on your calculator. Diagnostics provide the correlation coefficient, *r*, and the coefficient of determination, r^2. To activate the Diagnostics, press **2nd CATALOG** and scroll down to **DiagnosticOn**. Once you have pressed **2nd CATALOG**, your calculator is in Alpha mode; this allows you to simply press **D**, and the calculator will quickly scroll to the "D" area of the catalog.

Calculator Screen

With the arrow at **DiagnosticOn**, press **ENTER**; **DiagnosticOn** will be pasted to the home screen. Press **ENTER** again and the Diagnostics will be turned on. After pressing **ENTER**, the calculator will respond with **DONE**. Now, when the regression is run, the correlation coefficient and the coefficient of determination will be provided as part of the output.

To run a regression, press **STAT** and then **CALC**. Under **CALC**, either 4 or 8 will provide a least-square regression line for the data. The example below uses 8, which gives the regression line in the form of $\hat{y} = a + bx$. (This is the more common form used in statistics.) Press **ENTER** and **LinReg (a+bx)** will be pasted to the home screen. Now specify where your data is stored, *x* first and *y* second. The location of the variables must be separated with a comma.

You might find it useful to have the calculator paste the regression equation in the **Y** = window so that you can graph the line. To do this, place a comma after the location of the *y* data and then press **VARS**. Select **Y-VARS** from the top of the screen using your left arrow. Now select **1: FUNCTION**. This will bring up a screen that lists Y_1 to Y_8. Select where your want your regression equation to be pasted, and press **ENTER**. Now you are ready to run the regression. Press **ENTER**.

Calculator Screen

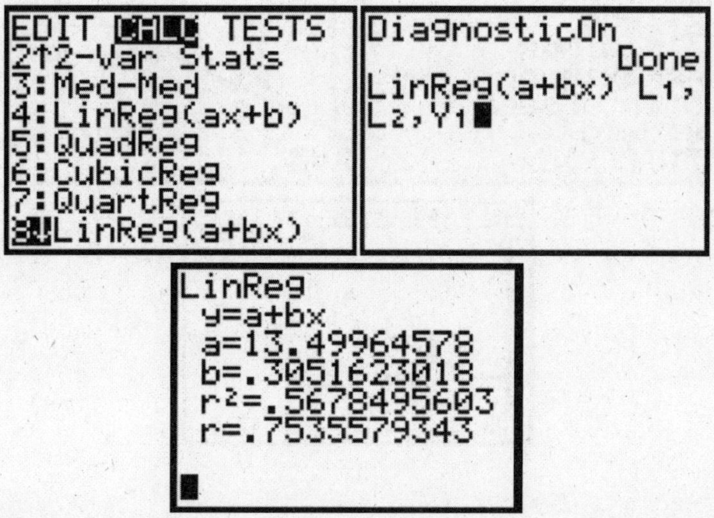

Notice your Y = window. The regression equation is there. Press **GRAPH**. With the equations graphed over the scatterplot, you get a good visual of the fit of the data.

Calculator Screen

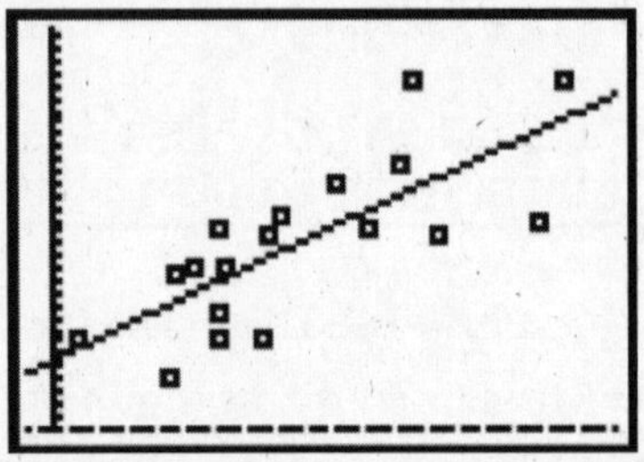

You can also calculate predicted values from the graph. Say we want to predict the amount of sales that could be expected if $3,200 is spent on advertising. Press **2nd CALC** and select **1: value** and press **ENTER**. This will bring you back to the graphed regression line, with an **X** = in the bottom left corner. Type in your value—in this case, 32 (to match the previous data, in which "hundreds of dollars" is assumed). Be careful to check that the data is rounded in the same manner as the data used to develop your regression line. The calculator will then provide you the predicted value, or the *y*-value for your given *x*.

A note of caution: If the *x*-value that you specify is larger than the window, you will get the message **ERR: INVALID**. Simply reset your window to include the *x*-value.

Calculator Screen

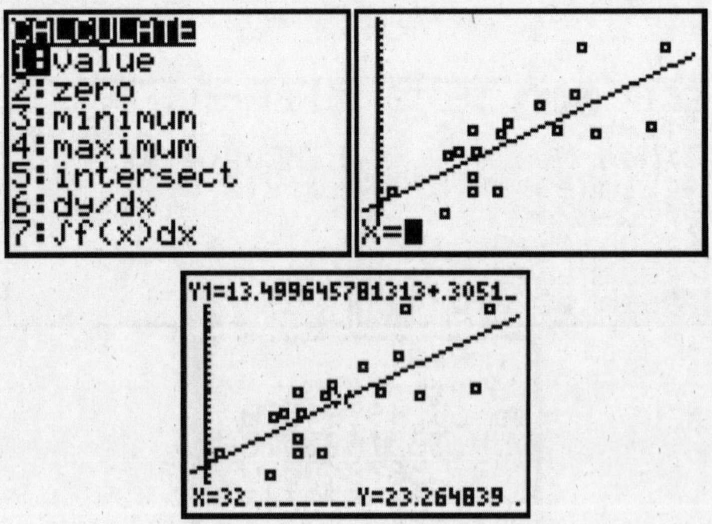

Residuals

With regression lines, you will always want to check the residuals. These are calculated automatically when you run the regression, and are stored in a list called **RESID**. To examine a plot of the residuals, you will want to turn off your current plot of the data and the regression line. To turn the regression line off, place the cursor over the equal sign and press **ENTER**. This way you do not erase the equation in case you need to use it again.

To turn the current scatterplot off, press **2nd STAT PLOT** and then press **ENTER**. Place the cursor over **OFF** and press **ENTER**.

To plot your residuals, press **2nd STAT PLOT** and then press **ENTER**. Select any empty plot from the available list. Turn the plot **ON** and select the scatterplot under **Type**. For the **Xlist**, put the location where your original *y* data was stored (in this case, L_2). For the **Ylist**, you want the residual data. This will give you a plot of the observed data versus the predicted data. With your cursor on the **Ylist**, press **2nd LIST** and select **NAMES**. Scroll through this list and you will find **RESID**. With your cursor over **RESID**, press **ENTER** and it will be pasted as your choice for **Ylist**. Select the mark you would like to use and press **ZOOM 9**.

Calculator Screen

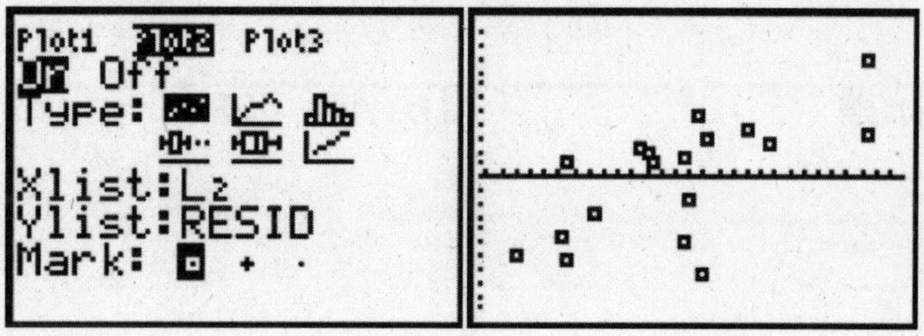

CALCULATOR TIPS FOR PROBABILITY

The Basics

Your calculator has several built-in functions that are useful in probability problems. To access them, press **MATH**, and from the top of the screen select **PRB**.

Calculator Screen

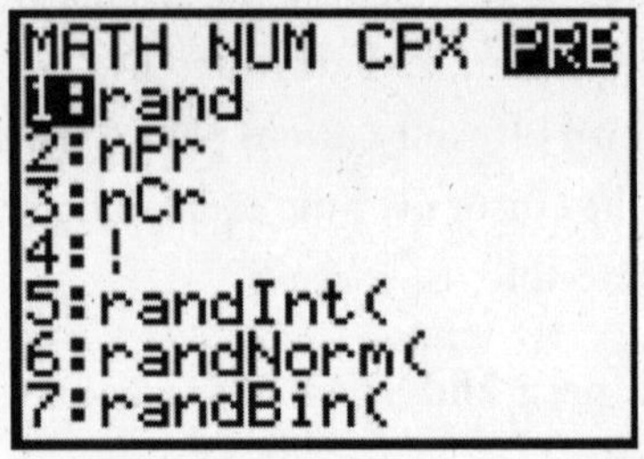

The difference between **rand** and **randInt** is that **rand** will give you a value between 0 and 1; **rand (n)** will give you *n* number of random numbers; **randInt** provides random integers. **randInt (lower bound, upper bound, n)** allows you to specify the smallest integer (lower bound) and the largest integer (upper bound) you want included, as well as the total number (*n*) of random integers you want to generate.

The function **nPr** yields permutations, and the function **nCr** yields combinations. When using the permutation and combination functions, you must first specify *n*. For example, the number of permutations of 10 values taken 3 at a time would be as shown below.

Calculator Screen

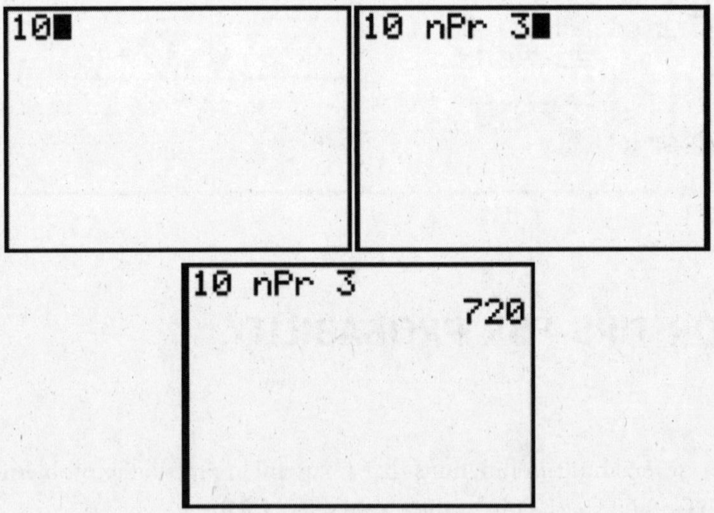

SIMULATION

Simulation on the calculator is best demonstrated with an example. A small commuter airline in Canada knows from past experience that, on average, 90% of its ticketed passengers show up for any flight. With the increase in fuel and other operating costs, it would like to double-book seats to ensure as full a flight as possible. If the plane holds 36 passengers, what is the probability that the company would overbook and agitate customers if they sold 40 seats for each flight?

Because 90% of the ticketed passengers show up for the flight, let the integers 0 to 8 represent the passengers that show up for the flight and 9 represent the passengers that do not show up. We can now generate a number of trials and count how many 9s occur, and thus determine the chances of too many passengers showing up for the flight.

Because the company plans to sell 40 seats, we will generate 40 random integers between 0 and 9 as follows: **randInt (0, 9, 40)**.

To facilitate tallying the number of 9s that occur, we can store the 40 random numbers to a list: **randInt (0, 9, 40) STO** $\rightarrow L_1$

Calculator Screen

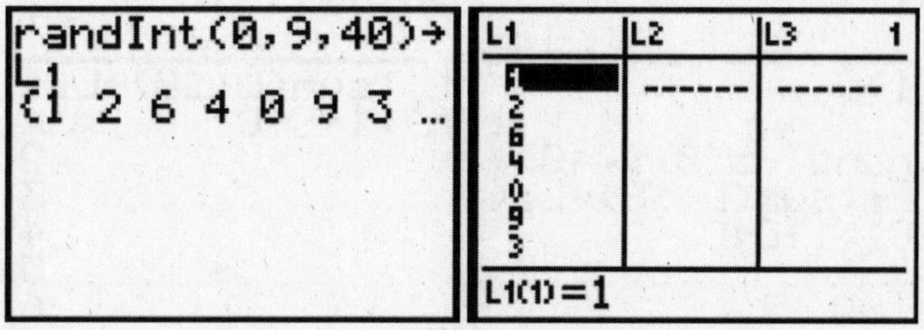

Now, to tally the number of 9s that occurred, the command **sum** $(L_1 \leq 8)$ will sum the number of values 0 to 8 that occurred in the 40 random numbers. In our example, there were 35. Thus, 35 of the 36 seats would be filled even though 40 seats were sold. The **sum** command is under **2nd LIST** and then **MATH**. The ≤ sign is under **2nd TEST**.

Calculator Screen

This simulation can be run multiple times, saving the number of seats that are filled per flight. To do this, you first need to set a counter. This is done by pressing **1 STOR $\rightarrow$ C**. This sets the value of C at 1. The C is just the alpha C, which can be accessed by pressing **ALPHA C**. Once this is done, input the following:

$$\textbf{randInt (0, 9, 40) STOR} \rightarrow \boldsymbol{L_1}\textbf{:sum} \ (L_1 \leq \textbf{8}) \ \textbf{STOR} \rightarrow \boldsymbol{L_2} \ \textbf{(C) :1 + C STOR} \rightarrow \textbf{C}$$

A colon tells the calculator that what follows is a new command. It functions in much the same way as a period at the end of a sentence.

The screen shots below show what your home screen looks like once you type in the formula/program. Then, each time you press **ENTER** and increase the counter, the program is run and a new value is stored in $\boldsymbol{L_2}$. Note that each time you press **ENTER**, your $\boldsymbol{L_1}$ is replaced with a new set of 40 random integers between 0 and 9.

Calculator Screen

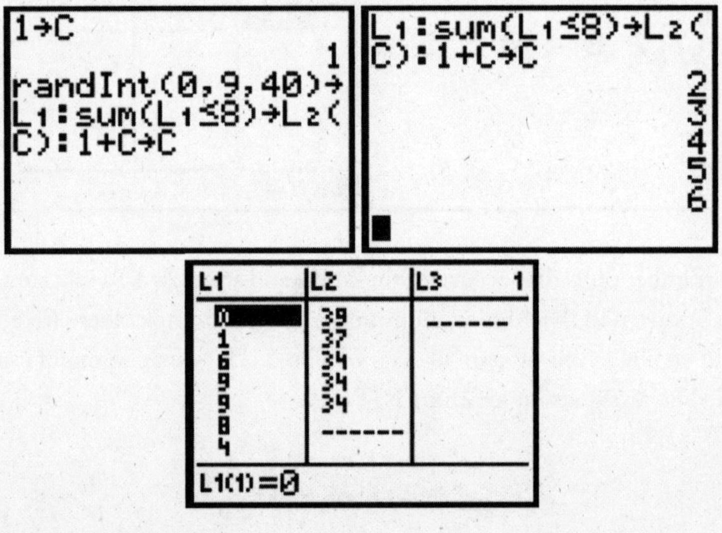

The Normal Distribution

Press **2nd DIST**. The first three menu items relate to the normal distribution. Given the mean and standard deviation of a normal population, you can calculate the probability (area under the curve) of the specific value or values greater than or less than the given value. For **normalpdf**, the arguments required are the specific value you are testing, the mean, and standard deviation of the population. If you do not give the mean and standard deviation, it is assumed they are 0 and 1, respectively. The values are entered as **normalpdf (x, μ, s)**.

For example, given the mean score for the math portion of a placement examination is 83 with a standard deviation of 5.2, we can calculate the probability of a student scoring 90 on the exam.

Calculator Screen

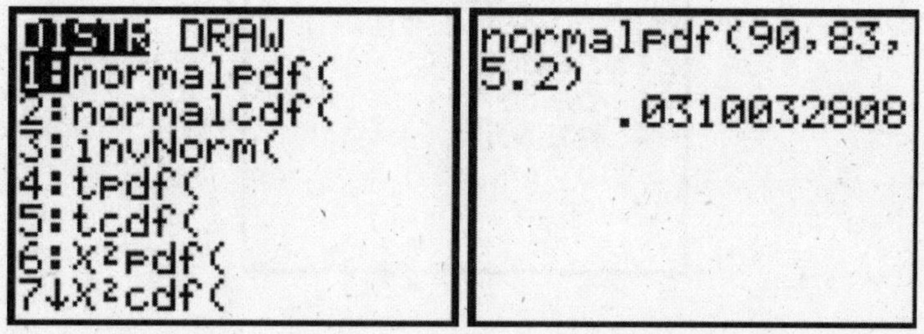

If the question is not the probability of a specific value but of a range of values, we would use the cumulative normal distribution function, or **normalcdf**. The arguments necessary here are the lower limit, upper limit, population mean, and standard deviation. Again, if the mean and standard deviation are not specified, it is assumed that the values have been standardized with a mean of 0 and a standard deviation of 1. These values are input as **normalcdf (lower bound, upper bound, μ, σ)**.

Using the same example, let's calculate the probability of achieving a score of 90 or better.

Calculator Screen

The calculator also gives us the option to draw this situation so that we can visualize what is happening. Notice on the screen listing the various distributions that the second option at the top of the screen is **DRAW**.

If we select that, we get the following screen.

Calculator Screen

ShadeNorm draws the normal density function specified by μ and σ and shades the area between the lower and upper bounds. It is entered as **ShadeNorm (lower bound, upper bound, μ, s)**. For our example, that would be: **ShadeNorm (90, 100, 83, 5.2)**. To activate the draw, press **ENTER**. Make sure you have adjusted your window and turned off all other plots and graphs.

Calculator Screen

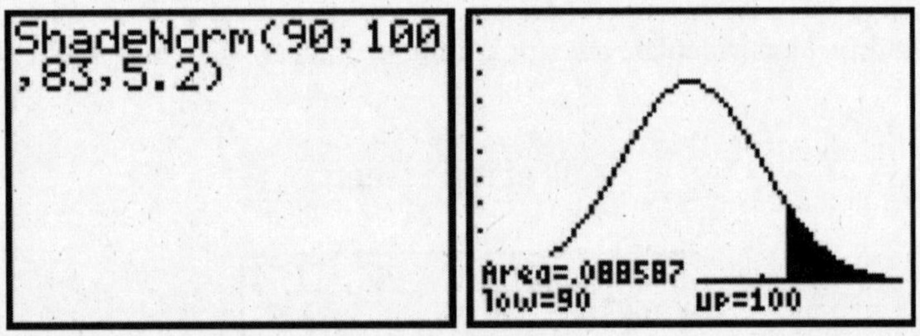

An easy way to set your window is as follows:

$X \text{ min} = \mu - 4\sigma$

$X \text{ max} = \mu + 4\sigma$

$xscl = \sigma$

$Y \text{ min} = 0$

$Y \text{ max} = normal\ pdf\,(\mu,\mu,\sigma)$

$yscl = 1$

Simply replace μ and σ based on the specifics of the problem.

Sometimes we know the probability of something occurring and need to work back to the specific value that would give that probability. The **invNorm** function computes the inverse cumulative normal distribution function for the given area and the specified mean and standard deviation.

That is, it computes the z-score for the standardized normal distribution, or the actual score/value that would give that area for any other normal distribution. The values for this function are input as **invNorm (probability, μ, σ)**.

Using our example from above, say we want to know the score required to be in the top 10% of those taking the exam. This means that 90% of the exam scores are lower than your score, so the input values are **invNorm (0.90, 83, 5.2)**.

Calculator Screen

ASSESSING NORMALITY OF DATA

Frequently, we need to determine if the data is normal. As you recall, the normal probability plot will aid you in assessing the normality of data. The calculator will perform the normal probability plot for you. It is found under **STAT PLOT**. Press **2nd STAT PLOT**, turn the plot on, and select the last of the six graphic options, and then specify where your data is located and whether you prefer the data to be shown on the x- or y-axis. Then press **ZOOM 9**.

Calculator Screen

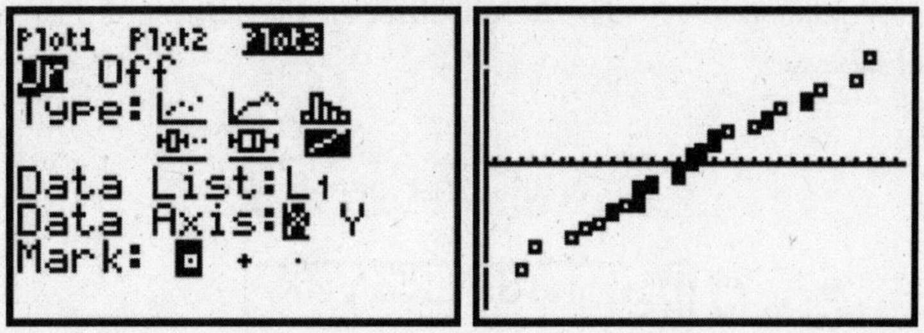

Because the normal probability plot is fairly linear, the data is approximately normal.

BINOMIAL PROBABILITY DISTRIBUTIONS

The question answered by the binomial distribution is, what is the probability of getting x success in n trials? Your calculator provides two binomial distribution functions: (1) the binomial probability density function, or **binompdf**; and (2) the cumulative binominal density function, or **binomcdf**. The **binompdf** will give the probability at the value specified. For example, if we know that the rate of defective field hockey sticks from a certain manufacturer is 7%, what is the probability that in a shipment of 15 sticks, 5 will be defective? The arguments required by the calculator are the number of trials, the probability, and the number of "successes" desired (keeping in mind that "success" in this case means finding a defective field hockey stick). These values are input as **binompdf (number of trials, P, X)**. In our example, the values would be **binompdf (15, 0.07, 5)**.

The probability density functions are found under **DISTR** on your calculator. Press **2nd DIST**. The **binompdf** is option 0.

Calculator Screen

The probability of getting 5 defective sticks in a shipment of 15 is 0.0024.

Let's change the question. What is the probability that, in a shipment of 15 sticks, we would get *at most* 5 defective sticks? Here we want $P(X=0) + P(X=1) + P(X=2) + P(X=3) + P(X=4) + P(X=5)$. This is done easily with the cumulative density function, **binomcdf**. (In the list of distributions on your calculator, this is choice A.) The calculator sums the probabilities up to the specified value. The arguments required by the calculator are the number of trials, the probability, and the number of successes.

Calculator Screen

Now imagine a different question. What is the probability that, in a shipment of 15 sticks, we would get *at least* 5 defective sticks? That would be $P(X = 5) + P(X = 6) + P(X = 7) + \ldots +$ $P(X = 13) + P(X = 14) + P(X = 15)$. This is also $1 - P$ (at most 4 defective sticks). We can calculate this probability as shown below.

Calculator Screen

Geometric Probability Distributions

In the case of a binomial random variable, the number of trials is fixed and the binomial variable X counts the number of successes in that fixed number of trials. Situations in which the goal is to determine a fixed number of trials before the first success are addressed using the geometric probability distribution. We want to determine the number of trials needed to obtain the first success.

As in the case of the binomial distribution, the calculator provides you two choices: (1) the geometric probability density function, **geometpdf**; or (2) the cumulative geometric probability density function, **geometcdf**. For both of these distributions, the arguments required by the calculator are the probability and the trial number of the first success.

Example:

If 7% of the field hockey sticks are defective, what is the probability that the first defective stick in a shipment will be the fourth one examined?

Solution:

The geometric density functions are found under **DISTR** on your calculator. Press **2nd DIST**. The **geometpdf** is choice D. To answer the question above, we would use **geometpdf (0.07, 4)**.

Using the cumulative geometric probability density function, we could answer the question, "What is the probability that the first defective stick is found in the first four that are examined?" This would be **geometcdf (0.07, 4)**.

Calculator Screen

CALCULATOR TIPS FOR STATISTICAL INFERENCE

Your calculator can be used to perform many of the operations and tests necessary to draw conclusions or inferences from the data you are given. Just remember: *While the calculator can crunch the numbers, it is up to* **you** *to interpret the results.*

Press **STAT** and select **TESTS**. You will see the various tests of significance and confidence interval calculations that are built into your calculator.

Calculator Screen

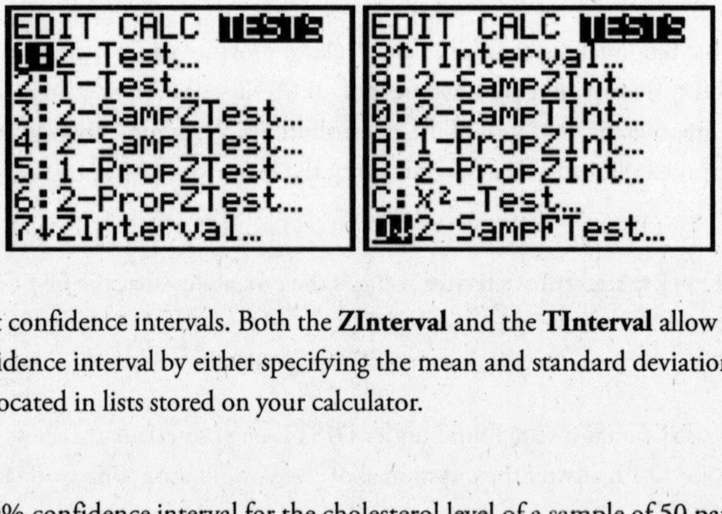

Let's first look at confidence intervals. Both the **ZInterval** and the **TInterval** allow you to calculate the desired confidence interval by either specifying the mean and standard deviation or by using the data that is located in lists stored on your calculator.

Example:

Calculate the 90% confidence interval for the cholesterol level of a sample of 50 patients at a local hospital, given that $\overline{x} = 178$ and $\sigma = 12$.

Solution:

Press **STAT** and select **TESTS**. Under **TESTS**, select 7: **ZInterval**. Specify the statistics from the problem and press **ENTER**.

Calculator Screen

If your data is in a list, you have the option of specifying the location of the data and the value of σ, and letting the calculator do the rest.

Example:

Data for the intelligence quotients of 15 randomly selected students taking AP Statistics at Central County High School has been placed in L_1. For this particular test, $\sigma = 8$. Calculate a 95% confidence interval about the mean score.

Solution:

Calculator Screen

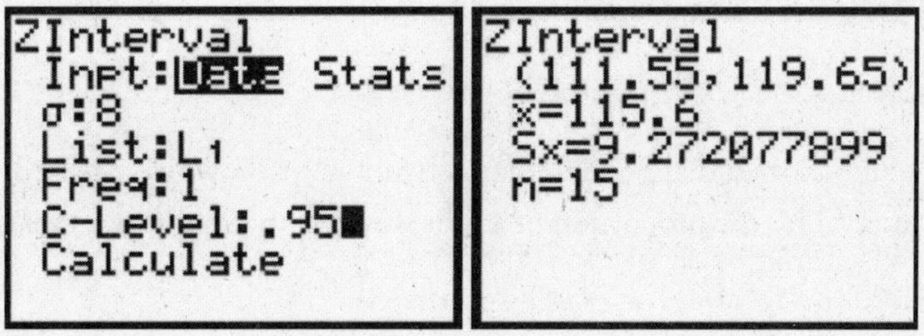

The **TInterval** works in exactly the same manner as the **ZInterval**. Recall that you use the **ZInterval** if σ (the standard deviation for the population) is known, and **TInterval** if it is not known.

For comparing two samples, we can use **2-SampZInt** or **2-SampTInt**. Again, which one you use depends on whether or not σ_1 and σ_2 are known. You still have the option of using statistics from the samples or the data, if it is entered into lists on the calculator.

Notice that you are prompted for information on both samples.

Calculator Screen

CONFIDENCE INTERVALS FOR PROPORTIONS

The calculation for determining the confidence intervals for proportions requires that the actual counts be entered into the calculator and not the proportions. The calculator will then compute the proportion.

Calculator Screen

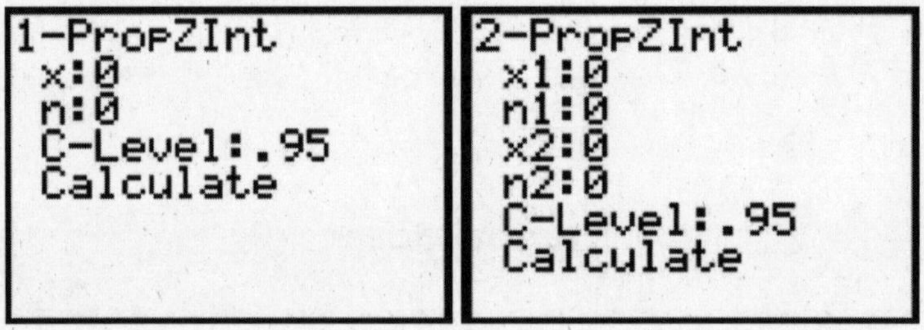

Example:

A sample of 550 adults showed that 387 felt that network news shows were biased. Calculate a 95% confidence interval for the proportion of adults who feel that network news is biased.

Solution:

Once the data is entered and the cursor is highlighting the option **Calculate**, press **ENTER**.

Calculator Screen

Remember, you do not enter the proportion of the sample when estimating a confidence interval about a proportion; you must enter the actual count.

TEST OF SIGNIFICANCE FOR THE MEAN OF ONE SAMPLE

The **Z-Test** and **T-Test** functions allow you to test a specified null hypothesis.

Notice that the setup on the calculator is the same for both tests. Again, you have the option of providing the summary statistics or using the data as entered into a list on your calculator.

The alternative hypothesis you are testing is specified in the second to last line.

The available choices are H_a: $\mu \neq \mu_0$, H_a: $\mu < \mu_0$, and H_a: $\mu > \mu_0$.

Calculator Screen

Also note that you are given the option of calculating the test or drawing the test.

Suppose we know from a large sample of adult men monitored over a two-year period that the distribution of their systolic blood pressure is approximately normal, with $\overline{x} = 131$. A random sample of 78 adult men from New York State is taken, and their mean systolic blood pressure is found to be 127 with a standard deviation of 21. Does this prove that, on average, adult men in New York State have a lower systolic blood pressure?

Enter the appropriate values on the **T-Test** screen, highlight **Calculate**, and press **ENTER**. The resulting screen gives you a statement of your alternative hypothesis, the *t*-statistic, and the probability that the *t*-statistic would be less than the given *t*. Based on your stated level of significance, you can now analyze the results and answer the question. If $\alpha = 0.05$, then the given *p*-value is less than α and we would have evidence to reject the null hypothesis in favor of the alternative. In other words, the systolic blood pressure of adult men in New York State is lower than that of adult men as a whole. Remember: *It is important to state your results within the context of the question.*

Calculator Screen

Selecting the **Draw** option would give you the following graph. The area less than $t = -1.6822$ is shaded. It is the size of this area that we are interested in analyzing. Below the diagram, you are given the *t*-statistic and the *p*-value.

Calculator Screen

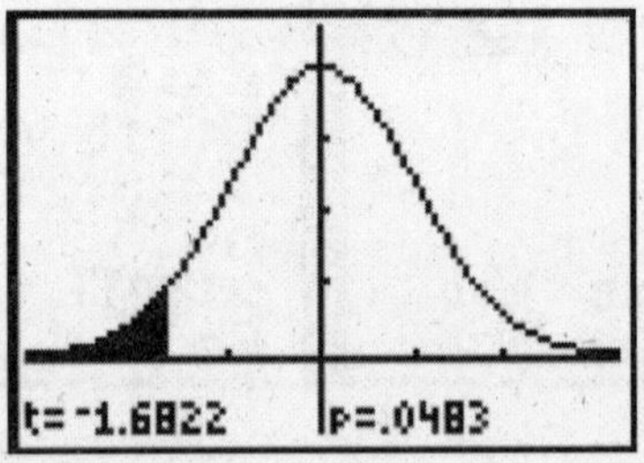

Test of Significance for the Mean of Two Samples

The options for a two-sample test of significance are much like those of a one-sample test. Information must be provided for both samples. Note that the hypothesis is stated as the relationship between μ_1 and μ_2.

With the two-sample *t*-test, you are also given the option of pooling. Remember: Do not pool unless you know (or have strong reason to believe) that the population variances are equal.

Calculator Screen

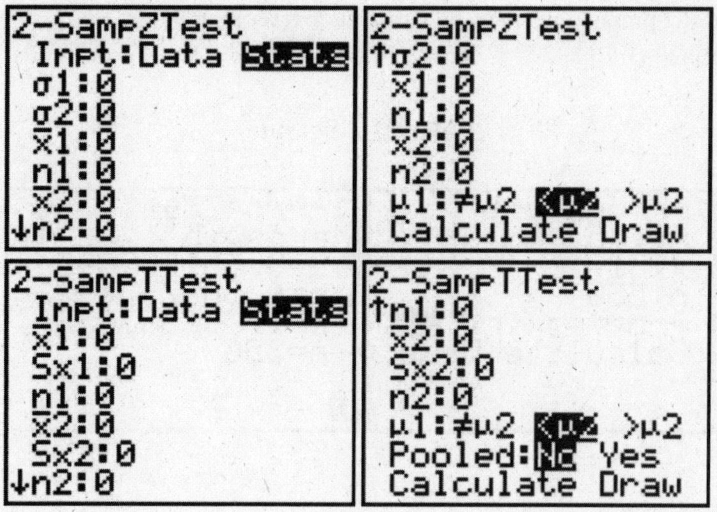

Tests of Significance for Proportions

Tests of significance for proportions are very straightforward. For a one-sample test of significance, you must provide the p_0 that you want to test against, as well as the actual *x* value (the count from the sample that meets the specified criteria) and sample size *n*. As with the tests for the mean, you have the option of either calculating the probability or drawing it.

Calculator Screen

Example:

A group representing a network news station states that less than 50% of adults feel that their news is biased. Based on a sample of 550 adults, you find that 387 adults feel the network's news is biased. Does the sample provide evidence that more than 50% of adults feel the network's news is biased?

Solution:

Here we are testing the proportion from the sample against the claim that $p_0 = 0.5$. Specifically, we are testing that the proportion from the sample is greater than 0.5.

Calculator Screen

The extremely small *p*-value gives strong evidence that the proportion of adults who feel that the network's news is biased is higher than 50%.

Now let's look at a graph of this same information by selecting **Draw**.

Calculator Screen

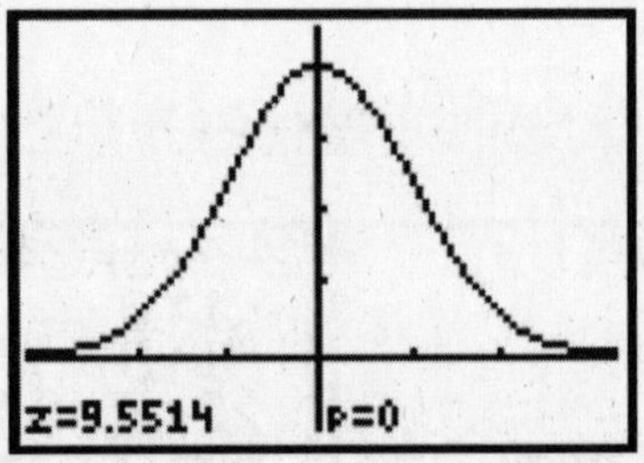

Notice that you cannot see any shading. This is because the *z*-value is outside the window range and *p*-value is so small. Also, the probability is given as zero at the bottom of the graph, as opposed to the value of 6.509757×10^{-22} that we received using the **Calculate** option.

The **2-PropZTest** operates in the exact same manner as the **1-PropZTest**, except that you are comparing the proportion from the two samples to each other and not to a population proportion. Thus, you are only required to input the values for x_1, n_1, x_2, and n_2.

Calculator Screen

CHI-SQUARE TEST OF SIGNIFICANCE

On graphing calculators, the chi-square test of significance is also called the chi-square test for independence, or the homogeneity of proportion. The data must be entered into the calculator as a matrix.

Calculator Screen

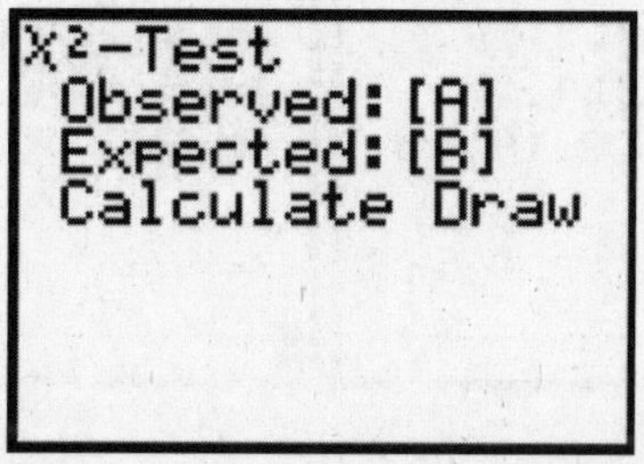

We will use an example from chapter 18 to see how this test can be done on the calculator. In a study of exercise habits in men working in the health care profession in Chicago, researchers classified the 356 sampled employees according to the level of education they completed and their exercise habits. The researchers wanted to determine if there is an association between the level of education completed and exercise habits at the 5% significance level. The data from the study are provided in the following table.

Observed Counts of Education Level Completed

Exercise Habits	*College*	*Some College*	*HS*	*Total*
Regularly	51	22	43	116
Occasionally	92	21	28	141
Never	68	9	22	99
Total	211	52	93	356

Enter the observed counts in a matrix. Do NOT include the TOTAL columns. To do this, select **2nd MATRIX** and then select **EDIT**. First, you must specify the dimensions of your matrix. In this case, it is a three-by-three matrix. Now enter your observed data. You do not have to calculate the expected data; the calculator will do that for you and paste it into Matrix [B]. Once your observed data is entered, you can calculate the χ^2 test. The output screen gives the χ^2 value, the probability, and the degrees of freedom. You can also look at Matrix [B] to see the expected counts. This is important to ensure that all of the conditions for the χ^2 test are met.

Calculator Screen

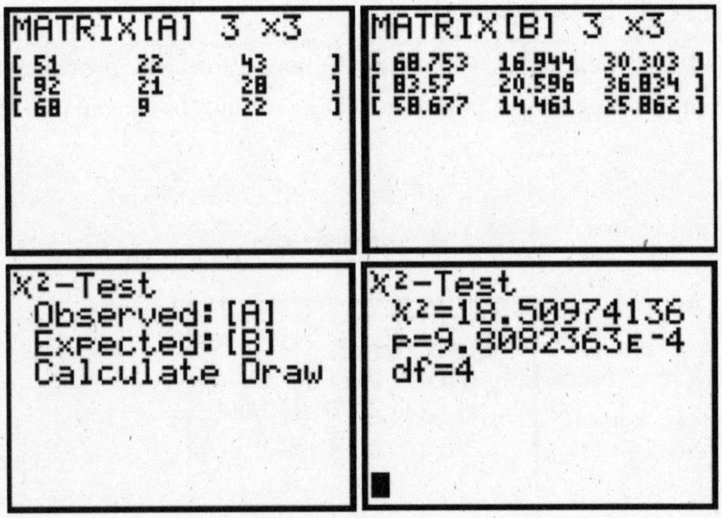

If we had specified **Draw** instead of **Calculate**, we would have the following diagram.

Calculator Screen

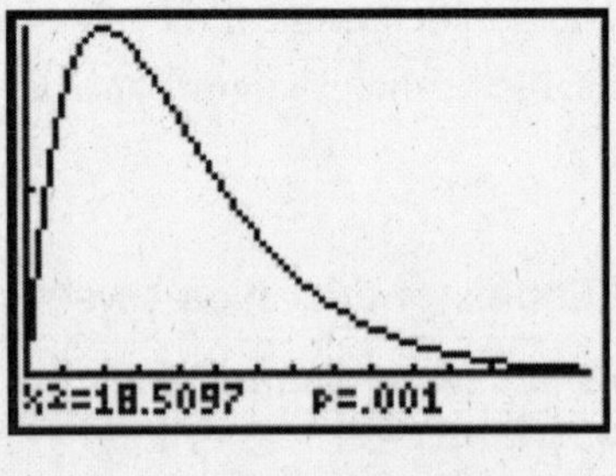

Note that because the *p*-value is so small, the shading is all but impossible to see. Nevertheless, the *p*-value and χ^2 value are provided at the bottom of the diagram. You can compare this to the results on page 312.

Because the χ^2 test for homogeneity of proportion is calculated in a comparable manner, you can use the same procedure to carry out the test.

To facilitate the calculations of the goodness of fit test, we enter our observed values in one list and the expected values in another. Below is a set of data with observed data in L_1 and the expected data in L_2. This is the first example of the goodness of fit test in chapter 18 dealing with the genetic alterations of lobsters on page 307.

Calculator Screen

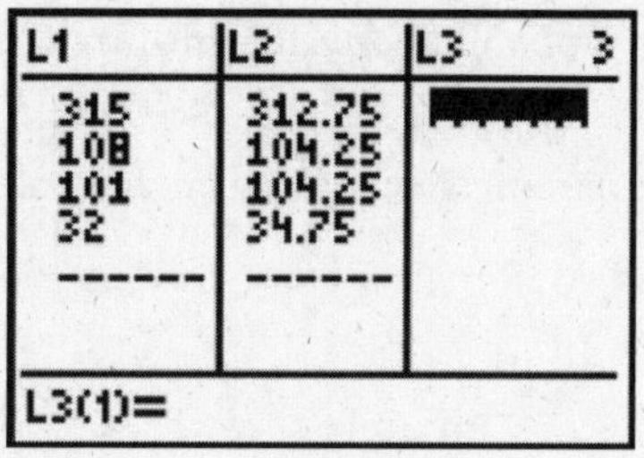

Recall the formula $\chi^2 = \sum_{i=1}^{n} \frac{(\text{observed value} - \text{expected value})^2}{\text{expected value}}$.

We can have the calculator do each of the individual calculations by placing the cursor at the top of the L_3 and entering the following formula: $(L_1 - L_2)^2/L_2$. The calculator will then take each row and calculate the (observed value – expected value)^2 and divide it by the expected value, and store the results in L_3. Press **ENTER**.

Calculator Screen

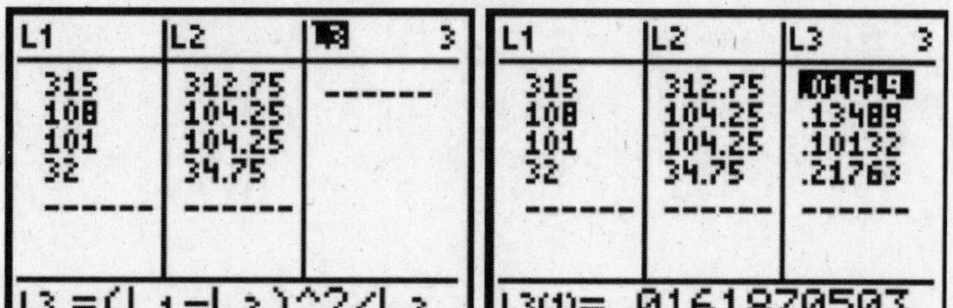

To calculate the χ^2 value, press **2nd QUIT**. This will return you to the home screen. To sum the values in L_3, press **2nd LIST** and select **MATH**. Under **MATH**, select **sum**. This will paste sum to the home screen; you now specify that you want the sum of $\mathbf{L_3}$, and press **ENTER**.

Calculator Screen

With the χ^2 you can calculate the probability of value greater than that observed by using the distributions on the calculator. Press **2nd DISTR** and select $\chi^2 cdf$. Press **ENTER** and χ^2 *cdf* will be pasted to the home screen. The parameters required are lower limit, upper limit, and degrees of freedom.

Calculator Screen

This gives us the *p*-value for the test. From here, you can do the analysis to assess whether you reject or fail to reject the null hypothesis. In this case, there is not sufficient evidence to reject the null hypothesis.

QUICK REFERENCE LIST OF CALCULATOR COMMANDS

randInt (a, b, n)	Generates random integers, **n** at a time, between (and including) **a** and **b**.
randBin (a, p, n)	Generates **n** binary numbers (0 or 1) with **a** occurring with a probability of **p**. *Example: randBin (1, 0.5, 100) will generate 100 random numbers, either 0 or 1, where the occurrence of 1 has the probability of 0.5. This example simulates the toss of a coin.*
nPr	Permutations: How many ways can you pick **r** things out of **n**, where order is important. That is, (1,2,3) is different from (2,3,1).
nCr	Combinations: How many ways can you pick **r** things out of **n**, where order is not important. That is, (1,2,3) is the same as (3,2,1).
normalpdf (x, μ, σ)	Computes the normal pdf at **x** for the given mean and standard deviation. Leaving out μ and σ, it assumes standardized normal.
normalcdf (lower, upper, μ, σ)	Computes the normal pdf between the given lower bound and upper bound for the specified mean and standard deviation.
invNorm (area, μ, σ)	Computes the inverse cumulative normal distribution function for the given area and the specified mean and standard deviation. That is, it computes the ***z***-score for the standardized normal or the actual score/value that would give that area for any other normal distribution.
tpdf (x, df)	Computes the pdf for the student *t*-distribution probability at a specified **x** with the given degrees of freedom.
tcdf (lower, upper, df)	Computes the *t*-distribution probability between the given lower bound and upper bound for the specified degrees of freedom.
χ^2pdf (x, df)	Computes the pdf for the χ^2 distribution at a specified x-value for the given degrees of freedom, **df**.
χ^2cdf (lower, upper, df)	Computes the χ^2 distribution probability between the given lower and upper bounds for the specified degrees of freedom.
binompdf (# of trials, p, x)	Computes a probability at **x** for the discrete binomial distribution with the specified number of trials and probability **p** of success on each trial.
binomcdf (# of trials, p, x)	Computes a cumulative probability at **x** for the discrete binomial distribution with the specified number of trials and probability p of success on each trial.
geometpdf (p, x)	Computes a probability at **x**, the number of the trial on which the first success occurs, for the discrete geometric distribution with probability of success **p**.
geometcdf (p, x)	Computes a cumulative probability at **x**, the number of the trial on which the first success occurs, for the discrete geometric distribution with probability of success **p**.
ShadeNorm (lower, upper, μ, σ)	Draws the normal density function specified by μ and σ and shades the area between the lower and upper bounds.
Z-Test	Use when σ is known, normal distribution or $n \geq 30$.
T-Test	Use when σ is unknown, normal population or $n \geq 30$.
2-SampZTest	Use when σ_1 and σ_2 are known, normal distributions, independent samples.
2-SampTTest	Use when σ_1 and σ_2 are unknown, normal populations or $n_1 + n_2 > 40$, independent samples. If $\sigma_1 = \sigma_2$, use pooled test. If $\sigma_1 \neq \sigma_2$, use unpooled test.
1-PropZTest	Use when $np > 10$, and $n(1 - p) > 10$.
2-PropZTest	Use when $n_1p_1 > 10$, $n_1(1 - p_1) > 10$, $n_2p_2 > 10$, $n_2(1 - p_2) > 10$, independent samples. If $p_1 = p_2$, use test of equal variances. If $p_1 \neq p_2$, use test of unequal variances.
ZInterval	Use when σ is known, SRS, population is normal or $n \geq 30$.
TInterval	Use when σ is unknown, SRS, population is normal or $n \geq 30$.
2-SampleZInt	Use when σ_1 and σ_2 are known, normal populations or $n_1 + n_2 > 40$, independent samples.
2-SampleTInt	Use when σ_1 and σ_2 are unknown, normal populations or $n_1 + n_2 > 40$, independent samples.
1-PropZInt	Use when $np > 10$, and $n(1 - p) > 10$, SRS, large population.
2-PropZInt	Use when $n_1p_1 > 10$, $n_1(1 - p_1) > 10$, $n_2p_2 > 10$, $n_2(1 - p_2) > 10$, SRS, large populations, independent samples.
χ^2 – Test	Computes χ^2 test for association on the two-way table of counts in Observed Matrix. Null hypothesis for two-way table is: no association exists between row variables and column variables.

For $-\infty$ as a lower bound, use –1EE99; for ∞, use 1EE99. On the TI-83, EE is the second key (yellow) next to the comma key.

Part Four

PRACTICE TESTS

HOW TO TAKE THE PRACTICE TESTS

The next section of this book consists of practice tests. Taking a practice AP exam gives you an idea of what it's like to answer these test questions for a longer period of time, one that approximates the real test. You'll find out which areas you're strong in, and where additional review may be required. Any mistakes you make now are ones you won't make on the actual exam, as long as you take the time to learn where you went wrong.

The two full-length practice tests in this book each include 40 multiple-choice questions and 6 free-response (essay) questions. You will have 90 minutes for the multiple-choice questions, a 10-minute reading period, and 90 minutes to answer the free-response questions. Before taking a practice test, find a quiet place where you can work uninterrupted for 3 hours. Time yourself according to the time limit at the beginning of each section. It's okay to take a short break between sections, but for the most accurate results you should approximate real test conditions as much as possible. Use the 10-minute reading period to plan your answers for the free-response questions, but don't begin writing your responses until the 10 minutes are up.

As you take the practice tests, remember to pace yourself. Train yourself to be aware of the time you are spending on each question. Try to be aware of the general types of questions you encounter, as well as being alert to certain strategies or approaches that help you to handle the various question types more effectively.

After taking a practice exam, be sure to read the detailed answer explanations that follow. These will help you identify areas that could use additional review. Even when you've answered a question correctly, you can learn additional information by looking at the answer explanation.

Finally, it's important to approach the test with the right attitude. You're going to get a great score because you've reviewed the material and learned the strategies in this book.

HOW TO COMPUTE YOUR SCORE

Now for the exciting part of the exam—computing your score! The score you obtain on the practice tests in this book approximate your performance on the actual test. The questions on the actual test are not the exact same questions offered in this book, but they are similar in both form and content.

There are five possible final scores on the AP Exam:

5. Extremely well-qualified
4. Well-qualified
3. Qualified
2. Possibly qualified
1. No recommendation

Think of a 5 as an A, a 4 as a B, a 3 as a C, a 2 as a D, and a 1 as an F (or no score). Most schools accept a score of 4 or 5; some schools will accept a score of 3.

The practice tests are composed of multiple-choice questions and free-response questions. The multiple-choice questions are scored by an electronic scanner, while a team of trained reviewers scores the free-response questions by hand. These questions are scored during the month of June and exam scores are sent out to students and colleges in July. If you haven't received your score by September, contact the College Board.

SCORING THE MULTIPLE-CHOICE QUESTIONS

To compute your score on the multiple-choice portion of the two practice tests, calculate the number of questions you got wrong on each test and then deduct $\frac{1}{4}$ of that number from the number of correct answers you got on each test. For example, if you got 6 multiple-choice questions wrong, you would deduct 1.5 ($6 \times \frac{1}{4}$) from the number of questions you answered correctly. For example, on a 100-question exam, you would subtract 1.5 from 94 (the number of questions answered correctly), and your score would then be a 93 (92.5 rounded up), or a 5 for the multiple-choice portion of the exam.

SCORING THE FREE-RESPONSE QUESTIONS

The essay reviewers have specific points that they are looking for in each free-response question. Frame your answer in complete, coherent sentences. Make sure that you present the various components of your answer in the right order, and in an order that will be intelligible to your readers. In addition to these basic structural concerns, reviewers will be seeking specific pieces of information in your answer. Each piece of information that they are able to find and check off in your answer is a point toward a better score.

To figure out your approximate score for the free-response questions, look at the key points found in the sample response for each question. For each key point you miss, subtract a point. For each key point you include, add a point. Figure out the number of key points there are in each question, then add up the number of key points you did include out of the total number of key points. Set up a proportion equal to 100 to obtain your approximate numerical score.

Calculating Your Composite Score

Your score on the AP exam is a combination of your score on the multiple-choice portion of the exam and the free-response section. Sometimes the multiple-choice section is weighted more heavily than the free-response section and vice-versa, but you can still calculate an approximate score based on your performance on the sample tests in this book.

Add together your score on the multiple choice portion of the exam and your approximate score on the free-response section of the exam. If your score is a decimal, then round up to a whole number. Divide by two to obtain your approximate score for each full-length exam.

If your score falls between 80 and 100, you're doing great. Keep up the good work! If your score is lower than 79, there's still hope. Keep studying and you will be able to obtain a much better score on the exam before you know it.

Best of luck!

Practice Test 1 Answer Grid

1.	Ⓐ Ⓑ Ⓒ Ⓓ Ⓔ	11.	Ⓐ Ⓑ Ⓒ Ⓓ Ⓔ	21.	Ⓐ Ⓑ Ⓒ Ⓓ Ⓔ	31.	Ⓐ Ⓑ Ⓒ Ⓓ Ⓔ
2.	Ⓐ Ⓑ Ⓒ Ⓓ Ⓔ	12.	Ⓐ Ⓑ Ⓒ Ⓓ Ⓔ	22.	Ⓐ Ⓑ Ⓒ Ⓓ Ⓔ	32.	Ⓐ Ⓑ Ⓒ Ⓓ Ⓔ
3.	Ⓐ Ⓑ Ⓒ Ⓓ Ⓔ	13.	Ⓐ Ⓑ Ⓒ Ⓓ Ⓔ	23.	Ⓐ Ⓑ Ⓒ Ⓓ Ⓔ	33.	Ⓐ Ⓑ Ⓒ Ⓓ Ⓔ
4.	Ⓐ Ⓑ Ⓒ Ⓓ Ⓔ	14.	Ⓐ Ⓑ Ⓒ Ⓓ Ⓔ	24.	Ⓐ Ⓑ Ⓒ Ⓓ Ⓔ	34.	Ⓐ Ⓑ Ⓒ Ⓓ Ⓔ
5.	Ⓐ Ⓑ Ⓒ Ⓓ Ⓔ	15.	Ⓐ Ⓑ Ⓒ Ⓓ Ⓔ	25.	Ⓐ Ⓑ Ⓒ Ⓓ Ⓔ	35.	Ⓐ Ⓑ Ⓒ Ⓓ Ⓔ
6.	Ⓐ Ⓑ Ⓒ Ⓓ Ⓔ	16.	Ⓐ Ⓑ Ⓒ Ⓓ Ⓔ	26.	Ⓐ Ⓑ Ⓒ Ⓓ Ⓔ	36.	Ⓐ Ⓑ Ⓒ Ⓓ Ⓔ
7.	Ⓐ Ⓑ Ⓒ Ⓓ Ⓔ	17.	Ⓐ Ⓑ Ⓒ Ⓓ Ⓔ	27.	Ⓐ Ⓑ Ⓒ Ⓓ Ⓔ	37.	Ⓐ Ⓑ Ⓒ Ⓓ Ⓔ
8.	Ⓐ Ⓑ Ⓒ Ⓓ Ⓔ	18.	Ⓐ Ⓑ Ⓒ Ⓓ Ⓔ	28.	Ⓐ Ⓑ Ⓒ Ⓓ Ⓔ	38.	Ⓐ Ⓑ Ⓒ Ⓓ Ⓔ
9.	Ⓐ Ⓑ Ⓒ Ⓓ Ⓔ	19.	Ⓐ Ⓑ Ⓒ Ⓓ Ⓔ	29.	Ⓐ Ⓑ Ⓒ Ⓓ Ⓔ	39.	Ⓐ Ⓑ Ⓒ Ⓓ Ⓔ
10.	Ⓐ Ⓑ Ⓒ Ⓓ Ⓔ	20.	Ⓐ Ⓑ Ⓒ Ⓓ Ⓔ	30.	Ⓐ Ⓑ Ⓒ Ⓓ Ⓔ	40.	Ⓐ Ⓑ Ⓒ Ⓓ Ⓔ

PRACTICE TEST 1

SECTION I: MULTIPLE-CHOICE QUESTIONS

Time—90 Minutes 40 Questions

Directions: Section 1 of this examination contains all multiple-choice questions. Decide which of the suggested answers best suits the question. You may use the formulas and tables found in the Appendix on p. 431 on the test.

1. What are the quartiles of the distribution displayed in the stemplot below?

Stem	Leaf
3	112358
4	0133789
5	2235577
6	46
7	569
8	3
9	8

(A) 36 and 70
(B) 40 and 64
(C) 41 and 65
(D) 43 and 83
(E) 48 and 81

2. The volumes of cola in bottles of a certain soft drink brand are normally distributed with a mean of 16 ounces and a standard deviation of 0.1 ounces. Which of the following intervals represents the middle 40% of this distribution?

(A) 15.98 to 16.02 ounces
(B) 15.97 to 16.03 ounces
(C) 15.96 to 16.04 ounces
(D) 15.95 to 16.05 ounces
(E) 15.92 to 16.08 ounces

3. The prizes awarded to winning contestants on a long-running television quiz show have a mean of $9,400 and a standard deviation of $4,500 per episode. Nonwinners receive nothing. For each winner, there is one nonwinner. The producers of the show are planning to double the dollar values for winning contestants and also award an extra $500 to each contestant on every episode. Under this new plan, what will the mean and standard deviation be for the prizes awarded per episode?

(A) Mean $18,800 and standard deviation $4,500
(B) Mean $18,800 and standard deviation $9,000
(C) Mean $22,800 and standard deviation $4,500
(D) Mean $22,800 and standard deviation $9,000
(E) Mean $22,800 and standard deviation $9,500

GO ON TO THE NEXT PAGE

4. The students in a college all take a standardized test scored on a scale from 0 to 100 points. The mean and standard deviation are 53 and 16, respectively. Hugh's test result has a *z*-score of 1.19 and Neal's has a *z*-score of 2.06. Neal's test score is how many points higher than Hugh's?

(A) 0.87

(B) 14

(C) 19

(D) 33

(E) 46

5. The governor of a state claims that the proportion of rural voters who support her economic policies, p_r, is the same as the support among urban voters, p_u. One of her political advisors, however, suspects this might not be the case. If this advisor wants to conduct a study to challenge the governor's claim, which of the following hypotheses would be appropriate?

(A) $H_0: \hat{p}_r = \hat{p}_u, H_a: \hat{p}_r > \hat{p}_u$

(B) $H_0: p_r = p_u, H_a: p_r > p_u$

(C) $H_0: p_r = p_u, H_a: p_r < p_u$

(D) $H_0: \hat{p}_r = \hat{p}_u, H_a: \hat{p}_r \neq \hat{p}_u$

(E) $H_0: p_r = p_u, H_a: p_r \neq p_u$

6. A manufacturer wants to test a sample of the products his company produces. Products are produced in sets of 100, and each set of 100 is called a *lot*. The company produces hundreds of lots each day. The manufacturer decides to randomly select five lots from each day's production run and test every product in these five lots. Which of the following best describes this type of sampling?

(A) Cluster

(B) Stratified random

(C) Convenience

(D) Systematic

(E) Simple random

GO ON TO THE NEXT PAGE

7. A calculus teacher is investigating the relationship between his students' grades prior to the final exam (PRIOR) and their grade on the final exam itself (EXAM). He conducts a regression analysis of his students from last year with the following results.

Source	DF	Sum of Sq	Mean Squ	F-ratio
Regression	1	2290.3	2290.3	38.52
Residual Error	30	1783.5	59.5	

Predictor	Coef	SE Coef	T	P
Constant	8.63	10.08	0.86	0.399
PRIOR	0.8126	0.1309	6.21	0.000

S = 7.71046 R-Sq = 56.2% R-Sq(adj) = 54.8%

According to the regression line, which of the following is the predicted final exam grade for a student who has an average of 85 prior to the final?

(A) 21
(B) 48
(C) 78
(D) 79
(E) 85

8. A cookbook has recipes for 10 chicken dinners, 10 beef dinners, and 15 vegetarian dinners. Pat plans dinner by picking a recipe at random from the book. Every one of the 35 meals has an equal chance of being selected each time Pat plans a dinner. What is the probability Pat prepares exactly 3 vegetarian dinners over 7 consecutive days?

(A) $\frac{3}{7}$
(B) $\left(\frac{15}{35}\right)^3$
(C) $\left(\frac{15}{35}\right)^3\left(\frac{20}{35}\right)^4$
(D) $3\left(\frac{15}{35}\right)\left(\frac{20}{35}\right)$
(E) $\binom{7}{3}\left(\frac{15}{35}\right)^3\left(\frac{20}{35}\right)^4$

9. For a population that consists of a list of numerical data, which of the following sample statistics represents the best point estimator for the standard deviation of the population?

(A) $\sqrt{\sum(x_i-\bar{x})^2 p_i}$
(B) $\sqrt{np(1-p)}$
(C) $\sqrt{\frac{p(1-p)}{n}}$
(D) $\sqrt{\frac{1}{n-1}\sum(x_i-\bar{x})^2}$
(E) $\frac{\sqrt{\frac{\sum(y_i-\hat{y}_i)^2}{n-2}}}{\sqrt{\sum(x_i-\bar{x})^2}}$

GO ON TO THE NEXT PAGE

10. Each member of a four-person relay team swims 50 meters, one after the other. Past results have shown that the first swimmer has a mean time of 28.3 seconds and a standard deviation of 0.4 seconds. The second has a mean of 29.1 seconds and a standard deviation of 0.6 seconds. The third has a mean of 30.2 seconds and a standard deviation of 0.1 seconds. The fourth has a mean of 27.0 seconds and a standard deviation of 0.7 seconds. Assuming each swimmer's time is independent of the other swimmers, what is the standard deviation for the entire relay?

(A) 0.45 seconds
(B) 0.9 seconds
(C) 1.01 seconds
(D) 1.8 seconds
(E) 50.7 seconds

11. A manufacturer is trying to determine whether a new method for constructing skateboards, method A, is faster than the old method, method B. A well-designed experiment is conducted. The hypotheses are $H_0{:}\mu_A = \mu_B$ and $H_a{:}\mu_A < \mu_B$, where μ_A is the mean time it takes to construct a skateboard using method A, and μ_B is the mean time for method B. The analysis shows a *p*-value of 0.02. Which of the following must be true?

(A) If H_0 is true, there is a 2% chance of observing a difference in sample means at least as extreme as the difference in mean times observed in the sample.
(B) If H_a is true, there is a 2% chance of observing a difference in sample means at least as extreme as the difference in mean times observed in the sample.
(C) A 95% confidence interval for the difference in sample means will include 0.
(D) The power of the test is 98%.
(E) H_0 would be rejected at the 1% significance level but not at the 5% level.

12. Kumar tosses a fair coin six times and happens to get heads all six times. He knows that in the long run the coin will only come up heads half the time, so he figures that the next toss is due to come up tails. Which of the following is the best assessment of Kumar's reasoning?

(A) Kumar is wrong. Successive tosses of a fair coin are independent of each other. The six prior tosses are irrelevant to the next toss.
(B) Kumar is wrong. The fact that he got heads six times in a row shows that the coin is "locked" into a streak. It is likely the streak will continue.
(C) Kumar is right. The proportion of tosses that come up heads is too high after these six tosses, so the law of large numbers requires that tails start to show up in quantities that will allow the proportion to settle down to 0.5. Another heads will make this process even harder.
(D) Kumar is right. The probability of getting seven heads in a row is very small, so it is very likely he will get tails on the next toss.
(E) Kumar is right. It is likely he will get tails on most of the next four tosses. This way he will have only six or seven heads out of ten tosses, which is a reasonable number.

GO ON TO THE NEXT PAGE

13. A polling organization wants to spend less money collecting data. They look to do this by decreasing the sample sizes they use when conducting surveys. Assuming the organization maintains sound statistical practices in the way data is collected, how will this affect the reliability of the results?

   (A) Their statistics will be subject to less variability when using smaller samples if they also work with proportionately larger populations.
   (B) Their statistics will be subject to increased sampling bias, so their results will be less trustworthy.
   (C) Their statistics will be subject to more variability when using smaller samples.
   (D) Their statistics will have about the same variability if they also work with proportionately smaller populations.
   (E) Their statistics will have about the same variability when using smaller samples regardless of population size.

14. The distribution of the salaries of a company's employees is strongly skewed to the right. Which of the following correctly describes a consequence of this fact?

   (A) The mean is larger than the median.
   (B) The distribution is normal.
   (C) The distribution is chi-square.
   (D) The distribution has no outliers to the left.
   (E) The distribution has at most two or three clusters of data.

15. For the eight planets in the solar system, a list is made of the length of a year Y (measured in Earth years) and the mean orbital radius R (measured in millions of km). A regression analysis yields $\hat{Y} = -13 + 0.0408R$, $r^2 = 0.98$, and the following residual plot.

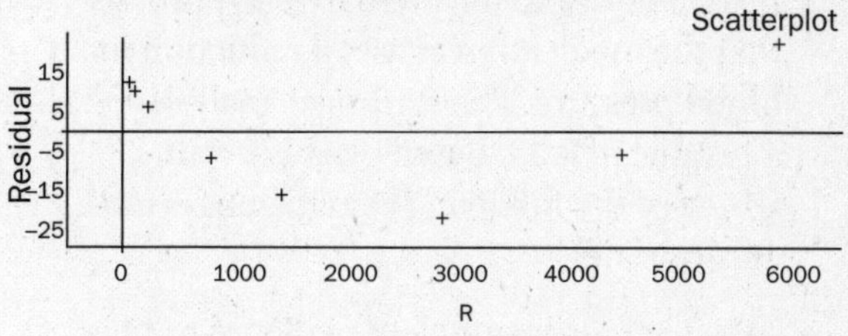

Which of the following is the most accurate statement about the regression results?

   (A) The value of r^2 shows there is a strong linear relationship between length of year and mean orbital radius.
   (B) The residual plot shows there is a strong linear relationship between length of year and mean orbital radius.
   (C) The slope of the regression line shows there is a negative association between length of year and mean orbital radius.
   (D) The regression line underpredicts the length of a year for a planet with mean orbital radius of 100 million km.
   (E) The regression line makes a very accurate prediction for the length of a year for a planet with mean orbital radius of 100 million km.

GO ON TO THE NEXT PAGE

16. A researcher designs and executes an experiment to determine whether a certain medication will decrease blood pressure in patients with hypertension. The experiment was well-designed with a null hypothesis that blood pressure is unaffected by the medication, and an alternative hypothesis that the medication causes a reduction in blood pressure. The statistical analysis yields a *p*-value of 0.33. Based on this *p*-value, which of the following conclusions is most appropriate?

(A) There is insufficient evidence to reject the null hypothesis.
(B) There is sufficient evidence to accept the null hypothesis.
(C) There is insufficient evidence to reach a conclusion about either the null or alternative hypothesis.
(D) There is insufficient evidence to accept the alternative hypothesis.
(E) There is sufficient evidence to reject the alternative hypothesis.

17. For a set of ten boxes, seven contain one blue marble and one red marble, while the other three contain one blue marble and three red marbles. You select a box at random and then select a marble from that box at random. Which of the following is the probability that the marble is blue?

(A) $\frac{10}{26}$
(B) $\frac{0.5+0.25}{2}$
(C) $(0.7+0.5)(0.3+0.25)$
(D) $(0.7)(0.5)(0.3)(0.25)$
(E) $(0.7)(0.5)+(0.3)(0.25)$

18. The correlation between two scores *X* and *Y* is −0.71. Which of the following is a correct statement about the relationship between the values of *X* and *Y*?

(A) 71% of the variation in *Y* is explained by the least-squares regression of *Y* on *X*.
(B) There is a moderate tendency for an increase in the value of *X* to cause a decrease in the value of *Y*.
(C) There is a moderate tendency for an increase in the value of *X* to cause an increase in the value of *Y*.
(D) Among the values of *X* and *Y*, there is a moderate tendency for decreasing values of *X* to correspond to decreasing values of *Y*.
(E) Among the values of *X* and *Y*, there is a moderate tendency for decreasing values of *X* to correspond to increasing values of *Y*.

19. Which of the following is a valid reason for performing a *z*-test rather than a *t*-test when performing a significance test for a population mean?

(A) The distribution of the sample is symmetric.
(B) The sample mean is reasonably close to μ_0, the mean under the null hypothesis.
(C) The population standard deviation is known.
(D) The sample size is small.
(E) The sample was chosen in an unbiased manner.

GO ON TO THE NEXT PAGE

20. For a large table of random numbers, which of the following is a correct statement?

    (A) The sequence 00 will not appear more than once.
    (B) Entries in one part of the table are independent of entries in another part.
    (C) A row of 40 digits will include each digit 0 through 9 exactly four times.
    (D) The sequence 05 will occur more often than the sequence 15 since the sequence 05 represents both the two-digit number 05 and the one-digit number 5.
    (E) The sequence 0123456789 is likely to occur in the first 1,000 digits.

21. Sherry conducts an experiment, collects data, and computes a *z*-test statistic from the data. She uses a two-sided test and finds that the test statistic is statistically significant. What would Sherry have concluded if she had conducted a one-sided test?

    (A) It would be significant since the *p*-value for a one-sided test is half the *p*-value for a two-sided test.
    (B) It would be significant since the *p*-value for a one-sided test is twice the *p*-value for a two-sided test.
    (C) It would not be significant since the *p*-value for a one-sided test is half the *p*-value for a two-sided test.
    (D) It is impossible to tell since the *p*-value for a one-sided test is twice the *p*-value for a two-sided test. We do not know the *p*-value or the significance level.
    (E) It is impossible to tell since we do not know on which side the test is conducted.

22. The boxplots below summarize data sets I and II. Which of the following statements is a correct conclusion that can be drawn from these boxplots?

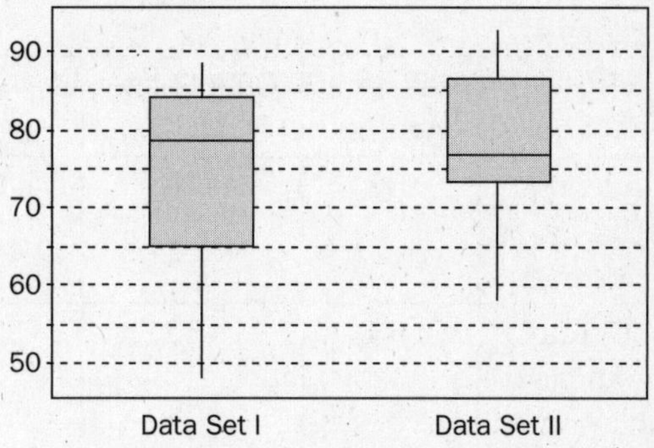

    (A) The 40th percentile of data set II is smaller than the 60th percentile of data set I.
    (B) Data set I has a smaller interquartile range.
    (C) About 25% of the values in data set I are between 65 and 84.
    (D) Data set I contains more data points.
    (E) Data set II has a larger range.

23. When designing an experiment, why is it important to protect against confounding?

    (A) Confounding interferes with randomization.
    (B) Confounding increases the likelihood of bias.
    (C) Confounding makes it impossible to determine the actual cause of any observed change in the response variable.
    (D) Confounding decreases the strength of the association between the explanatory variable and the response variable.
    (E) Confounding makes it difficult to double blind an experiment.

GO ON TO THE NEXT PAGE

24. The following table classifies the students in the senior class of Johnson High School by whether they attended Johnson as freshmen, and where they expect they will live when they are 30 years old (in-state or out-of-state).

	In-State at Age 30	Out-of-State at Age 30	Total
Attended Johnson as freshman	302	118	420
Did not attend Johnson as a freshman	48	32	80
Total	350	150	500

Which of the following is the marginal distribution of where the seniors expect to live when they are 30?

(A) 84% were at Johnson as freshmen, 16% were not.

(B) Of the students expecting to live out-of-state, 79% were at Johnson as freshmen and 21% were not.

(C) Of the students who attended Johnson as freshmen, 72% expect to be in-state and 38% expect to be out-of-state.

(D) 70% expect to live in-state, 30% expect to be out-of-state.

(E) Of the students who did not attend Johnson as freshmen, 60% expect to be in-state and 40% expect to be out-of-state.

25. Chelsea designs an experiment to measure the difference in subjects' pulse rates while listening to two recordings of the same piece of music, one with a strong drum track and one with the drum track removed. She works with 50 subjects, measuring each subject's pulse rates in response to each recording. The order in which each subject listens to the two tracks is randomized, and a period of time is allowed between trials. Which of the following would be the most appropriate statistical test in this situation?

(A) Matched pairs *t*-test

(B) One-sample proportion *z*-test

(C) Chi-square test for homogeneity of proportions

(D) One-sample *t*-test

(E) Two-sample *t*-test

26. A random sample of 500 likely voters is polled just prior to a city election. Of these voters, 280 say they will vote to re-elect the mayor. Which of the following is a 95% confidence interval for the proportion of votes the mayor will receive in the election?

(A) $0.56 \pm 1.645\sqrt{\frac{0.56(1-0.56)}{500}}$

(B) $0.56 \pm 1.645\sqrt{0.56(1-0.56)}\sqrt{\frac{1}{280}+\frac{1}{220}}$

(C) $0.56 \pm 1.96\sqrt{\frac{0.56(1-0.56)}{500}}$

(D) $0.56 \pm 1.96\frac{0.56}{\sqrt{500}}$

(E) $0.56 \pm 1.96\sqrt{0.56(1-0.56)}\sqrt{\frac{1}{280}+\frac{1}{220}}$

GO ON TO THE NEXT PAGE

27. Two fair, six-sided dice are rolled and the numbers are added together. Chelsea performs this experiment until she rolls a sum of 7. She records the number of rolls this takes, and repeats the process until she has recorded 100 counts for the number of rolls necessary to see a roll of 7. A cumulative relative frequency diagram of her results is shown below.

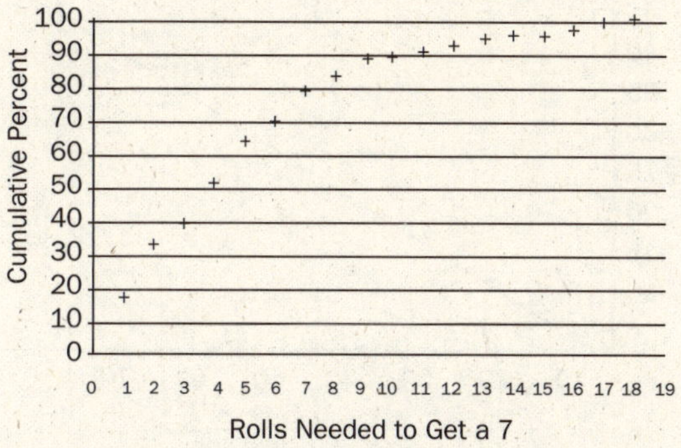

According to this diagram, about what percentage of the time did it take three or more rolls to get a 7?

(A) 6%
(B) 33%
(C) 39%
(D) 61%
(E) 67%

28. Which of the following is NOT a valid reason for blocking when designing an experiment?

(A) Blocking controls the effects of an outside variable by bringing that variable into the experiment to form the blocks.
(B) Blocking allows the researcher to isolate variability due to a factor other than the explanatory variable under investigation.
(C) Blocking can diminish confounding.
(D) Blocking reduces the need for replication.
(E) Blocking reduces variation.

29. Which of the following statements is (are) true about normal distributions?

I. The mean is exactly the same as the median.
II. A normal distribution with standard deviation of 2 has a peak that is higher than a normal distribution with standard deviation of 1.
III. A normal distribution is completely described by giving its mean and standard deviation.

(A) I only
(B) II only
(C) III only
(D) I and III
(E) I, II, and III

GO ON TO THE NEXT PAGE

30. A well-designed experiment is conducted in order to estimate the difference between two population means. A 98% confidence interval for the difference in population means is 2 ± 7 units. Which of the following statements must be true?

(A) The results of the experiment are unlikely to occur if the difference in population means is 3.

(B) The results of the experiment are consistent with a difference in population means of 0.

(C) There is a 98% chance that a second run of the experiment would have a difference of sample means between –5 and 9.

(D) There is a 98% chance that the difference in population means is between –5 and 9.

(E) There is a 0.98 probability that the result of the experiment correctly estimates the difference of population means.

31. The scatterplot below displays runs allowed versus hits allowed for 11 of the 12 pitchers for a baseball team. Also included is a plot of the least-squares regression line. What would be the effect if the 12th pitcher, who allowed 75 hits and 16 runs, was included?

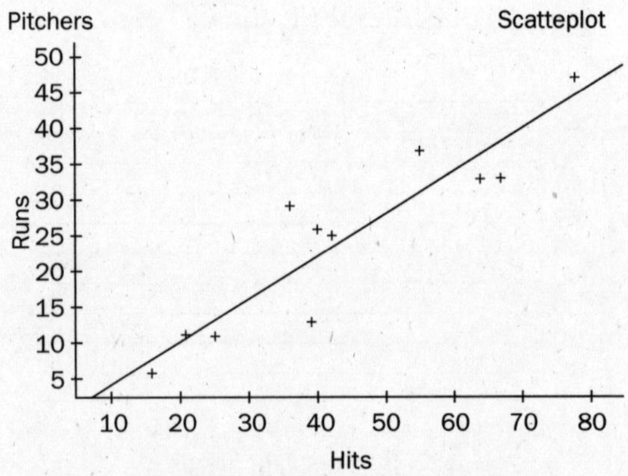

Runs = 0.594 Hits – 1.4; $r^2 = 0.84$

(A) The slope would decrease and the association would strengthen.

(B) The slope would decrease and the association would stay approximately the same.

(C) The slope would decrease and the association would weaken.

(D) The slope would increase and the association would weaken.

(E) The slope would increase and the association would strengthen.

GO ON TO THE NEXT PAGE

32. Seamus is deciding whether to use a 95% confidence interval or a 99% confidence interval for the population mean when analyzing a large sample he has randomly collected from a population. Which of the following statements about the intervals is true?

(A) The margin of error for the 95% confidence interval is 5% of the sample mean.

(B) The margin of error for the 99% confidence interval is 2.576.

(C) If the sample size is at least 1,000, the margin of error for the 99% confidence interval will be much larger than the margin of error for the 95% confidence interval.

(D) The 99% confidence interval will be wider than the 95% confidence interval regardless of the size of Seamus's sample.

(E) The width of the 99% confidence interval will be smaller than the margin of error for the 95% confidence interval.

33. A correlation and linear regression analysis is performed on two separate scatterplots, A and B. Both have identical slopes and intercept values, but the correlation coefficient *r* for scatterplot A is 0.75 and 0.50 for scatterplot B. What conclusion can be drawn concerning the total deviation of the points from the corresponding regression line?

(A) The sum of the deviations for A is less than the sum of the deviations for B.

(B) The sum of the squares of the deviations for A is less than the sum of the squares of the deviations for B.

(C) The ratios of the deviations differ by $0.75 - 0.50 = 0.25$.

(D) The ratios of the squares of the deviations differ by $0.75 \div 0.50 = 1.5$.

(E) None of the above answers can be determined from this information.

GO ON TO THE NEXT PAGE

34. An executive at an advertising firm must decide whether to continue with the current ad campaign for a product or move on to a new campaign. The firm tested the two campaigns with focus groups, using a null hypothesis that there is no difference in the campaigns, and an alternative hypothesis that the new campaign is superior at motivating consumers to shop at the store. Which of the following best describes the consequences of a type II error?

    (A) The firm will continue with the old campaign even though the new campaign is better.
    (B) The firm will switch to the new campaign even though the old campaign is better.
    (C) The firm will continue with the old campaign even though there is no difference in the two campaigns.
    (D) The firm will switch to the new campaign even though there is no difference in the two campaigns.
    (E) The firm will switch to the new campaign even though the evidence is inconclusive.

35. All the first grade students in a large school system take an intelligence test. The scores are approximately normally distributed with a mean of 100 and a standard deviation of 10. Two of the test-takers are selected at random. What is the probability that their scores will be within 5 points of each other?

    (A) 0.14
    (B) 0.28
    (C) 0.38
    (D) 0.5
    (E) 0.68

36. A researcher is trying to find a 90% confidence interval for the proportion of human pregnancies that run longer than 270 days. The researcher wants a margin of error no larger than 0.02. Of the following sample sizes, which is the smallest that will accomplish this goal?

    (A) 2,402
    (B) 1,692
    (C) 1,028
    (D) 625
    (E) 133

37. A lumber yard produces wooden beams that are approximately normally distributed with a mean length of 4 meters and a standard deviation of 0.03 meters. A construction project uses 50 of these beams. Which of the following represents the probability that the mean length of the beams used in the project is at least 3.99 meters?

    (A) $P\left(z > \dfrac{4.00 - 3.99}{\dfrac{0.03}{50}}\right)$
    (B) $P\left(z > \dfrac{3.99 - 4.00}{\dfrac{0.03}{50}}\right)$
    (C) $P\left(z > \dfrac{3.99 - 4.00}{0.03}\right)$
    (D) $P\left(z > \dfrac{4.00 - 3.99}{\dfrac{0.03}{\sqrt{50}}}\right)$
    (E) $P\left(z > \dfrac{3.99 - 4.00}{\dfrac{0.03}{\sqrt{50}}}\right)$

GO ON TO THE NEXT PAGE

38. A consumer products researcher wants to design an experiment testing the durability of two different types of roofing tiles, brand X and brand Y. Ten test roofs are built, roofs 1 through 10, which will be subjected to simulated weather in a controlled environment. Roofs 1 through 5 are covered with brand X tiles, and roofs 6 through 10 are covered with brand Y. One of the tile brands is exposed to cold and snow, and the other is subjected to heat and rain, with the assignment made at random. After the simulation has ended, the water resistance of the two types of tiles is compared. Which of the following is a valid observation of the experiment?

(A) The experiment is well-planned and should serve its purpose.
(B) The sample size of 10 is too small.
(C) The treatments in this experiment are the tiles applied to the roofs.
(D) Tile brands were randomly assigned to roofs.
(E) There is a flaw in the design. Each brand of tile should be exposed to both sets of weather conditions.

39. A random number generator is designed to pick a random number from the set $\{1, 2, 3, 4\}$. The random number generator is tested by having it produce 200 random numbers. The number of instances of each result is shown in the table below.

Result	1	2	3	4
Count	61	50	44	45

A chi-square test for goodness of fit will be used to determine whether these counts are consistent with equal probabilities of each result. Which of the following is the value of the chi-square test statistic for this experiment?

(A) 0
(B) 0.0728
(C) 3.64
(D) 4.11
(E) 182

40. Which of the following gives the mean μ_X and variance σ_X^2 for the following random variable X?

X	100	300	500	1000
Probability	0.6	0.2	0.1	0.1

(A) $\mu_X = 270$ and $\sigma_X^2 = 76{,}100$
(B) $\mu_X = 270$ and $\sigma_X^2 = 111{,}875$
(C) $\mu_X = 270$ and $\sigma_X^2 = 615{,}600$
(D) $\mu_X = 475$ and $\sigma_X^2 = 111{,}875$
(E) $\mu_X = 475$ and $\sigma_X^2 = 149{,}167$

IF YOU FINISH BEFORE TIME IS CALLED, YOU MAY CHECK YOUR WORK ON THIS SECTION ONLY. DO NOT TURN TO ANY OTHER SECTION IN THE TEST. **STOP**

SECTION II: FREE-RESPONSE QUESTIONS

Time—90 Minutes 6 Questions

Directions: Section II of this exam contains six questions: five free-response questions and one investigative task. The investigative task is the last question and will comprise 25% of your score for Section II. You should spend about 25 minutes on it. The other five questions comprise 75% of your score for this section. You should spend about 13 minutes each on these.

Clearly show the methods used as you write out your answers. You will be scored on the soundness of your methods and reasoning, and on finding the correct answers.

Part A: Questions 1–5
Spend about 65 minutes on this part of the exam.

1. The histogram below shows the snowfall, in inches, for each snow season in Minneapolis, Minnesota, from 1884 to 2005.

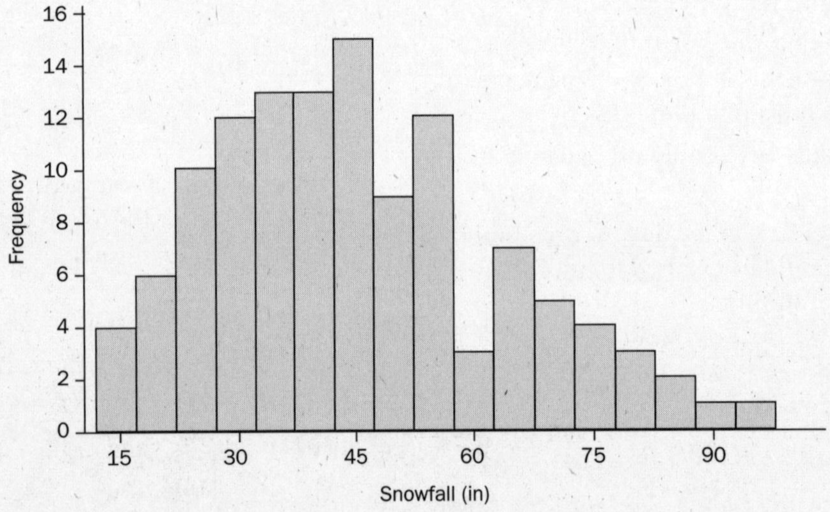

(a) Write a few sentences describing this distribution.

(b) The meteorologist at a local radio station wants to report a center of this distribution. He believes that his listeners will want it to sound like Minneapolis gets a lot of snow. Which measure of center, mean or median, would help him give his listeners this impression? Explain.

(c) Given that there are 121 observations represented in this histogram, estimate Q_1 as accurately as the histogram allows.

GO ON TO THE NEXT PAGE

2. The editor of a city's newspaper has recently increased the paper's coverage of the high technology industry, and she wants to find out if the newspaper's readers approve. The editor will use the newspaper's Website, which is available only to newspaper subscribers, as a method for administering the survey. The Website will administer the brief survey to every 50th visitor before the visitor can proceed to the news content of the site. The newspaper's Website is configured so that no visitor will be given the survey more than once.

   (a) What is the population of interest for the survey? Identify a potential source of bias for this method of selecting a sample of the population. Explain.
   (b) Describe how the bias in sample selection identified in the previous question might affect the estimate of the proportion of the population who approve of the increased coverage.
   (c) Is this a simple random sample of visitors to the newspaper's Website? Explain.

3. Tom and Viv host an annual party at their home. Past experience shows that if they invite 35 guests, the following table summarizes the number of guests who will actually make it to the party.

Number who make it to the party	29	30	31	32	33	34	35
Probability	0.18	0.23	0.27	0.15	0.09	0.05	0.03

Assume they invite 35 guests.

   (a) Tom and Viv are only providing enough food for 32 guests. What is the probability that there will be enough food?
   (b) The food for the party costs Tom and Viv $224. If they have 33 guests or more, they will send out for more food at a flat cost of $40. What is the expected value of the amount of money they will spend on food?
   (c) Given that Tom and Viv have to send out for extra food, what is the probability that all 35 guests make it to the party?

GO ON TO THE NEXT PAGE

4. Moore's Law claims that the number of transistors that can be fit on an integrated computer chip grows exponentially. The number of transistors per chip (Transistors) is recorded for selected years (Year) from 1971 to 2003. Computer output for a regression analysis of log(Transistors) versus Year is given below.

The log used is base 10.

	Regression Analysis: log(Transistors) versus Year			
Predictor	Coef	SE Coef	T	P
Constant	−300.29	10.71	−28.03	0.000
Year	0.154018	0.005388	28.59	0.000

S = 0.220768 R-Sq = 98.7% R-Sq(adj) = 98.6%

(a) According to this model, what will be the number of transistors per chip in 2010?

(b) Interpret r^2.

The following residual plot was produced as part of the regression analysis.

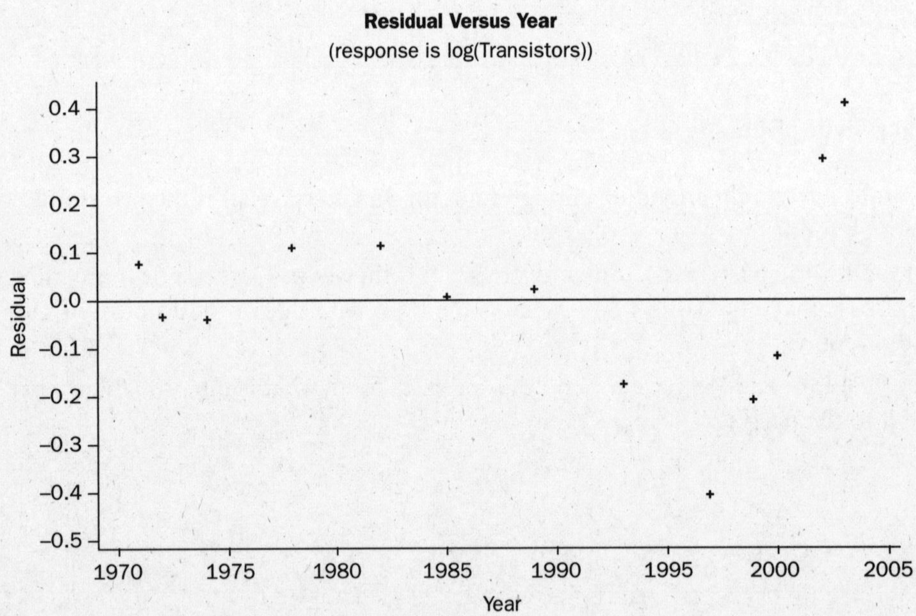

(c) Using this residual plot, comment on the ability of this regression model to make a good prediction for the number of transistors per chip in 2010.

GO ON TO THE NEXT PAGE

5. A math professor suspects a connection between the mean time a student studies each night and the student's term grade for the 300 students in a large class he teaches. He finds the mean daily study time for each student in the class. He notes the term grades for a random sample of six students from those who study an average of at least 30 minutes per night, and he notes the term grades for a random sample of five students from those who study less than 30 minutes per night. The grades are listed below.

Mean Minutes Study Time per Night	Term Grade					
≥30 minutes	85	71	93	79	74	86
<30 minutes	74	58	68	80	70	

(a) Is there evidence that the mean term grade for students who study for a mean of at least 30 minutes per night is higher than the mean term grade for those who study less? Perform an appropriate statistical test using $\alpha = 0.05$ to answer this question.

(b) Suppose that a study with this design found that the difference in mean term grades for the two groups is statistically significant. Can we conclude that studying an average of at least 30 minutes per night causes students to earn higher term grades?

GO ON TO THE NEXT PAGE

Part B—Question 6

Spend about 25 minutes on this part of the exam.

6. A test prep course is under consideration by a large high school to help its students prepare for a standardized college admissions test. A random sample of 10 students from the school is selected to see if the course will improve their scores. The students are given versions of the test before taking the course and after, with results summarized below. In addition, this table has a column with entries labeled "+" or "–" depending on whether the student's score increased or decreased after taking the course.

Student	Before Course	After Course	Improvement?
1	20	25	+
2	30	29	–
3	15	14	–
4	23	24	+
5	18	20	+
6	19	26	+
7	22	24	+
8	14	19	+
9	18	20	+
10	21	22	+

A matched pairs test could be used with this design, but there is an alternative using only the plus and minus signs.

(a) Let p represent the proportion of students in the population who will show improvement in test scores from taking the course. Write null and alternative hypotheses in terms of p for a test to determine whether or not the course improves scores on the standardized test.

(b) If your null hypothesis is true, use the binomial distribution formula to find the probability that a randomly selected sample of 10 students will show exactly 8 students whose scores showed improvement and 2 whose scores got worse. Assume that no scores remain the same.

(c) Find the p-value for this significance test.

(d) Is there sufficient evidence to reject your null hypothesis?

IF YOU FINISH BEFORE TIME IS CALLED, YOU MAY CHECK YOUR WORK ON THIS SECTION ONLY. DO NOT TURN TO ANY OTHER SECTION IN THE TEST.

STOP

Practice Test 1: Answer Key

1.	B	11.	A	21.	E	31.	C
2.	D	12.	A	22.	A	32.	D
3.	D	13.	C	23.	C	33.	B
4.	B	14.	A	24.	D	34.	A
5.	E	15.	D	25.	A	35.	B
6.	A	16.	A	26.	C	36.	B
7.	C	17.	E	27.	E	37.	E
8.	E	18.	E	28.	D	38.	E
9.	D	19.	C	29.	D	39.	C
10.	C	20.	B	30.	B	40.	A

PRACTICE TEST 1

ANSWERS AND EXPLANATIONS

SECTION I

1. B

There are 27 numbers in the distribution. Using the $\frac{n+1}{2}$ rule, the median will be the $\frac{27+1}{2}$, or the 14th number of the list. Since there are 13 numbers above the median and 13 below, the quartiles will be the $\frac{13+1}{2}$, or the 7th numbers from the top and bottom of the list. The 7th number from the top is 40, and the 7th number from the bottom is 64, so those are the quartiles. You could also determine this by entering all the numbers in your calculator and having it find the quartiles for you. Be sure to look for quartiles and nothing else. Otherwise you might be confused by observing that the mean $\bar{x}$ is 53.07, and the standard deviation s_x is about 17.42. Choice (A) gives you the numbers that are one standard deviation above and below the mean, but these have nothing to do with quartiles. You might also perform a more involved erroneous calculation, finding the quartiles of a normal distribution with a mean of 53.07 and a standard deviation of 17.42 to get the numbers in choice (C). These values are wrong because the given distribution is clearly nonnormal due to its skewness. Choice (D) makes no sense either, but for less elaborate reasons. These numbers can be obtained by finding the middle stem values, 40 and 80, and then taking the middle leaf from each of these. This method could give approximately correct answers for a uniform distribution, but this is unimodal and skewed. Finally, 48 and 81, the numbers from choice (E), are 25% and 75% of the way from the minimum to the maximum numbers in the distribution. These are not quartiles, but they could give a decent approximation for a uniform distribution. As in (D), this is not uniform.

2. D

You can approach this problem directly from the question or by working backwards from the choices. Working backwards, simply check each interval to see whether it corresponds to a 40% probability. Working directly, the middle 40% will run from the 30th to the 70th percentile. Working from a table or with a calculator, these values are about 0.525 standard deviations above and below the mean. That makes the endpoints of the interval $16 \pm (0.525)(0.1)$ or about 15.95 to 16.05. You get choices (B) or (E) if you follow this process but use the wrong percentiles. Choice (A) ignores the fact this is a normal distribution and simply creates an interval with width 0.04, which has a 4 in it but does not capture the middle 40% of the distribution. Choice (C) is a different take on ignoring normality; 40% of the standard deviation 0.1 is 0.04, so this interval is 16 ± 0.04. This computation has nothing to do with the middle 40%.

3. D

The mean amount of money going to the winner will double, and the producers will award $4,000 to the nonwinners, so the mean prizes per episode will total $22,800. That rules out (A) and (B). Doubling the prize money will double any measure of spread, but adding on $4,000 will not affect the spread. The new standard deviation would then become $9,000, which makes (D) the correct answer.

4. B

Hugh's test is 1.19 standard deviations above the mean, so he got a score of $53 + 1.19 \cdot 16$ or about 72. Similarly, Neal earned $53 + 2.06 \cdot 16$ or about 86. That makes a difference of 14 points, so

(B) is correct. Hugh's score of 72 is 19 points above the mean, and Neal's 86 is 33 points above the mean. This might make choices (C) or (D) appealing if you were comparing their scores to the mean instead of to each other. You might be tempted to work directly with the difference in z-scores, 2.06 – 1.19, or 0.87. This number matches wrong answer (A). You can use this method as long as you realize that 0.87 is the number of standard deviations between test scores and not the difference in the scores themselves. A difference of 0.87 standard deviations is $0.87 \cdot 16$, or 14 points, which matches correct answer (B). If you multiply 0.87 by 53 instead of the 16, you get incorrect answer (E). This choice is designed to trap you if you proceed haphazardly instead of with careful understanding.

5. E

The hypotheses in a hypothesis test should always be statements about parameters, not statistics. This rules out choices (A) and (D). Nothing in the question indicates this should be a one-sided test, so you are left with the only two-sided test among the remaining alternatives, (E).

6. A

The sampling described in the problem is an excellent example of cluster sampling. Clusters of products are randomly selected rather than individual products, and each cluster can reasonably be expected to resemble every other cluster. Stratified random sampling would require splitting the population into strata so that the strata are designed to be different from each other. Convenience sampling involves no randomization, so that's not right for this problem. Systematic sampling is a design that takes, for example, every tenth item from a list, which is not at all what is described here. Finally, if this were simple random sampling then the manufacturer would randomly select 500 products from the entire population of products, not select random lots.

7. C

Source	DF	Sum of Sq	Mean Sq	F-ratio
Regression	1	2290.3	2290.3	38.52
Residual error	30	1783.5	59.5	

Predictor	Coef	SE Coef	T	P
Constant	8.63	10.08	0.86	0.399
PRIOR	0.8126	0.1309	6.21	0.000

S = 7.71046 R-Sq = 56.2% R-Sq(adj) = 54.8%

The only part of the output that matters for this question is boxed above. The regression equation is EXAM = 8.63 + 0.8126 PRIOR. If PRIOR = 85, then the predicted final exam score is 77.7, or 78, choice (C). Choices (A) and (D) use incorrect versions of the regression equation, which you might guess at from other parts of the table. Choice (B) is 56.2% of 85, which is an inappropriate use of r^2 (listed in the table as R-Sq). Choice (E) simply uses the PRIOR grade of 85 as a guess for the EXAM grade.

8. E

This probability can be found using a binomial distribution with $n = 7$ and $p = \frac{15}{35}$. In general, $P(X = k) = \binom{n}{k} p^k (1 - p)^{n-k}$, so for this problem $P(X = 3) = \binom{7}{3}\left(\frac{15}{35}\right)^3\left(\frac{20}{35}\right)^4$. The other choices provide seemingly plausible but incorrect alternatives to a binomial approach. You might arrive at choice

(A) either by using the fact that $\frac{3}{7}$ of the meals in the book are vegetarian, or the fact that the question mentions that 3 of the 7 days will be vegetarian. Regardless, this answer does not address the overall question under consideration. Choice (B) ignores the fact that 4 days must be nonvegetarian. Choice (C) remedies this, but it ignores the fact that the 3 vegetarian days and 4 nonvegetarian days could occur in many different orders. Choice (D) takes this into consideration, but there are more than 3 different orderings of the meals. That brings you back to (E), which correctly assesses that there are $\binom{7}{3}$ possible orderings.

9. D

Choice (D) gives the correct formula for sample standard deviation s_x, the sample statistic used to estimate population standard deviation. The other choices all give formulas for standard deviations, but from other contexts altogether. The formula given in (A) is for σ_X, the standard deviation for a discrete probability distribution. Choice (B) is the standard deviation σ_x for a binomial distribution with parameters n and p, and choice (C) is the standard deviation for $\sigma_{\hat{p}}$ for the sampling distribution of proportions for the same type of binomial distribution. Choice (E) gives the formula for s_{b_1}, the standard deviation for the slope of a least-squares regression line.

10. C

For the sum of independent random variables, standard deviations do NOT add up, but variances do (variance is the square of the standard deviation). The sum of the four variances is 1.02, so the standard deviation is the square root of 1.02, or about 1.01 seconds. The wrong answer choices are all instances of misapplying formulas or making plausible but incorrect guesses about the formulas. If you add the standard deviations you get 1.8, choice (D). If you divide this by 4, you get choice (A), 0.45. If you try to divide the sum of 1.8 by $\sqrt{4}$ you get (B), 0.9. Choice (E) is absurd, but you might arrive at this answer if you treat the means as values of a random variable and the standard deviations as probabilities. The answer 50.7 is what you would get if you find the "mean" of this distribution.

11. A

Choice (A) gives a standard explanation for the meaning of a p-value, and it is correct. Choice (B) is wrong because it begins with the assumption that H_a is true instead of H_0. Choice (C) would be right if it were a 99% confidence interval instead of 95%, but it's wrong as stated. Calculations of power require a specific alternative and require several more steps than (D) suggests, so this choice is incorrect. Choice (E) is wrong as well. It could be corrected if the 1% and 5% were reversed.

12. A

Choice (A) states the truth of the matter clearly and correctly. Choice (B) is not supported by any statistical reasoning. Choice (C) is wrong because the effect of the six heads on the long-range proportion of heads does not require a bunch of tails to balance things out. A few hundred tosses with equal numbers of heads and tails will reduce the impact of the original six heads just fine. Don't be fooled by the fact that (C) is the only choice to identify by name the key concept for this question, the law of large numbers. It's still wrong. Choice (D) ignores the fact that the first six heads already happened, so it is irrelevant to any upcoming probability. You aren't contemplating seven heads in a row, you are looking to see if one toss comes up heads. Choice (E) is wrong because it isn't crucial to make the results of ten tosses seem plausible, not when we already know the first six tosses were unlikely. Ten tosses do not constitute "in the long run" as mentioned in the question.

13. C

As long as sound statistical procedures are used, sampling variability increases when sample size decreases, and that's why (C) is correct and (E) is not. Working with proportionately larger or smaller populations does not change this fact, so choices (A) and (D) are wrong. As for (B), sampling bias will not be present as long as sound statistical practices are followed.

14. A

Normal distributions aren't skewed, so choice (B) cannot be right. Chi-square distributions are skewed to the right, but not all right-skewed distributions are chi-square. This rules out choice (C). Outliers can occur on either side of a distribution, even strongly skewed ones, so (D) is wrong, too. Skewed distributions can have many clusters, so that rules out (E). Choice (A) is true by elimination, but it is true on its own merits as well. The median of a distribution is a resistant measure, but the mean is not. The mean will lie farther to the right than the median for a distribution that is skewed to the right.

15. D

The curved pattern in the residual plot shows conclusively that the relationship is nonlinear, so that rules out choices (A) and (B). The slope of the regression line is a positive number, 0.0408, so the association is positive. That eliminates (C) as an answer. That leaves (D) and (E). You might not be able to sort out which point in the residual plot corresponds to a mean orbital radius of 100 million km, but all the possibilities show a positive residual. Since the formula for a residual is $Y - \hat{Y}$, a positive residual means that the actual value Y is more than the predicted value $\hat{Y}$. The regression line underpredicts, so (D) is correct.

16. A

A p-value of 0.33 is not significant at any reasonable significance level, so there is insufficient evidence to reject the null hypothesis. No other conclusion is supported by the analysis. When you compute a p-value, you assume that the null hypothesis is true and then see how odd the observed results would seem to be under this assumption. A p-value of 0.33 means that the observed result would not be odd at all, so the sample results are consistent with the assumption that the null hypothesis is true. The null hypothesis may or may not be true, but the analysis does not give you cause to reject it. At no point are you assuming the alternative hypothesis is true, so you have no grounds to reach a conclusion about it.

17. E

The experiment can be outlined using the following tree diagram.

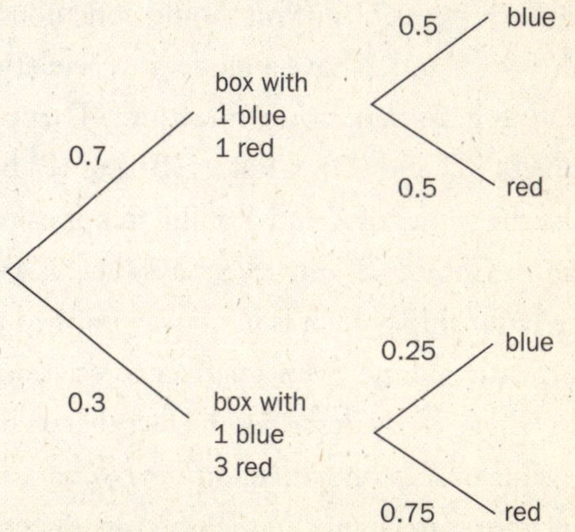

The probability of picking a box of the first type and then a blue from that box is (0.7)(0.5). The probability of picking a box of the second type and then a blue from that box is (0.3)(0.25). Thus the overall probability of picking a blue marble is (0.7)(0.5) + (0.3)(0.25). Choice (C) reflects this line of thinking but reverses addition and multiplication. You should recognize that it gives you the wrong answer since it multiplies out to a probability greater than 50%, which is not a reasonable answer. Choice (D) also uses this table but multiplies all the individual probabilities together. This choice gives a probability that is unreasonably small, so you should recognize that it cannot be right. Choice (A) ignores the boxes and simply finds the number of blue marbles out of the total number of marbles. This is incorrect since the use of boxes makes it a little more likely that you might draw a blue marble (7 of the 10 boxes give you a 50-50 chance of drawing a blue marble). Finally, choice (B) ignores the fact that there are 7 boxes of one type and 3 of the other. The answer for (B) would be correct if there were an equal number of each type of box, but there aren't.

18. E

Choice (A) is a textbook interpretation of r^2, but that's not the statistic the question provides. The correlation $r = -0.71$, so you should conclude that the values of X and Y have a moderate correlation and tend to head in opposite directions. That rules out choices (C) and (D). Choices (B) and (E) both describe the values of X and Y going in opposite directions. Choice (B), however, says there is a cause-effect relationship, which is in no way implied by any correlation alone, even a perfect correlation. Choice (E) correctly describes the fact that a change in the value of X in one direction *corresponds* to a tendency for Y to change in the opposite direction.

19. C

One of the main reasons for using a t-test is to allow inference even though the population standard deviation is unknown. You would use a z-test if you knew the population standard deviation. The symmetry of the sample matters in relation to sample size, but it is irrelevant as far as whether to use a z-test or not. Choices (B), (D), and (E) are irrelevant as well.

20. B

A significant aspect of tables of random numbers is that entries are independent of each other, so (B) is correct. Choice (A) is incorrect because a table of random numbers contains all two-digit entries 00 through 99 equal numbers of times. On average, every five lines of 40 digits each will contain 00. Choice (C) is incorrect because the equal occurrence of entries is true only on average, so it is not necessarily true that each block of 10 or 20 or 40 digits will contain uniform counts of each digit. Choice (D) is incorrect because 05 does not in fact stand for the one-digit number 5. The single digit 5 stands for 5. Choice (E) is wrong because the sequence 0123456789 is unlikely to occur in the table at all unless the table is extraordinarily large. For the table to include all 10-digit sequences it would have to be at least 10 times 10,000,000,000, or 100 billion digits long. The first 1,000 digits of the table only make up a tiny fraction of this number.

21. E

If p is the p-value for the two-sided test, the p-value for a one-sided test would be either $\frac{1}{2}p$ or $1-\frac{1}{2}p$ depending on which side the one-sided test is conducted. That's what makes it impossible to tell. Choice (A) seems plausible until you get to choice (E). Choice (C) begins with the same reasoning as (A), but then reaches the wrong conclusion. Choices (B) and (D) give the wrong formula for the p-value of a one-sided test.

22. A

Recall that the quartiles and the median divide a data set into segments, each containing about 25% of the data. The 60th percentile of data set I is clearly above 78, and the 40th percentile of data set II is clearly below 77. That makes (A) correct. As for the other choices, (B) is wrong because the interquartile range (IQR) is the height of each box. With this in mind, it's easy to see that data set I has a larger IQR. For (C), the span from 65 to 84 covers about 50% of the values, not 25%. Choice (D) is wrong because boxplots do not show how many data points are in a data set. Finally, choice (E) is wrong because it is data set I that has the larger range.

23. C

This question demands that you know what confounding means, and there is no context to help you figure it out if you don't know. When variables in an experiment are confounded, there is no way to distinguish which is the true cause of the observed results; that is why (C) is correct. The other answers, though, all sound like they refer to things that would be a problem in the design of an experiment, and so they may lure you to guess them. (A) is incorrect because randomization is a cure for confounding. (B) is incorrect because confounding is unrelated to bias; bias is a systematic tendency to favor certain outcomes. Confounding can be present in the complete absence of bias. (D) is incorrect because there can be a very strong association between the explanatory and response variables even when the relationship is confounded with some third variable. As for choice (E), double blinding can guard against a particular instance of confounding. Just as with (A), the answer reverses the true relationship between confounding and a measure that combats it.

24. D

The question is asking for the proportion of seniors expecting to live in-state at age 30, which is $\frac{350}{500}$ or 70%, and the proportion expecting to live out-of-state, which is $\frac{150}{500}$ or 30%. The other choices are all proportions to be found from the table but they do not answer the question asked. Choice (A) is the marginal distribution of where Johnson seniors spent their freshman year. Choice (B) is the conditional distribution of where seniors spent their freshman year given that they expect to live out-of-state at age 30. Choice (C) is the conditional distribution of where seniors expect to live given that they attended Johnson as freshmen. Choice (E) is the conditional distribution of where seniors expect to live given that they did not attend Johnson as freshmen.

25. A

Since the goal is to measure the difference in pulse rates for individual subjects, and since paired data is available, a matched-pairs t-test is the best statistical test here. Choice (E) is an option, but then you lose the ability to measure individual differences. Choice (D) ignores the fact that there are two sets of data collected. Choices (B) and (C) make no sense for this context.

26. C

For this question $n = 500$, $\hat{p} = \frac{280}{500} = 0.56$, the critical value z^* for a 95% confidence interval is 1.96, and the standard error is $\sqrt{\frac{\hat{p}(1-\hat{p})}{n}}$. This makes (C) the correct answer. If you use a garbled version of the standard error formula for the mean, you get (D), and if you use a garbled version of the standard error formula for the difference of sample proportions, you get (E). Choice (A) is wrong because it uses z^* from a 90% confidence interval, and (B) is wrong for combining the mistakes of choices (A) and (E).

27. E

A cumulative relative frequency diagram shows the percentage of trials in which a given number or less was observed. The value above 1 is about 17%, so 17% of the trials resulted in a 7 on the first roll. The value above 2 is about 33%, so 33% of the trials resulted in a 7 on the first or second roll. That's answer choice (B), so it isn't the right answer. In all the other trials, 67% of them, it took at least three rolls to get a 7. That's choice (E), the right answer. The other answer choices are based on various mistakes in reading the graph. For (A), 6% of the trials took exactly three rolls to get a 7. Choice (C) tempts you because it took three *or fewer* rolls 39% of the time. Finally, it took *more than* three rolls 61% of the time, and that's choice (D).

28. D

Replication is crucial for any experiment regardless of other features of the design. Choices (A) and (B) are wrong because they describe exactly what blocking is all about. Choices (C) and (E) also describe the main motivations for blocking, but they may confuse you because they are so concise. Don't let that throw you. They hit the nail on the head as far as describing why researchers block experiments.

29. D

Statement I is true since normal distributions are perfectly symmetric. A sample from a normal distribution won't be perfectly symmetric, but a normal distribution itself will be. Statement II is false; the larger the standard deviation, the more spread-out the distribution and the lower the peak will be. A normal distribution with standard deviation 2 will have a *lower* peak than a normal distribution with standard deviation 1. Statement III is true. Recall that normal distributions are often written in the form $N(\mu,\sigma)$. In addition, at least one other question on this exam describes a normal distribution by giving its mean and standard deviation.

30. B

Since the number 0 occurs well within the 98% confidence interval, you can fail to reject at the 2% significance level the hypothesis that the difference of population means is 0. In other words, the results are consistent with the possibility that the difference in population means is 0. This same type of analysis can be used to show that the observed results are consistent with a difference of population means equal to 3, so choice (A) is wrong. Choices (C), (D), and (E) are all wrong because the 98% part of a 98% confidence interval does not refer to a probability in the way these choices describe.

31. C

A point at (75,16) would be influential, since it is far from the other points and because it is off to one side of the distribution of the explanatory variable *hits*. This rules out choice (B). Like a see-saw, the line would tilt towards the influential point, decreasing the slope. That rules out (D) and (E). This new point would weaken the moderately strong linear association. That makes (C) the correct answer.

32. D

Intervals with higher confidence levels must be wider. That's because you need to look in a wider interval if you are going to be more confident your interval contains the population mean. The margin of error for a confidence interval for the mean is $z^* \cdot \frac{\sigma}{\sqrt{n}}$, so choices (A), (B), and (C) don't make much sense. Choice (E) is an unusual comparison of margin of error for one interval against the width of the other, but a little consideration shows that this choice could only be true if the 95% confidence interval is wider than the 99% confidence interval, which it is not.

33. B

Since r for scatterplot A is closer to 1 than scatterplot B, its points are typically closer to the regression line. This can be quantified by taking the difference of each point from the corresponding point on the line, squaring it, and summing the values for each point. If the differences are not squared, as in (A), the points above the line will have a positive difference, the points below the line will have a negative difference, and the sum will be zero. No other conclusions can be drawn concerning the deviations from the information provided.

34. A

A type I error occurs when H_0 is true but it is rejected anyway. A type II error occurs when H_0 is false but it is accepted anyway. In this context, it means that the old campaign will be used, which knocks out choices (B), (D), and (E), even though H_0 is false, which knocks out choice (C). That leaves (A) as the correct answer.

35. B

Since the individual scores themselves are approximately normally distributed, the difference in scores for two randomly selected children will also be approximately normally distributed with mean 0 and standard deviation $\sqrt{10^2 + 10^2}$, or about 14.14.

The probability that the difference is between –5 and 5 is then about 0.28, which is the correct answer. If you only look at differences between 0 and 5, you get choice (A), 0.14. Choice (C) represents the probability that a single test-taker gets a score between 95 and 100, and would only be attractive if you decided that the first test-taker might get 100, so the second would have to get from 95 to 105. This sort of approach is attractive to students not comfortable working with the difference between random variables, but it always leads to the wrong answer. Choices (D) and (E) are also incorrect "shortcut" answers. Choice (D) simply divides the 5-point difference by the standard deviation of 10 to get 0.5, and choice (E) is the familiar proportion of observations that lie within 1 standard deviation of the mean for a normal distribution.

36. B

The margin of error for any confidence interval is given by (critical value)(standard error of statistic). Here, the critical value is $z^* = 1.645$ (actually 1.644854), and the standard error is $\sqrt{\frac{\hat{p}(1-\hat{p})}{n}}$, where $\hat{p}$ is the sample proportion and n is the sample size. For a given sample size, the standard error is largest when $\hat{p} = 0.5$, so we will use this value to determine the appropriate sample size. The work proceeds as follows.

$$1.645\sqrt{\frac{0.5(1-0.5)}{n}} \le 0.02$$

$$1.645 \cdot \frac{0.5}{\sqrt{n}} \le 0.02$$

$$\frac{1.645 \cdot 0.5}{0.02} \le \sqrt{n}$$

$$1691.3 \le n.$$

Thus, (B) is the correct answer. If the more precise value of z^* were used, it turns out that 1,691 would be a large enough sample size. Of the given choices, though, $n = 1{,}692$ is still the smallest that will accomplish the researcher's goal.

You could also figure out this answer by working backwards from the answer choices. Notice that the answers are arranged in decreasing order, so you will make fastest progress if you try (C) first and see if that answer is a large enough sample size. It isn't, so then all you have to do is decide whether (A) or (B) is the better answer. This method still requires you to know that you need to use $\hat{p} = 0.5$.

Of the wrong answers, (A) and (C) correspond to incorrect values of z^*. If you use z^* for a 95% confidence interval, you get (A), and if you use z^* for an 80% confidence interval, you get (C). Choice (D) is the answer you would get if you wanted the standard error to be less than 0.02 instead of the margin of error. Choice (E) is the result of using $\hat{p} = 0.02$ instead of $\hat{p} = 0.5$ in your calculations.

37. E

The sampling distribution of the sample mean of n items from a population with approximate distribution $N(\mu_X, \sigma_X)$ is approximately $N\left(\mu_X, \frac{\sigma_X}{\sqrt{n}}\right)$. Thus, the sampling distribution for the mean length of 50 beams has mean 4 m and standard deviation $\frac{0.03}{\sqrt{50}}$. That means the z-score of 3.99 m is $z = \frac{3.99 - 4.00}{\frac{0.03}{\sqrt{50}}}$, and so the probability that the mean length is at least 3.99 m is

$$P\left(z > \frac{3.99 - 4.00}{\frac{0.03}{\sqrt{50}}}\right).$$

The wrong answer choices all use the same solution method but include mistakes in various ways, either in using the wrong sampling standard deviation or reversing the subtraction in the z-score numerator.

38. E

Answer (E) is the correct answer. Each brand of tile should have been randomly assigned and exposed to both weather types.

Answer (A) is incorrect because there was not randomization in the assignment of the tiles to the roofs, because each type of tile did not receive both treatments, and because the experimental units (the tiles) were not assigned to the treatments (the weather conditions) randomly. Answer (B) is incorrect. A sample size of 10 is sufficient to allow for replication by placing more than one tile of each type in each treatment group. Answer (C) is incorrect because the treatments in this experiment are the weather conditions applied to the tiles. Answer (D) is incorrect because the tiles were not randomly assigned to the roofs. Roofs 1–5 were covered in brand X tiles and roofs 6–10 were covered in brand Y tiles.

39. C

The chi-square test statistic is calculated as follows.

$$x^2 = \frac{(61-50)^2}{50} + \frac{(50-50)^2}{50} + \frac{(44-50)^2}{50} + \frac{(45-50)^2}{50} = 3.64.$$

If you forget to square, you get choice (A), if you square the denominators, you get (B), and if you forget to even have denominators, you get (E). Choice (D) is the chi-square ($df = 3$) critical value for a tail probability of 0.25, which is not helpful in the calculation of the test statistic for the given data.

40. A

The formulas are

$$\mu_X = \sum x_i p_i \quad \text{and} \quad \sigma_X^2 = \sum (x_i - \overline{x})^2 p_i, \text{ so}$$

$$\mu_X = (100)(0.6) + (300)(0.2) + (500)(0.1) + (1000)(0.1) = 270$$

$$\text{and } \mu_X = \sum x_i p_i \quad \text{and} \quad \sigma_X^2 = \sum (x_i - \overline{x})^2 p_i$$

$$\sigma_X^2 = (100 - 270)^2(0.6) + (300 - 270)^2(0.2) + (500 - 270)^2(0.1) + (1000 - 270)^2(0.1) = 76{,}100.$$

Choice (B) is wrong because it computes σ_X^2 the same way as choice (D) (see below). Choice (C) is wrong because it uses $\sum (x_i - \overline{x})^2$ for σ_X^2. Choice (D) is wrong because its mean and variance are computed from the list of numbers 100, 300, 500, and 1,000, without using the given probabilities at all. This would only be correct if the distribution were uniform. Choice (E) is wrong because its values are computed as in (E), and because it gives the value of s_X^2 rather than σ_X^2 for the list of numbers.

SECTION II ANSWERS

1. (a) The distribution is centered at about 40 to 45 inches, and has a range from about 10 inches up to about 100 inches of snowfall in a year. The graph is unimodal and skewed right. There are no gaps and there appear to be no outliers.

   (b) The meteorologist should report the median snowfall per year. Since the graph is skewed right, the mean will be larger than the median. That is because the median is a resistant measure, while the mean is nonresistant. In other words, the mean is affected by skewness more than the median.

   (c) The bars of the histogram cover the intervals 12.5 to 17.5, 17.5 to 22.5, 22.5 to 27.5, and so on. The first bar contains 4 observations, so 17.5 is approximately the $\frac{4}{121}$ or 3.3rd percentile. The first four bars include 32 observations, so 32.5 is approximately the $\frac{32}{121}$ or 26.4th percentile. It is a reasonable estimate that Q_1, the 25th percentile, is about 32.

2. (a) The population of interest is the readership of the newspaper. A potential source of bias is undercoverage. Not everyone in the newspaper's readership has easy Internet access, so they are systematically excluded from the survey.

   Alternate answer:

   Other sources of bias include nonresponse (Website visitors may choose not to take the poll) and response bias (survey subjects are using a computer while being asked about coverage of high technology).

   (b) Readers who have better access to the Internet will be overrepresented in the sample, and these survey subjects will be more likely to approve of increased coverage of technology.

   Alternate answer:

   For nonresponse as a source of bias, those who choose to respond may be more likely to have strong opinions than those who do not. Survey subjects who are using a computer and who have strong opinions are more likely to approve of increased coverage. For response bias as a source of bias, survey subjects are again more likely to approve of increased coverage of technology because of the circumstances in which the survey is being conducted (i.e., through the means of high technology).

   (c) This is a systematic sample of visitors to the Website and not a simple random sample. In a simple random sample every individual has the same chance of being selected and every sample has the same chance of being selected. That is not the case here, since consecutive visitors to the site cannot both be included in the sample.

3. (a)

   $$P(\text{guests} \le 32) = 0.18 + 0.23 + 0.27 + 0.15 = 0.83.$$

   (b) The following table shows the probability distribution for the expenditure on food.

Food expenditure	\$224	\$264
Probability	0.83	0.17

The expected amount of money spent on food is (224)(0.83) + (264)(0.17) = \$230.80.

Alternatively, the expected amount is \$224 + (\$40)(0.17) = \$230.80.

(c) $P(35 \text{ guests and at least 33 guests make it})$

$$= \frac{P(35 \text{ guests and at least 33 guests make it})}{P(\text{at least 33 guests make it})}$$

$$= \frac{0.03}{0.17} \approx 0.18.$$

4. (a) $\log(\text{Transistors}) = -300.29 + 0.154018 \text{ Year}$

$$\log(\text{Transistors}) = -300.29 + 0.154018(2010) = 9.28618.$$

$$\text{Transistors} = 10^{9.28618} \approx 1{,}933{,}000{,}000.$$

(b) r^2 is 98.7%. This means that 98.7% of the variation in the value of log(Transistors) is explained by the least-squares regression of log(Transistors) on Year.

(c) First of all, using this model to predict the number of chips per circuit in 2010 is an extrapolation. Extrapolation is risky for any model. For this particular model, the residual plot is showing that the model is a pretty good fit for data prior to 1995 since there is no clear pattern to the residuals. Since 1995, however, there is a clear upward sloping linear pattern to the residuals, which suggests that the prediction for 2010 is too low.

5. (a) **Step 1:** $\mu_{\geq 30}$ = mean term grade for the population of students whose mean study time is at least 30 minutes per night, $\mu_{<30}$ = mean term grade for the population of students whose mean study time is less than 30 minutes per night.

H_0: $\mu_{\geq 30} = \mu_{<30}$ OR
H_0: $\mu_{\geq 30} - \mu_{<30} = 0$
H_a: $\mu_{\geq 30} > \mu_{<30}$ H_a: $\mu_{\geq 30} - \mu_{<30} > 0$

Step 2: Conduct a two-sample *t*-test.

Conditions met?

i. The samples are independent simple random samples from their respective populations.

ii. The combined sample size is 11, which is small. Note that a dot plot for each sample shows each is reasonably symmetric with no gaps or outliers. Both populations are plausibly normal.

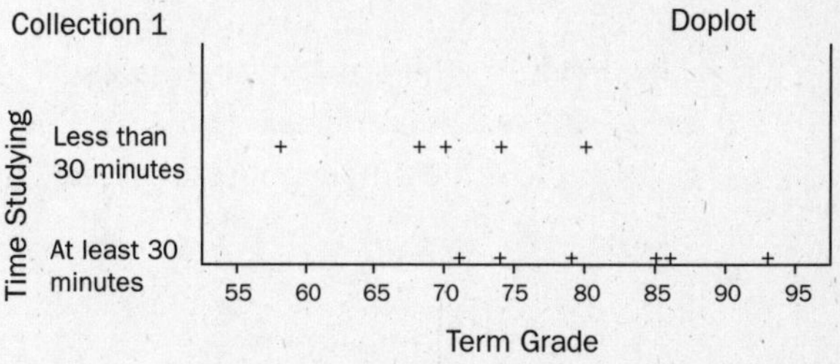

Step 3: The sample statistics are $n_{\geq 30} = 6$, $\bar{x}_{\geq 30} = 81.33$, $s_{\geq 30} = 8.21$, $n_{<30} = 5$, $\bar{x}_{<30} = 70$, and $s_{<30} = 8.12$.

$$\text{Test statistic: } t = \frac{(\bar{x}_{\geq 30} - \bar{x}_{<30}) - 0}{\sqrt{\frac{s^2_{\geq 30}}{n_{\geq 30}} + \frac{s^2_{<30}}{n_{<30}}}} = 2.29$$

with $df = 8.68$ for a *p*-value of 0.024.

Step 4: Because the *p*-value is less than $\alpha = 0.05$, the difference in grades is statistically significant. We reject the null hypothesis that the mean term grade is the same for both populations.

(b) No cause-and-effect relationship can be established because this is not an experiment. There is no assignment of subjects to treatment groups. The effect of study time could be confounded with other factors, such as the desire to do well in the course that prompts this extra study time. It may be that an unmotivated student who studies more would not get anything out of that extra study time.

6. (a) $H_0: p = 0.5$ and $H_a: p > 0.5$. The null hypothesis says that the proportion of all students whose scores show improvement from taking the course is 0.5, so the course neither helps nor hurts. The alternative hypothesis says that more than half of all students will improve their scores by taking the course.

(b) The binomial probability with $p = 0.5$ and $n = 10$ is

$$P(X=8) = \binom{10}{8}(0.5)^8(0.5)^2 = 0.044.$$

(c) The p-value is the probability of seeing results at least as extreme as those seen in the sample if the null hypothesis is true.

$$\begin{aligned} p\text{-value} &= P(X \geq 8) \\ &= \binom{10}{8}(0.5)^8(0.5)^2 \\ &\quad + \binom{10}{9}(0.5)^9(0.5)^1 \\ &\quad + \binom{10}{10}(0.5)^{10}(0.5)^0 \\ &= 0.0547. \end{aligned}$$

(d) At the 5% significance level, there is insufficient evidence to reject the null hypothesis, that half of all students would see their scores improve from taking this course. If students were just as likely to see their scores go up as go down after taking the course, then there is a 0.0547 probability of seeing results at least as extreme as seen here. This is not rare enough to reject the null hypothesis at the 5% significance level since 0.0547 is greater than 0.05.

Practice Test 2 Answer Grid

1. Ⓐ Ⓑ Ⓒ Ⓓ Ⓔ	11. Ⓐ Ⓑ Ⓒ Ⓓ Ⓔ	21. Ⓐ Ⓑ Ⓒ Ⓓ Ⓔ	31. Ⓐ Ⓑ Ⓒ Ⓓ Ⓔ
2. Ⓐ Ⓑ Ⓒ Ⓓ Ⓔ	12. Ⓐ Ⓑ Ⓒ Ⓓ Ⓔ	22. Ⓐ Ⓑ Ⓒ Ⓓ Ⓔ	32. Ⓐ Ⓑ Ⓒ Ⓓ Ⓔ
3. Ⓐ Ⓑ Ⓒ Ⓓ Ⓔ	13. Ⓐ Ⓑ Ⓒ Ⓓ Ⓔ	23. Ⓐ Ⓑ Ⓒ Ⓓ Ⓔ	33. Ⓐ Ⓑ Ⓒ Ⓓ Ⓔ
4. Ⓐ Ⓑ Ⓒ Ⓓ Ⓔ	14. Ⓐ Ⓑ Ⓒ Ⓓ Ⓔ	24. Ⓐ Ⓑ Ⓒ Ⓓ Ⓔ	34. Ⓐ Ⓑ Ⓒ Ⓓ Ⓔ
5. Ⓐ Ⓑ Ⓒ Ⓓ Ⓔ	15. Ⓐ Ⓑ Ⓒ Ⓓ Ⓔ	25. Ⓐ Ⓑ Ⓒ Ⓓ Ⓔ	35. Ⓐ Ⓑ Ⓒ Ⓓ Ⓔ
6. Ⓐ Ⓑ Ⓒ Ⓓ Ⓔ	16. Ⓐ Ⓑ Ⓒ Ⓓ Ⓔ	26. Ⓐ Ⓑ Ⓒ Ⓓ Ⓔ	36. Ⓐ Ⓑ Ⓒ Ⓓ Ⓔ
7. Ⓐ Ⓑ Ⓒ Ⓓ Ⓔ	17. Ⓐ Ⓑ Ⓒ Ⓓ Ⓔ	27. Ⓐ Ⓑ Ⓒ Ⓓ Ⓔ	37. Ⓐ Ⓑ Ⓒ Ⓓ Ⓔ
8. Ⓐ Ⓑ Ⓒ Ⓓ Ⓔ	18. Ⓐ Ⓑ Ⓒ Ⓓ Ⓔ	28. Ⓐ Ⓑ Ⓒ Ⓓ Ⓔ	38. Ⓐ Ⓑ Ⓒ Ⓓ Ⓔ
9. Ⓐ Ⓑ Ⓒ Ⓓ Ⓔ	19. Ⓐ Ⓑ Ⓒ Ⓓ Ⓔ	29. Ⓐ Ⓑ Ⓒ Ⓓ Ⓔ	39. Ⓐ Ⓑ Ⓒ Ⓓ Ⓔ
10. Ⓐ Ⓑ Ⓒ Ⓓ Ⓔ	20. Ⓐ Ⓑ Ⓒ Ⓓ Ⓔ	30. Ⓐ Ⓑ Ⓒ Ⓓ Ⓔ	40. Ⓐ Ⓑ Ⓒ Ⓓ Ⓔ

PRACTICE TEST 2

SECTION I: MULTIPLE-CHOICE QUESTIONS

Time—90 Minutes 40 Questions

Directions: Section 1 of this examination contains all multiple-choice questions. Decide which of the suggested answers best suits the question. You may use the formulas and tables found in the Appendix on p. 431 on the test.

1. Which of the following would NOT be a good display to show the clusters in a distribution?

   (A) Stem and leaf plot
   (B) Histogram
   (C) Dotplot
   (D) Boxplot
   (E) Cumulative frequency plot

2. A simple random sample of 30 ears of corn is collected from a field. The ears of corn have a mean weight of 0.21 kg with a standard error of the mean of 0.006. Which of the following should be used to compute an 80% confidence interval for the mean weight of all the field's ears of corn?

   (A) $0.21 \pm 0.854(0.006)$
   (B) $0.21 \pm 0.854\left(\frac{0.006}{\sqrt{30}}\right)$
   (C) $0.21 \pm 1.311(0.006)$
   (D) $0.21 \pm 1.311\left(\frac{0.006}{\sqrt{30}}\right)$
   (E) $0.21 \pm 1.96\left(\frac{0.006}{\sqrt{30}}\right)$

3. A movie studio releases 12 films in one year, earning a mean gross revenue of \$14.4 million per film. The next year, the studio releases 8 films with a mean gross of \$23.2 million per film. Which of the following is the mean gross revenue for the studio over these two years?

   (A) \$4.1 million
   (B) \$17.92 million
   (C) \$18.8 million
   (D) \$19.68 million
   (E) \$37.6 million

4. Continuous random variables X and Y are independent, with parameters $\mu_X = 30$, $\mu_Y = 40$, $\sigma_X = 15$, and $\sigma_Y = 5$. Which of the following gives the correct values for μ_{X+Y} and σ_{X+Y}?

   (A) $\mu_{X+Y} = 70$ and $\sigma_{X+Y} = 14.1$
   (B) $\mu_{X+Y} = 70$ and $\sigma_{X+Y} = 15.8$
   (C) $\mu_{X+Y} = 70$ and $\sigma_{X+Y} = 20$
   (D) $\mu_{X+Y} = -10$ and $\sigma_{X+Y} = 14.1$
   (E) $\mu_{X+Y} = -10$ and $\sigma_{X+Y} = 15.8$

GO ON TO THE NEXT PAGE

5. A sports researcher is studying statistics for a set of professional baseball teams over a single season. The researcher performs a regression analysis on the total payroll (in millions of dollars) versus the number of wins. An output of the results is shown below.

Regression Analysis: Wins versus Payroll

Predictor	Coef.	SE	T	P
Constant	66.42	12.46	5.33	0.000
Payroll	0.2070	0.1673	1.24	0.240

S = 16.9082 R-Sq = 11.3% R-Sq(adj) = 3.9%

Which of the following is the best interpretation for the slope of the regression line?

(A) An addition of about \$5 million in payroll corresponds to 1 extra win.
(B) An addition of about \$17 million in payroll corresponds to 1 extra win.
(C) An addition of about \$66 million in payroll corresponds to 1 extra win.
(D) Five extra wins correspond to an addition of about \$1 million in payroll.
(E) One extra win corresponds to an addition of about \$17 million in payroll.

6. The residual for the least-squares regression line $\hat{y} = 3.422 + 5.551x$ at the data point $(9.227, y)$ is equal to 1.811. Which of the following is the value of y?

(A) 56.452
(B) 54.641
(C) 52.830
(D) 11.038
(E) 7.416

7. The standard deviation for a large population is 3. A simple random sample is taken from this population, and this sample is used to calculate a 96% confidence interval for the population mean. If this confidence interval has a margin of error of 0.283, what is the sample size?

(A) 3
(B) 112
(C) 344
(D) 432
(E) 474

8. Why is it possible to show a cause-and-effect relationship between two variables using an experiment and not an observational study?

(A) Unlike observational studies, experiments can be double-blinded.
(B) Unlike observational studies, experiments control for confounding variables.
(C) Unlike observational studies, experiments can be blocked.
(D) Experiments tend to be conducted with more care for details than observational studies.
(E) Experiments tend to use larger sample sizes than observational studies.

GO ON TO THE NEXT PAGE

9. A lottery has a winner every day. The winning number is chosen randomly each day, and the probability of winning on any one day is 0.03. Alice plans to buy a lottery ticket every day for 500 days, and Bob plans to buy a ticket every day for 5,000 days. Which of the following is the best assessment of how the law of large numbers applies to this situation?

   (A) Alice will probably have a higher proportion of wins than Bob.
   (B) Bob will probably have a higher proportion of wins than Alice.
   (C) The difference between Alice's proportion of wins and Bob's proportion of wins will probably be close to 0.03.
   (D) Alice will probably have a proportion of wins closer to 0.03 than Bob.
   (E) Bob will probably have a proportion of wins closer to 0.03 than Alice will.

10. Trucks are used in many configurations. Three common configurations are tractors with 1, 2, and 3 trailers. Accident statistics for these configurations in 2002 are listed below.

Fatal U.S. Truck Accidents in 2002 by Configuration

Truck Configuration	Number of Registered Vehicles	Number of Accidents
Tractor, 1 trailer	150,000	2,889
Tractor, 2 trailers	5,100	154
Tractor, 3 trailers	200	1

Assume these statistics represent simple random samples for each configuration. Which of the following would be the most appropriate test to use to determine whether the populations for each configuration have the same rate of fatal accidents?

   (A) One-sample proportion z-test
   (B) Two-sample proportion z-test
   (C) Chi-square test for independence
   (D) Chi-square goodness-of-fit test
   (E) Chi-square test for homogeneity of proportions

GO ON TO THE NEXT PAGE

11. Consider these three statements about *t*-distributions. Which is/are true?

    I. The center for every *t*-distribution is 0.

    II. The larger the degrees of freedom, the less a *t*-distribution resembles the normal distribution.

    III. The one-tail probability for any specific *df* is double the two-tailed probability.

    (A) I only
    (B) II only
    (C) III only
    (D) I and II
    (E) I and III

12. A mattress manufacturer wants to test a new line of mattresses to see if a disturbance by a sleeper on one side of a mattress will be noticeable to a sleeper on the other side. A random sample of 500 mattresses will be used for the test. A house of cards is set up on one side of each mattress, and then a bowling ball is dropped on the other side. This is done two more times, and the whole process is repeated with the sides switched. The initial side assignments are selected at random for each mattress. A mattress is considered to be resistant to disturbances if the house of cards collapses two times or fewer for the six trials on each mattress. The manufacturer will find it acceptable if more than 80% of all mattresses are resistant to disturbances. Which of the following best describes a shortcoming in the design of this study?

    (A) The sample size is insufficient.
    (B) This design will not permit the mathematical calculations needed for statistical inference.
    (C) The measurements are subject to response bias.
    (D) Convenience sampling is used in the design.
    (E) The design lacks realism.

13. The U.S. government tracks the number of employees in 22 different mining sectors, ranging from anthracite to uranium. The following histogram displays the number of employees in each of the 22 sectors.

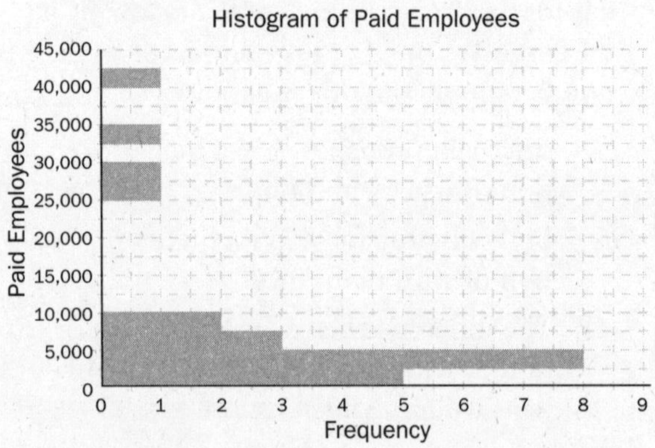

What percentage of mining sectors has 7,500 or fewer paid employees?

(A) 73%
(B) 59%
(C) 27%
(D) 16%
(E) 6%

GO ON TO THE NEXT PAGE

14. Measures x and y were taken on a set of 15 subjects. A scatterplot of these measures is shown below. Five points are labeled A, B, C, D, and E.

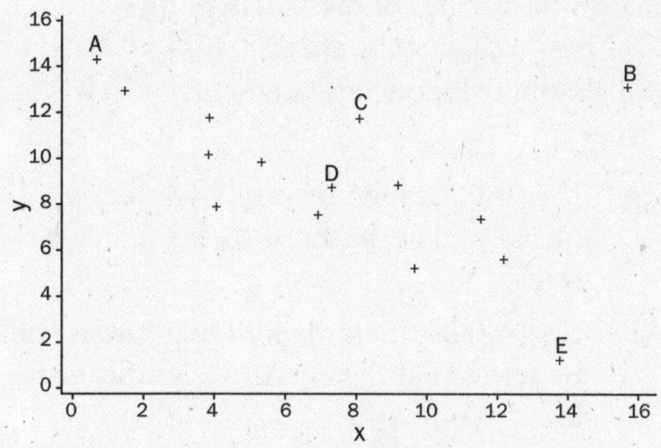

Which of the following is the label for the point that is most influential on the correlation coefficient r?

(A) A

(B) B

(C) C

(D) D

(E) E

15. A major manufacturer of marble tabletops sells them to some customers at wholesale prices and to others at retail prices. The following probability distribution shows estimates of next year's sales for wholesale customers.

Wholesale Sales

Number sold	1,000	5,000	10,000	20,000
Probability	0.25	0.30	0.40	0.05

The estimates of next year's retail sales are shown below.

Retail Sales

Number sold	500	1,000	5,000
Probability	0.15	0.65	0.20

If wholesale customers are charged $300 per tabletop and retail customers $600 per tabletop, what is the company's expected revenue for the next year?

(A) $8,475

(B) $2,732,142.86

(C) $2,850,000

(D) $3,060,000

(E) $3,813,750

16. The following are selected summary statistics for the women's long jump in the 2004 Olympic Games. Distances are measured in meters.

N	Mean	SE Mean	St Dev	Variance	Sum
37	6.4630	0.0439	0.2673	0.0715	239.1300

Minimum	Q1	Median	Q_3	Maximum	IQR
5.6400	6.3750	6.4600	6.6850	6.9500	0.3100

What interval contains, approximately, the middle 50% of the distribution?

(A) 6.375 to 6.685 m

(B) 6.1957 to 6.7303 m

(C) 6.15 to 6.77 m

(D) 5.91 to 6.84 m

(E) 5.64 to 6.46 m

GO ON TO THE NEXT PAGE

17. A certain genetic abnormality is present in 1% of all gerbils. A test for this abnormality correctly identifies its presence 98% of the time. It correctly identifies its absence 95% of the time. If a gerbil is selected at random from the population, what is the probability the test will show the presence of the abnormality?

(A) 0.98

(B) 0.95

(C) 0.0593

(D) 0.0495

(E) 0.01

18. A bag contains 6 red marbles and 4 blue marbles. Ralph randomly draws a marble from the bag, notes its color, and then replaces the marble in the bag. He remixes the marbles and repeats the process until he draws a blue marble. What is the probability this takes Ralph more than 3 draws?

(A) 0.36

(B) 0.216

(C) 0.167

(D) 0.144

(E) 0.0864

19. A consumer research firm reports that the 95% confidence interval for the proportion of consumers who prefer brand K tissue is 0.38 ± 0.04. If the research firm also conducts a significance test of the null hypothesis $H_0: p = p_0$ against the alternative $H_a: p \neq p_0$, which of the following statements must be true?

(A) If $p_0 = 0.35$, then the null hypothesis will be rejected at the $\alpha = 0.05$ significance level.

(B) If $p_0 = 0.35$, then the null hypothesis will be rejected at the $\alpha = 0.01$ significance level.

(C) If $p_0 = 0.38$, then the null hypothesis will be rejected at the $\alpha = 0.05$ significance level.

(D) If $p_0 = 0.38$, then the null hypothesis will be rejected at the $\alpha = 0.01$ significance level.

(E) If $p_0 = 0.43$, then the null hypothesis will be rejected at the $\alpha = 0.05$ significance level.

GO ON TO THE NEXT PAGE

20. A rancher wants to study two breeds of cattle to see whether or not the mean weights of the breeds are the same. Working with a random sample of each breed, he computes the following statistics.

Cattle Weights

Cattle Breed	*n*	Mean (lbs)	Std Dev (lbs)
X	45	1,247	120
Y	30	1,300	50

He conducts a two-sample *t*-test for the difference of means. Which of the following will give a correct value for the standardized test statistic?

(A) $\dfrac{1{,}247-1{,}300}{92\sqrt{\dfrac{1}{45}+\dfrac{1}{30}}}$

(B) $\dfrac{1{,}247-1{,}300}{92\sqrt{\dfrac{1}{44}+\dfrac{1}{29}}}$

(C) $\dfrac{1{,}247-1{,}300}{\dfrac{92}{\sqrt{75}}}$

(D) $\dfrac{1{,}247-1{,}300}{\sqrt{\dfrac{120^2}{45}+\dfrac{50^2}{30}}}$

(E) $\dfrac{1{,}247-1{,}300}{\sqrt{\dfrac{120^2}{44}+\dfrac{50^2}{29}}}$

21. All the students at colleges A and B take a standardized intelligence test. The scores at each college are approximately normally distributed with parameters listed below.

College	N	Mean	Std Dev
A	1,500	130	25
B	6,000	125	20

Samples are taken from each college, each consisting of the entire college. The sample means are computed $\overline{x}_A$ and $\overline{x}_B$. Which of the following gives the correct standard deviation of the sampling distribution for $\overline{x}_A-\overline{x}_B$?

(A) $\sqrt{\dfrac{25^2}{1{,}500}+\dfrac{20^2}{6{,}000}}$

(B) $\sqrt{\dfrac{25^2}{1{,}500}-\dfrac{20^2}{6{,}000}}$

(C) 0

(D) $\dfrac{45}{\sqrt{7{,}500}}$

(E) $22.5\sqrt{\dfrac{1}{1{,}500}+\dfrac{1}{6{,}000}}$

GO ON TO THE NEXT PAGE

22. A pair of measures x and y were taken on 20 experimental units. A regression analysis is performed on the values of $\log y$ versus $\log x$, resulting in a least-squares regression line. The residual plot for this regression is below.

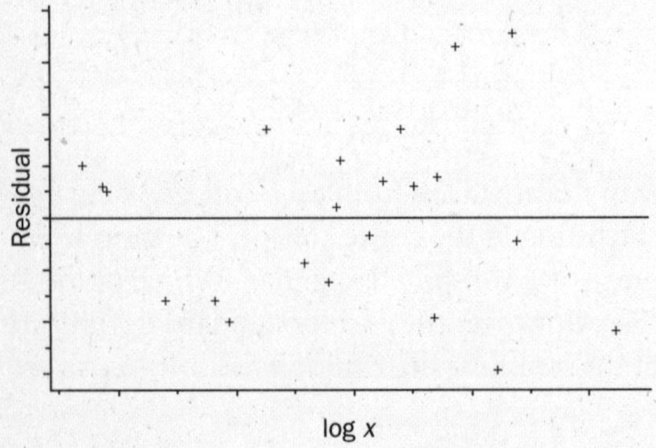

Which of the following conclusions is best supported by the residual plot?

(A) x has a linear relationship with log x.

(B) Log y has a linear relationship with x.

(C) Log y has a linear relationship with log x.

(D) y has a linear relationship with x.

(E) The correlation between y and x is approximately 0.

23. Scores on a standardized test are normally distributed with a mean of 510. Find the standard deviation for the test scores given that the probability of a score above 600 is 0.32.

(A) 42

(B) 90

(C) 120

(D) 132

(E) 192

24. A die is "loaded" so that the probability of rolling a result of 5 has a probability of 0.5 and rolls of 1, 2, 3, 4, and 6 each have a probability of 0.1. This die is to be rolled 100 times, and the mean of these results will be taken. The die will then be rolled another 100 times, and that mean will be taken. Repeated sets of 100 rolls will take place until 1,000 means have been collected. Which of the following best describes the distribution of these means?

(A) Approximately normal

(B) Exactly normal

(C) Approximately a chi-square distribution with 999 degrees of freedom

(D) Exactly a chi-square distribution with 999 degrees of freedom

(E) Exactly binomial

GO ON TO THE NEXT PAGE

25. Two sections of students take the same final exam. A back-to-back stem and leaf plot of the exam results are listed below.

Section 1		Section 2
8	4	
	5	3
8	5	8999
40	6	24
996	6	56
1	7	1234
9876	7	
44	8	44
8	8	
	9	14

Which of the following is a correct statement about the two distributions?

(A) Section 1 shows a larger range of scores.
(B) Section 2 has a higher median score.
(C) Section 1 has a larger interquartile range.
(D) The high score in section 2 is less than the high score in section 1.
(E) The largest gap in section 1 is larger than the largest gap in section 2.

26. A medical researcher suspects that right-handed subjects have longer right arms than left arms. He measures both arm lengths for a variety of right-handed subjects. Let $\overline{x}$ and μ_X represent the sample and population mean lengths of the right arms, $\overline{y}$ and μ_Y represent the sample and population mean lengths of the left arms, and $\overline{d}$ and μ_D represent the sample and population differences in arm length, right minus left, for each subject. Which of the following is the best pair of hypotheses for this experiment?

(A) $H_0: \mu_X = \mu_Y$ and $H_a: \mu_X > \mu_Y$
(B) $H_0: \overline{x} = \overline{y}$ and $H_a: \overline{x} > \overline{y}$
(C) $H_0: \mu_D = 0$ and $H_a: \mu_D > 0$
(D) $H_0: \mu_D = 0$ and $H_a: \mu_D \neq 0$
(E) $H_0: \overline{d} = 0$ and $H_a: \overline{d} \neq 0$

27. What does it mean to say that a statistic is an unbiased estimator of a parameter?

(A) The statistic is computed from a study that has no tendency to favor certain outcomes at the expense of others.
(B) The same value of the statistic obtained from one trial of a study will be obtained again when a second trial is conducted.
(C) The statistic is computed from a study that is double-blinded.
(D) The sampling distribution of the statistic is centered on the value of the parameter.
(E) The sampling distribution of the statistic is symmetric.

28. Tiffany is working with data that are normally distributed with a mean of 78 and a standard deviation of 12.4. In order to find the quartiles for this distribution, she uses table A, the table of standard normal probabilities. She looks up the value of z corresponding to an area of 0.75 to the left of z. The quartiles are then $78 \pm 12.4z$. Which of the following best assesses Tiffany's procedure?

(A) Tiffany's work is correct.
(B) Tiffany's work is incorrect because her formula should use the variance 12.4^2 instead of the standard deviation 12.4.
(C) Tiffany's work is incorrect because the quartiles are 0.25 standard deviations above and below the mean.
(D) Tiffany's work is incorrect because the quartiles are 0.75 standard deviations above and below the mean.
(E) Tiffany's work is incorrect because the quartiles are 1 standard deviation above and below the mean.

GO ON TO THE NEXT PAGE

29. An individual randomly places 3 X's on a blank tic-tac-toe board. What is the approximate probability that he gets the equivalent of tic-tac-toe (three in a row across, down, or diagonally)?

(A) 9.5%

(B) 1.2%

(C) 7.1%

(D) 33%

(E) 16%

30. For a given state, the following table breaks down the population of legislature's senators by gender and political party.

	Party A	Party B	Total
Male	6	19	25
Female	9	6	15
Total	15	25	40

Are these variables, gender and political party, independent?

(A) Yes, because the conditional distribution of political party is the same for male and female senators.

(B) Yes, because the marginal distribution of gender is the same as the marginal distribution of political party.

(C) No, because the conditional distribution of political party is not the same for male and female senators.

(D) No, because the marginal distribution of gender is not the same as the marginal distribution of political party.

(E) No, because there are different numbers of male and female senators.

31. The editor of the newspaper at a large university wants to measure student opinion on a proposed change in graduation requirements. The editor randomly selects 100 freshmen, 100 sophomores, 100 juniors, and 100 seniors to take a short survey. This plan is an example of what kind of sampling?

(A) Cluster sampling

(B) Simple random sampling

(C) Systematic sampling

(D) Stratified random sampling

(E) Convenience sampling

32. For a given large sample size *n*, each of the following confidence intervals for a population proportion are computed. Which has the largest margin of error?

(A) A 90% confidence interval with sample statistic $\hat{p} = 0.35$

(B) A 95% confidence interval with sample statistic $\hat{p} = 0.35$

(C) A 99% confidence interval with sample statistic $\hat{p} = 0.35$

(D) A 95% confidence interval with sample statistic $\hat{p} = 0.45$

(E) A 99% confidence interval with sample statistic $\hat{p} = 0.45$

GO ON TO THE NEXT PAGE

33. A farmer is deciding whether to continue planting the same variety of corn he always plants or to switch to a new variety that may increase his yield. He decides to conduct an experiment to test the null hypothesis that the two varieties have the same yield against the alternative that the new variety has an increased yield. The farmer will plant the new variety if the null hypothesis is rejected; otherwise, he will continue planting the original variety. Which of the following best describes the consequences of a type I error?

(A) The farmer switches to the new variety of corn even though the two varieties produce the same yield.

(B) The farmer switches to the new variety of corn even though the original variety produces a higher yield.

(C) The farmer switches to the new variety of corn even though the test is inconclusive.

(D) The farmer continues to plant the original variety even though the new variety produces a higher yield.

(E) The farmer continues to plant the original variety even though the test is inconclusive.

34. Johanna is conducting a significance test of $H_0: \mu = 50$ against the alternative $H_0: \mu > 50$. She works with 100 experimental units at a significance level of $\alpha = 0.01$. Johanna calculates the power of her test against the specific alternative $\mu = 51$. Which of the following changes in her work will result in a higher calculated power?

I. Increase the sample size to 200 and leave everything else unchanged.

II. Increase the significance level to $\alpha = 0.05$ and leave everything else unchanged.

III. Calculate power against the specific alternative $\mu = 50.5$ instead of $\mu = 51.0$ and leave everything else unchanged.

(A) I only

(B) I and II

(C) I and III

(D) II and III

(E) I, II, and III

35. For two independent events A and B, $P(A) = 0.4$ and $P(B) = 0.5$. Which of the following is the value of $P(A \cup B)$?

(A) 0.2

(B) 0.4

(C) 0.5

(D) 0.7

(E) 0.9

GO ON TO THE NEXT PAGE

36. Paired data for the percentage of men and women who smoke cigarettes is recorded for selected years from 1965 to 2002. The variable Male stands for the percentage of men who smoke in a given year; Female stands for the percentage of women. Summary statistics for these measurements are given below.

Descriptive Statistics: Male, Female

Variable	Mean	Std Dev
Male	30.38	6.89
Female	24.884	4.013

Correlation of Male and Female = 0.950

Formulas for the coefficients of a regression line $\hat{y} = b_0 + b_1 x$ are $b_1 = r\frac{s_y}{s_x}$ and $b_0 = \overline{y} - b_1\overline{x}$. Which of the following is an equation of the least-squares regression line for Female versus Male?

(A) $\widehat{\text{Female}} = 8.07 + 0.5533$ Male
(B) $\widehat{\text{Female}} = 16.61 + 0.5533$ Male
(C) $\widehat{\text{Female}} = 8.91 + 0.5257$ Male
(D) $\widehat{\text{Female}} = -10.21 + 1.6311$ Male
(E) $\widehat{\text{Female}} = -24.67 + 1.6311$ Male

37. A significance test is conducted using z procedures. When the test statistic is analyzed using one-sided procedures, it is significant at the $\alpha = 0.05$ level but not the $\alpha = 0.01$ level. Which of the following must be true about the test statistic if the analysis is conducted with a two-sided procedure?

(A) Significant at the $\alpha = 0.01$ level but not at the $\alpha = 0.10$ level
(B) Significant at the $\alpha = 0.10$ level but not at the $\alpha = 0.01$ level
(C) Significant at the $\alpha = 0.10$ level but not at the $\alpha = 0.05$ level
(D) Significant at the $\alpha = 0.05$ level but not at the $\alpha = 0.10$ level
(E) Significant at the $\alpha = 0.025$ level but not at the $\alpha = 0.005$ level

38. Which of these combinations is most effective at ensuring that the results of an experiment are not confounded by the placebo effect?

(A) Replication and blocking
(B) Blocking and control
(C) Control and blinding
(D) Blinding and randomization
(E) Randomization and replication

GO ON TO THE NEXT PAGE

39. A city newspaper reports that a referendum on funding a new mass transportation system for the city has the support of 51% of likely voters. The newspaper also notes that the poll results have a margin of error of 5%. Which of the following statements best describes what this means?

(A) There is a 5% chance that the poll results are wrong.
(B) There is a 95% confidence level that the poll results are correct.
(C) There is a 5% chance that the true level of support for the referendum among likely voters is not between 46% and 56%.
(D) The true level of support for the referendum among likely voters is likely to be between 46% and 56%.
(E) The true level of support for the referendum among likely voters must be between 46% and 56%.

40. A standardized college admissions test is designed to predict freshman year grade averages. A researcher suspects this is not the case, and that there is no relationship between scores on the admissions test and freshman year grades for those who take the test. The researcher wants to test the null hypothesis that the slope of the regression line for the population is 0 against the alternative that the slope is nonzero. A regression analysis for freshman grades, scored from 0 to 100, versus the admissions test scores, ranging from 0 to 1,000, is given below.

Dependent variable: freshman grade average

Source	DF	Sum of Sq	Mean Sq	F-ratio
Regression	1	79.59	79.59	1.59
Residual error	398	19,952.65	50.13	

Predictor	Coef	SE Coef	t-Ratio	Prob
Constant	79.018	2.515	31.42	0.000
Admissions score	−0.004447	0.003529	−1.26	0.208

Assume all conditions are met for a significance test for regression slope. Which of the following is a correct estimate for the *p*-value?

(A) Less than 0.01
(B) Between 0.01 and 0.05
(C) Between 0.05 and 0.10
(D) Between 0.10 and 0.20
(E) Greater than 0.20

IF YOU FINISH BEFORE TIME IS CALLED, YOU MAY CHECK YOUR WORK ON THIS SECTION ONLY. DO NOT TURN TO ANY OTHER SECTION IN THE TEST. **STOP**

SECTION II: FREE-RESPONSE

Time—90 Minutes 6 Questions

Directions: Section II of this exam contains six questions: five free-response questions and one investigative task. The investigative task is the last question and will comprise 25% of your score for Section II. You should spend about 25 minutes on it. The other five questions comprise 75% of your score for this section. You should spend about 13 minutes each on these.

Clearly show the methods used as you write out your answers. You will be scored on the soundness of your methods and reasoning, and on finding the correct answers.

Part A: Questions 1–5

Spend about 65 minutes on this part of the exam.

1. The following cumulative relative frequency histogram shows the distribution of the heights, in meters, of the principal mountains in each part of Canada.

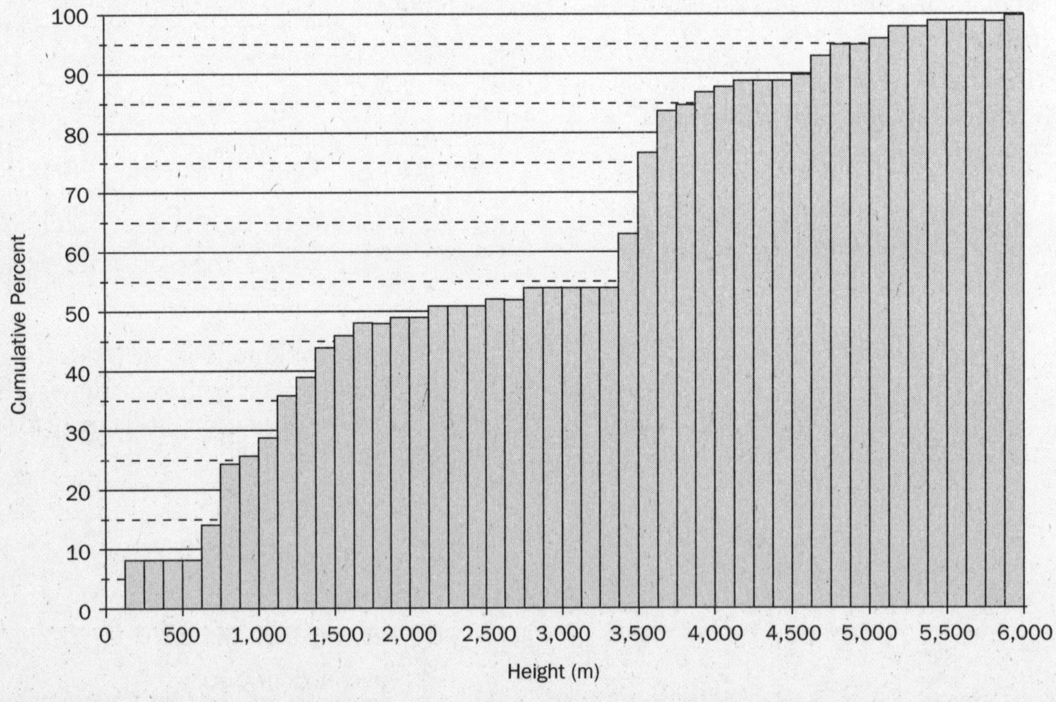

(a) Outlook Peak has a height of 2,210 m. About what percentile is this among this distribution of principal mountains of Canada?

(b) Find the interquartile range as accurately as the graph allows.

(c) Describe the clustering of the distribution of mountain heights. Explain your answer.

GO ON TO THE NEXT PAGE

2. People experiencing migraine headaches of the cluster variety experience sharp pain behind one eye. For people who get these headaches, their occurrence can be reduced by treatment with medicine X, which is administered in pill form. The effectiveness of medicine X is established by a randomized comparative experiment that measures the number of headaches experienced by the subjects in a six-month period.

   A new treatment, medicine Y, may do a better job. This new medicine is administered via a nasal spray. A study is to be conducted to compare these two treatments. There are 200 subjects available for the study.

   (a) Design a completely randomized experiment for this study.

   (b) Can this experiment be blinded? How?

   (c) The response to this treatment may differ for men and women. The subject pool is made up of 120 men and 80 women. Should you modify the design of your experiment? Explain. If you should modify the design, how?

GO ON TO THE NEXT PAGE

3. A monthly report on the visitors to a math reference Website is shown below.

Hour	Pages	Megabytes
0	1,846	38.63
1	1,362	26.46
2	934	19.38
3	7,412	92.41
4	843	17.95
5	2,109	34.29
6	920	19.15
7	1,419	27.28
8	1,168	22.35
9	1,862	39.88
10	1,814	36.52
11	2,956	60.04
12	2,231	44.78
13	1,566	34.13
14	2,037	43.22
15	2,317	51.8
16	2,534	56.69
17	2,100	45.58
18	1,720	34.88
19	2,247	47.78
20	2,482	49.12
21	2,556	53.18
22	2,150	45.54
23	1,977	40.89

For each hour of the day from hour 0 to hour 23, the Pages column shows the hourly breakdown of pages viewed that month. The Megabytes column shows the hourly breakdown of megabytes of information transferred. This information is what allows visitors to see the pages they visit.

Results from a regression analysis are shown below.

Response Variable: Megabytes

Predictor	Coef	SE Coef	T	P
Constant	15.805	2.576	6.13	0.000
Pages	0.011918	0.001056	11.29	0.000

S = 6.37185 R-Sq = 85.3% R-Sq(adj) = 84.6%

GO ON TO THE NEXT PAGE

(a) Find the residual for hour 12.

(b) What is the intercept of the least-squares regression line? Interpret this intercept in the given context.

(c) Find and interpret the value of the correlation coefficient.

4. Maurilio is conducting a taste test for two colas, C and P, at a large university. He gives the taste test to a random sample of students. The subjects first taste C, then taste P before stating a preference. The responses are shown below.

Preference		
C	P	Total
61	41	102

Maurilio then conducts a second survey using a separate random sample of students from the university. This time the subjects first taste P, then taste C before stating a preference. The responses are shown below.

Preference		
C	P	Total
45	60	105

Do Maurilio's data provide evidence that cola preference depends on the order in which the colas were tasted? Explain.

5. A golf ball manufacturer has come up with a less expensive way to produce the company's golf balls. The manufacturer wants to make sure the balls manufactured with the new method travel the same distance as the balls manufactured the old way. A machine is used to drive both types of golf balls a number of times, and the results are summarized below. Distances are in yards.

Manufacturing Method	*N*	Mean Drive	Std Dev	Std Error of the Mean
Old	50	258	25	3.536
New	50	246	40	5.657

(a) Find a 95% confidence interval for the mean difference in driving distances.

(b) Does your confidence interval provide evidence that balls manufactured by the two methods have the same driving distance? Explain.

GO ON TO THE NEXT PAGE

Part B: Question 6

Spend about 25 minutes on this part of the exam.

6. Guessing on a 40-question multiple-choice test can be simulated using a table of random numbers.

   (a) If each question has five choices, explain how you would use the random number table below to estimate the distribution of the number of questions guessed correctly on the entire test.

   (b) Use the excerpt from a table of random digits below to perform a simulation of 10 questions. Be sure to report the number of questions guessed correctly. Start at the leftmost digit of the first line of the table. Make sure your work is clear so that a grader can follow your work. You must label either directly on or above the table.

```
84392 36623 99640 35054 65254 04907 77876 85450

27957 26767 92610 86639 73450 51204 18178 01278

70586 19428 37154 53506 61260 20049 08585 10871
```

   (c) A simulation of guessing on a full 40-question multiple-choice test is conducted 500 times. Some of the results are shown below.

Questions guessed correctly	8	9	10
Simulations with this result	82	60	58

According to binomial probability calculations, out of the 500 simulations what are the *expected* number in which 8 questions will be guessed correctly? Repeat this for 9 and 10 questions.

   (d) Using the simulation results in (c), conduct a significance test to determine if the results of the simulation are consistent with the theoretical predictions.

IF YOU FINISH BEFORE TIME IS CALLED, YOU MAY CHECK YOUR WORK ON THIS SECTION ONLY. DO NOT TURN TO ANY OTHER SECTION IN THE TEST. **STOP**

Practice Test 2: **Answer Key**

1	D	11	A	21	C	31	D
2	C	12	E	22	C	32	E
3	B	13	A	23	E	33	A
4	B	14	B	24	A	34	B
5	A	15	D	25	B	35	D
6	A	16	A	26	C	36	A
7	E	17	C	27	D	37	B
8	B	18	B	28	A	38	C
9	E	19	E	29	A	39	D
10	E	20	D	30	C	40	E

PRACTICE TEST 2

ANSWERS AND EXPLANATIONS

SECTION I ANSWERS

1. D

Stem and leaf plots, histograms, and dotplots can do an excellent job of showing data clusters, although histograms can sometimes hide clustering if the bin width is too wide. Cumulative frequency plots can show clusters as well, but you may not be practiced in seeing this. A steep increase in a cumulative frequency plot corresponds to a data cluster, and a flat section corresponds to a gap. Boxplots, however, do a terrible job of showing data clusters. Only clusters of outliers show up in a boxplot. Other clusters are invisible.

2. C

A confidence interval for a mean is found using $\bar{x} \pm (z^*)$(standard deviation of $\bar{x}$) if the population standard deviation is known and $\bar{x} \pm (t^*_{n-1})$(standard error of $\bar{x}$) if it is not. In this case t^*_{29} is 1.311 using the *t*-table. We were given $\bar{x} \pm (t^*_{n-1})$(standard error of $\bar{x}$) in the question, so (C) is correct. If you looked up t^*_{29} incorrectly, you might get (A) as your answer. If you mistook the standard error of 0.006 for the standard deviation, you might get (D) as your answer. If you made both mistakes, you might get (B) or (E).

3. B

The mean gross revenue over two years is the two-year gross divided by the number of releases. The total gross in year 1 was 12 · 14.4, or $172.8 million, and in year 2 the total gross was $185.6 million. That's a total of 20 films grossing $358.4 million, or a mean gross of $17.92 million. That's (B).

Choice (A) will be your result if you compute $\frac{14.4}{12}$ plus $\frac{23.2}{8}$. Choice (C) is the mean of $14.4 million and $23.2 million with each figure receiving equal weight. You would get choice (D) if you reversed the number of films released each year, and (E) is the sum of the two mean gross revenues.

4. B

For independent random variables, $\mu_{X+Y} = \mu_X + \mu_Y$ and $\sigma^2_{X+Y} = \sigma^2_X + \sigma^2_Y$, so $\mu_{X+Y} = 70$ and $\sigma_{X+Y} = \sqrt{\sigma^2_X + \sigma^2_Y} = \sqrt{250} = 15.8$. The other answer choices reflect confusion between $X + Y$ and $X - Y$, and confusion over the formula for σ_{X+Y}.

5. A

The slope is 0.207 wins per million dollars of payroll. These are the units for slope since the formula is $r\frac{s_y}{s_x}$. This slope is equivalent to about 1 extra win per $5 million increase in payroll, so choice (A) is correct. Choices (B), (C), and (E) all use the wrong number for the slope, and choice (D) uses the right slope but uses the wrong units (millions of dollars of payroll per win).

Note that r^2 is 0.113, so the linear model is a poor predictor.

6. A

A residual is equal to $y - \hat{y}$, and since $x = 9.227$ we know that $\hat{y} = 3.422 + 5.551(9.227) = 54.641$. The residual is 1.811, so $\hat{y} = 56.452$. Choices (E) and (D) are the values you would get if you used either $y - x$ or $x - y$ as the residual formula. Choice (C) is what you would get if you used the value of $\hat{y} - y$ as the residual formula by mistake, and (B) is the correct value of $\hat{y}$ rather than y.

7. E

The sample size n can be found using the following calculations.

$$\text{margin of error} = z^* \frac{\sigma}{\sqrt{n}}$$

$$0.283 = 2.054 \cdot \frac{3.00}{\sqrt{n}}$$

$$n = \left(\frac{2.054 \cdot 3.00}{0.283}\right)^2 \approx 474$$

An algebra mistake in this calculation results in choice (A). You would get choice (B) for your answer if you left z^* out of your calculations. If you used z^* for a 92% confidence interval instead of 96%, you would get (C), and (D) results from using the commonly encountered $z^* = 1.96$.

8. B

It is control for confounding variables that allows experiments to establish a cause-and-effect relationship. This is done by means of comparison of treatments, replication, randomization, and sometimes with blocking. Choices (A) and (C) are correct statements, but they do not get to the central issue of establishing cause-and-effect. Choice (D) is not true; the care with which a study is conducted has to do with the diligence of the researcher and not the type of study being conducted. Choice (E) is incorrect because it is not true; there is no such tendency for experiments to use larger samples than observational studies.

9. E

The law of large numbers states that large samples tend to have a proportion of successes closer to the proportion in the population. Choice (E) best interprets this statement in the given context. Another way of thinking of this problem is that large samples tend to have less sampling variability, so again Bob is likely to have a proportion of wins closer to 0.03. Because of this, Alice's results are more likely to be extreme, so this might tempt you to answer (A), but Alice might just as well have an extreme result that is too small as too large. The same reasoning rules out (B). Regardless, neither (A) nor (B) relates to the law of large numbers. Choice (C) is simply wrong; the difference will probably be close to 0. Finally, choice (D) interprets the law of large numbers backwards.

10. E

You are being asked to compare the proportion of accidents per registered vehicle for each of the three configurations. In order to compare three proportions, the best test to use is the chi-square test for homogeneity of proportions. The table isn't set up properly to conduct the test, but it can easily be written in appropriate form, as shown below.

Fatal U.S. Truck Accidents in 2002 by Configuration

Truck Configuration	Safe Vehicles	Vehicles in Accidents	Total
Tractor, 1 trailer	147,111	2,889	150,000
Tractor, 2 trailers	4,946	154	5,100
Tractor, 3 trailers	199	1	200
Total	152,256	3,044	155,300

Let's consider the other answers. Choice (A) won't work because it only works for one proportion, and there are three here. Choice (B) could work, but it would require three different comparisons; choice (E) is more appropriate. Choice (C) is used to test for independence within one population, and the problem indicates there are three populations being studied here. Choice (D) would be right if the accident rates were being compared to a presupposed count of accidents for each configuration, but that isn't the case here.

11. A

Statement I is correct because all *t*-distributions are in fact centered at 0. Statement II is wrong because it has the degree of freedom relationship backwards; the larger the degrees of freedom, the *more* a *t*-distribution resembles a normal distribution. Statement III is also incorrect because the statement is reversed. Only statement I is correct, so (A) is the right answer.

12. E

The effect of a bowling ball on a house of cards is not an accurate way to determine if someone moving on one side of a mattress will be noticeable by someone on the other side. That makes (E) a good response, but the other choices must be considered to see if it is the best choice. Choice (A) is not a shortcoming; a sample of 500 mattresses is large for most purposes. Choice (B) is wrong since *z* procedures for proportions can be conducted using the sample proportion $\hat{p}$ of mattresses that are resistant to disturbances. Choice (C) makes no sense since response bias is only an issue with question-and-answer polling. Finally, choice (D) is not an issue here since the sample is selected randomly. You can now safely conclude that lack of realism is in fact the most glaring design problem among the answer choices because the design does not reflect the purpose of the study.

13. A

The histogram shows 16 of the 22 mining sectors have 7,500 or fewer paid employees, for 73% of the mining sectors, so (A) is correct. If you misread the table and only consider the sectors employing 5,000 or fewer people, you get choice (B). If you work with the sectors employing more than 7,500, you get 27%. The histogram table shows frequency, but if you misread it and believe it shows percentage instead, you will get choice (D). If you combine the mistakes of (C) and (D), you get (E).

14. B

This question can be approached visually or by considering the formula for r. For a visual approach, consider the appearance of the scatterplot if each of the points are removed. Only the omission of points B and E will have a significant impact on the look of the graph. To decide which is more influential, consider the look of the graph leaving out both B and E. Observe that E is more consistent with this pattern than B, so B is the most influential point on r. For a formula-based approach, consider the formula $r = \frac{1}{n-1}\sum\left(\frac{x_i - \bar{x}}{s_x}\right)\left(\frac{y_i - \bar{y}}{s_y}\right)$ from the official formula sheet. The value of r is derived exclusively from the z-scores of the x- and y-coordinates. A point (x_i, y_i) is influential if the product of the z-scores for each coordinate is as extreme as possible. As with the visual approach, the point that stands out the most is B.

15. D

The expected wholesale sales are 6,750 units, and the expected retail sails are 1,975 units, so the expected revenue is $6{,}750 \cdot 300 + 1{,}725 \cdot 600$, or \$3,060,000. The correct answer is (D). Choice (A) is incorrect because the expected counts are added together without considering the charge per tabletop. Choice (B) takes a few inappropriate shortcuts. The probabilities are ignored, and the seven sales categories are averaged. Then this average is multiplied by \$450, which is the average of the two prices. Choice (C) will be attractive if you use a rough estimate of sales, using 7,500 for wholesale and 1,000 for retail and then proceeding as in (D). It's not a bad estimate, but don't confuse it with the real answer. Choice (E) uses the correct expected sales counts, but then adds them together and multiplies by \$450.

16. A

The lower quartile Q_1, the median, and the upper quartile Q_3 split the distribution into segments that each comprise about 25% of the long jump distances. The interval between Q_1 and Q_3 contains, approximately, the middle 50% of the distribution. That makes answer choice (A), 6.375 to 6.685 m, the correct choice. Choice (E) contains 100% of the distribution, since it is the interval from the minimum to the maximum. Choice (D) is the interval from $Q_1 - 1.5$ IQR to $Q_3 + 1.5$ IQR. This is an interval commonly used to determine whether an observation is an outlier, but it is unrelated to the middle 50% of the distribution. Choice (C) is the interval of values within one IQR of the median, which again is unrelated to the middle 50%. Choice (B) is the interval representing long jumps within one standard deviation of the mean. If the distribution were normal, this would be the middle 68% of the distribution. Regardless, (B) is wrong.

17. C

The following probability tree outlines the probabilities involved with the genetic testing.

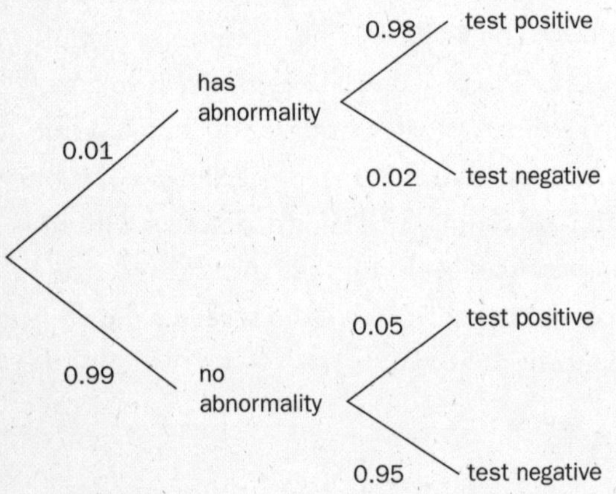

The probability that a randomly selected gerbil has the abnormality and tests positive is $0.01 \cdot 0.98$, or 0.098. The probability that the gerbil doesn't have the abnormality and tests positive is 0.0495. Thus, the probability that a randomly selected gerbil tests positive is $0.098 + 0.0495$, or 0.0593. Choice (A) is the probability that a gerbil tests positive given that it has the abnormality. Choice (B) is the probability that a gerbil tests negative given that it doesn't have the abnormality. Choice (D) is the probability that a randomly selected gerbil doesn't have the abnormality and tests positive. Choice (E) is the probability that a randomly selected gerbil has the abnormality.

18. B

There are two approaches to this problem. The first is to recognize that this problem is geometric, so the probability that it takes more than n draws before the first success is $(1 - p)^n$, where p is the probability of success. This means that the probability it takes Ralph more than 3 draws is $(1 - 0.4)^3$, or 0.216. The other approach is to realize that the scenario is equivalent to drawing 3 red marbles in a row, which would be 0.6^3, or 0.216. Choice (A) is the probability that Ralph draws 2 reds in a row. Choice (C) is the probability that it takes Ralph more than 3 draws if he does not replace the marble after each draw. Choice (D) is the probability he draws 2 reds then blue, and (E) is the probability he draws 3 reds then blue.

19. E

Values of p_0 that lie outside the confidence interval will lead to a rejection of the null hypothesis at $\alpha = 0.05$, so (E) must be true. Choices (A) and (C) are certainly false by this reasoning, and since 0.38 is within any confidence interval from this sample, then (D) is false as well. Choice (B) is also wrong. Since 0.35 can't be rejected at the 0.05 significance level, the p-value is more than 0.05. Therefore, 0.35 can't be rejected at the 0.01 level, either. Another approach is to reconstruct the values of $\hat{p}$ and n from the confidence interval to find that $\hat{p} = 0.38$ and $n =$ 566. This will allow you to find the p-values for each of the values of p_0 in the answer choices. This also leads to (E) as the correct answer.

20. D

The standardized test statistic for a z-test is given by $\frac{\text{statistic} - \text{parameter}}{\text{standard deviation of the statistic}}$. For a t-test, standard error replaces standard deviation. Here, the statistic is the difference of sample means $\bar{x} - \bar{y}$, or $1247 - 1300$, and the parameter from the null hypothesis is 0. The standard error for a difference of means problem is $\sqrt{\frac{s_X^2}{n_X} + \frac{s_Y^2}{n_Y}}$, or $\sqrt{\frac{120^2}{45} + \frac{50^2}{30}}$. Put this all together and the test statistic matches choice (D). You would get choice (A) if you erroneously used the pooled formula for standard error. This is certainly not appropriate since the standard deviations for the samples do not resemble each other. Choices (B) and (E) are similar to choices (A) and (D), but the sample sizes are all reduced by 1, probably because of confusion over degrees of freedom. Choice (C) is wrong because it makes inappropriate use of the standard error formula for the one-sample *t*-test using a pooled value of the standard deviation.

21. C

Since the samples consist of the entire population, there is no variability. Each time you sample an entire population, you get the exact same sample mean. The formula used in choice (A), $\sqrt{\frac{\sigma_1^2}{1,500}+\frac{\sigma_2^2}{6,000}}$, for the standard deviation of the difference of sample means only applies when the samples are much smaller than the populations. The remaining choices are a variety of different ways to misapply this formula or to misuse other standard deviation formulas.

22. C

The regression is performed on $\log y$ versus $\log x$, so the only linear relationship that could be established is between $\log y$ and $\log x$. That rules out choices (A), (B), and (D). Because the residual plot shows no strong pattern, there is in fact a linear relationship between $\log y$ and $\log x$. As for (E), there is a power relationship between y and x since the relationship between $\log y$ and $\log x$ is linear. That means there is a nonzero correlation between y and x. If the correlation were approximately 0, then knowing the value of x would be useless in predicting the value of y.

23. E

If X represents the test score random variable, then

$$P(X > 600) = 0.32$$
$$P(X < 600) = 0.68$$
$$P\left(z < \frac{600-510}{\sigma_X}\right) = 0.68$$
$$\frac{600-510}{\sigma_X} = 0.4677$$
$$\sigma_X \approx 192.$$

This value can also be found from the answer choices by using, the guess-and-check method. Choice (A) results from an algebra mistake in the final step of this process. You might arrive at wrong answer (B) if you mistakenly thought that a right-tail area of 0.32 corresponds to 1 standard deviation above the mean. This is true for a combined tail area of 0.32, which is not the case here. Choice (C) results from using the right method but using the normal table backwards. Choice (D) is the result if you use 0.68 instead of 0.4677 in the work above.

24. A

The central limit theorem guarantees that for large sample sizes the distribution of sample means is approximately normal. Even though a sample size of 1,000 is very large, the distribution of sample means won't be exactly normal for any discrete distribution. The chi-square distribution is unrelated to sample means, and so is the binomial distribution.

25. B

Variable	N	Minimum	Q_1	Median	Q_3	Maximum	IQR
Section 1	15	48.00	64.00	71.00	79.00	88.00	15.00
Section 2	17	53.00	59.00	66.00	79.00	94.00	20.00

The table above lists the five-number summary and the IQR for the two sections. It shows that statement (B) is the only correct choice among (A) through (D). Choice (E) is incorrect because the largest gap is the same for the two distributions: from 48 to 58 in section 1, and from 74 to 84 in section 2.

26. C

The question being asked is whether individuals have right arms that are longer than left arms. The relevant statistic is the $\bar{d}$, the mean difference in arm lengths for the subjects, and the relevant parameter is μ_D, the population difference in arm lengths. That rules out choices (A) and (B). Proper hypotheses are about parameters, not statistics, so that rules out (E). The researcher suspects that right-handed subjects have longer right arms, so that makes (C) the right answer. Another way of assessing this experiment is to realize that this is a one-sided matched pairs scenario, which also results in (C) as the correct answer.

27. D

Bias, in its general sense, is any systematic tendency to favor certain outcomes. When you talk about an unbiased *statistic*, however, you mean that the statistic does not show any systematic tendency to stray from the value of the parameter it is approximating. That is, the sampling distribution of the statistic is centered on the value of the parameter. Choice (E) also mentions the sampling distribution of the statistic, but it is not true that an unbiased statistic must have a symmetric sampling distribution. The other choices do not mention the sampling distribution of the statistic at all. Instead, they get unbiased *statistic* mixed up with the concept of an unbiased *study*.

28. A

Tiffany's work is in fact correct, and she will find that $z = 0.674$ for quartiles at 69.6 and 86.4. Quartiles are the 25th and 75th percentiles of a distribution. Choice (B) is wrong because variance is not a part of this calculation. Choice (C) will produce the 40th and 60th percentiles. Choice (D) produces the 23rd and 77th percentiles. The 16th and 84th percentiles result from choice (E).

29. A

A tic-tac-toe board has 9 boxes, and 3 boxes are marked at a time. The number of combinations is

$$\binom{9}{3} = \frac{9!}{3!(9-3)!} = \frac{9 \bullet 8 \bullet 7 \bullet 6!}{3 \bullet 2 \bullet 6!} = 3 \bullet 4 \bullet 7 = 84.$$

The odds of randomly achieving any set of 3 boxes is therefore $\frac{1}{84} \approx 0.012 = 1.2\%$. There are exactly 8 combinations that will produce tic-tac-toe, so the odds of achieving tic-tac-toe is $\frac{8}{84} = \frac{2}{21} \approx 0.095$, or 9.5%.

30. C

Two categorical variables are independent if the conditional distribution of one variable is the same for every category of the other variable. The conditional distribution of political party for men is 24% party A, and 76% party B. The conditional distribution of political party for women is 60% party A, 40% party B. These aren't even close, so the variables are not independent. Choice (C) is correct. The marginal distributions may contain the same numbers (62.5% male and 37.5% female, versus 62.5% party B and 37.5% party A), but this is unrelated to independence. Choice (E) is incorrect because independence is a property determined using proportions, not counts.

31. D

Stratified random sampling is a scheme that divides a population into strata and then randomly selects a sample from each stratum. The members of each stratum should resemble each other, and this is exactly what is going on in this scenario. If this were cluster sampling, then each cluster should resemble the population and the clusters themselves would be chosen randomly, not members of the clusters. Simple random sampling gives every possible sample the same chance of being selected, so that is not occurring here. If this were systematic sampling, then the sample would be taken by selecting, for example, every tenth student from a list of all students. Convenience sampling usually isn't randomized; it consists of selecting a sample that is convenient to collect rather than one that resembles the population.

32. E

Higher levels of confidence create wider confidence intervals, so that eliminates choices (A), (B), and (D). A confidence interval for a population proportion is wider when $\hat{p}$ is closer to 0.5, so the answer is (E). Another option is to use the formula $z^* \sqrt{\frac{\hat{p}(1-\hat{p})}{n}}$ for the margin of error for the values given in each choice, using the same value of n for each.

33. A

A type I error occurs when the null hypothesis is rejected even though it is true. Here, that would mean that the farmer switches to the new variety even though the two varieties produce the same yield. Choice (B) sounds plausible, but neither hypothesis mentions a higher yield for the new variety. Choice (D) is wrong because it describes a type II error. Choices (C) and (E) are wrong because the technical description of type I and II errors do not refer to inconclusive tests. What's more, inconclusive test results do not necessarily result in errors.

34. B

The power of a test is the probability that the test will correctly reject the null hypothesis when the null hypothesis is in fact false. Increasing the sample size will improve a test's ability to discriminate between alternatives, so statement I is true. As for statement II, increasing the significance level will decrease the probability of a type II error. The power of a test is one minus the probability of a type II error, so statement II is true. Statement III is false. It is easier for a test to distinguish between alternatives that are farther apart, so the change indicated in statement III will decrease the calculated power.

35. D

$P(A \cup B) = P(A) + P(B) - P(A \cap B)$ and, for independent events, $P(A \cap B) = P(A)P(B)$, so $P(A \cup B) = P(A) + P(B) - P(A)P(B)$, or 0.7. Choice (*A*) is $P(A \cap B)$, choice (*B*) is $P(A)$, and choice (C) is $P(B)$. Choice (E) would be the correct answer for $P(A \cup B)$ if the events were mutually exclusive instead of independent.

36. A

Here, $\bar{x} = 30.38$, $\bar{y} = 24.884$, $s_x = 6.89$, $s_y = 4.013$, and $r = 0.950$. As a result, $b_1 = 0.5533$ and $b_0 = 8.07$. The other choices all correspond to various mistakes in labeling summary statistics.

37. B

Because one-sided procedures show the test statistic to be significant at the $\alpha = 0.05$ level but not the $\alpha = 0.01$ level, the tail probability is between 0.05 and 0.01. For a two-sided test, the *p*-value will include the areas of both tails, so the *p*-value will be between 0.10 and 0.02. That means the statistic is significant at the $\alpha = 0.10$ level but not at the $\alpha = 0.01$ level. Choice (B) is correct. You might consider choice (C); after all, the *p*-value might be more than 0.05, in which (C) would be true. The question asks, though, which choice must be true. Choice (C) could be true, but it is not automatically true.

38. C

The placebo effect occurs when the subject(s) show the effect of a treatment they believe they are receiving, whether there is a genuine treatment or not. Control, or comparison of treatments, generally prevents confounding. In order for a control group to be treated the same as a treatment group, however, the experiment must be blinded. The subjects must not know which treatment they are receiving. The other elements of experimental design are less important in preventing confounding by the placebo effect.

39. D

The margin of error for a confidence interval gives an interval of values in which the population parameter is likely to be found. This is what choice (D) says, restated for the context of the problem. A margin of error of 5% is not a probability, so choices (A) and (C) are wrong. No confidence level or sample sizes are mentioned in the problem, so the assumption that the confidence level is 95% is unwarranted. That rules out (B). Confidence intervals do not always contain the population parameter, so (E) is incorrect as well.

40. E

The *p*-value for this significance test is found using a *t*-test statistic using 398 degrees of freedom. The test statistic is not given as part of the regression results, but it is easily obtained from this output.

$$t = \frac{b_1}{s_{b_1}} = \frac{\text{slope of the regression line}}{\text{standard error of the slope}} = \frac{-0.004447}{0.003529}$$

$= -1.26$. Consulting the *t*-table shows a tail area greater than 0.10. Since this is a two-tail test, the *p*-value is greater than 0.20. Alternatively, you can have your calculator compute the actual *p*-value of 0.208.

SECTION II ANSWERS

1. (a) The plot shows that 2,210 m corresponds to a percentile just below 50. Outlook Peak is at about the 48th percentile.

   (b) The plot shows Q_1, the 25th percentile, at about 875 m. Q_3, the 75th percentile, appears to be at about 3,500 m. That makes the interquartile range about 3,500 – 875, or 2,625 m.

   (c) Steep parts of this graph show clusters and completely flat parts show gaps. There are two principal clusters here, one from 625 m to 1,625 m and another from 3,375 m to 3,750 m.

2. (a) The following diagram shows the design.

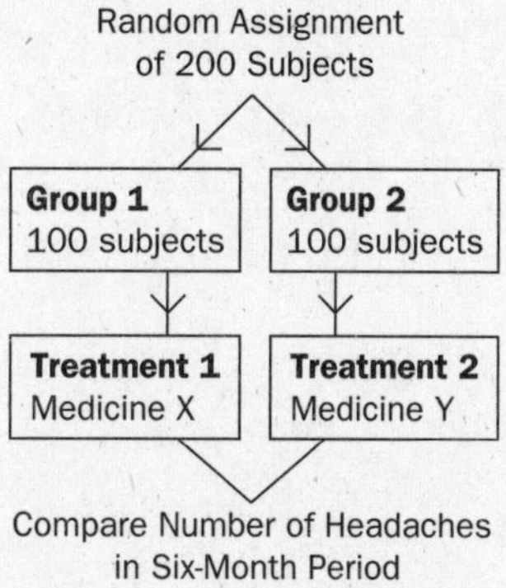

   (b) This experiment can be blinded by having the subjects in Group 1 take medicine X in pill form and a placebo nasal spray. Subjects in Group 2 take medicine Y as a nasal spray and a placebo pill. This way, neither group knows which active medicine is being administered.

   (c) The experiment should be blocked by gender so that gender does not confound the results. Block 1 contains the 120 men and Block 2 the 80 women. The design above is then followed for each block. The groups in Block 1 will contain 60 subjects each and the groups in Block 2 will have 40 subjects each.

3. (a) The regression equation is $\widehat{\text{Megabytes}} = 15.805 + 0.011918$ Pages. In hour 12 Pages = 2231, so $\widehat{\text{Megabytes}} = 42.39$. Thus:

   Residual = Megabytes – $\widehat{\text{Megabytes}} = 44.78 - 42.39 = 2.39$ megabytes of information.

   (b) The intercept is 15.805 megabytes of information. The regression line predicts that an hour that has 0 pages viewed corresponds to a transfer of 15.805 megabytes of information.

   (c) The correlation coefficient is positive because the slope is positive. Thus, $r = \sqrt{.853} = 0.924$. This means that the variables have a strong positive linear correlation. Large values of Pages have a strong tendency to associate with large values of Megabytes.

4. **Step 1:** Hypotheses.

Using p_1 for the proportion of students preferring C when C was tasted first and p_2 for the proportion of students preferring C when P was tasted first, the hypotheses are:

$H_0: p_1 = p_2$ and $H_a: p_1 \neq p_2$

This test will be conducted at the $\alpha = 0.05$ significance level.

Step 2: Check requirements for the two-sample proportion z-test.

(i) Each sample is an independent simple random sample of the population.

(ii) All of the following values are at least 10:

$$n_1\hat{p}_1 = 102 \cdot \frac{61}{102} = 61,\ n_1(1-\hat{p}_1) = 102 \cdot \frac{41}{102} = 41$$

$$n_2\hat{p}_2 = 105 \cdot \frac{45}{105} = 45,\ n_2(1-\hat{p}_2) = 105 \cdot \frac{60}{105} = 60$$

(iii) The population is at least 10 times each sample size. This is a large university, so it will have at least 1,020 and 1,050 students.

Step 3: Calculations.

The pooled sample proportion is $\hat{p} = \dfrac{61+45}{102+105} = \dfrac{106}{207} = 0.512.$

$$z = \frac{\hat{p}_1 - \hat{p}_2}{\sqrt{\hat{p}(1-\hat{p})}\sqrt{\dfrac{1}{n_1} + \dfrac{1}{n_2}}} = 2.44$$

p-value $= 0.015$.

Step 4: Conclusion.

The p-value of 0.015 is less than the significance level $\alpha = 0.05$. This test shows we should reject the null hypothesis, that the proportion of the population preferring C is the same regardless of which cola is tasted first, in favor of the alternative, which states that the order of presentation does matter.

Note: This work could be conducted a number of different ways. There are several choices of proportions that could be compared. The test could also have been conducted in a fundamentally different way. A chi-square test for homogeneity of proportions is also an appropriate method.

5. (a) This is a two-sample t-interval for the difference of means. The conservative estimate for df is 49, which is just about 50, so use $t^* = 2.009$ from the t-table. The confidence interval is thus

$$(\bar{x}_{old} - \bar{x}_{new}) \pm t^* \sqrt{\frac{s_{old}^2}{n_{old}} + \frac{s_{new}^2}{n_{new}}}$$

$$(258 - 246) \pm 2.009\sqrt{\frac{25^2}{50} + \frac{40^2}{50}}$$

$$12.0 \pm 13.4 \text{ yds, or } -1.4 \text{ to } 25.4 \text{ yds.}$$

Note: If a calculator is used, you may get –1.3 to 25.3 yds.

(b) This confidence interval can be used to test the null hypothesis that the two types of golf balls have the same mean driving distance against the alternative hypothesis that the mean driving distances differ. This test is conducted at the 5% significance level because we are using a 95% confidence interval. Because 0 is in the confidence interval, there is not enough evidence to reject the null hypothesis at the 5% level. The results are consistent with the hypothesis that the balls have the same mean driving distance, but this is by no means a proof of this claim.

6. (a) To simulate a question, select a digit from the table. If the digit is 1 or 2, the question has been guessed correctly. Any other digit means the question has been guessed incorrectly. Repeat this for 40 questions to simulate how many were guessed correctly on the test. Perform this 40-question simulation many times to simulate the distribution of the number of questions guessed correctly.

(b) Only the first 10 digits of the table are needed. The results for each question are labeled N for a wrong answer and Y for a right answer. This simulation shows two right answers for the 10 questions simulated.

N N N N Y   N N N Y N
8 4 3 9 2   3 6 6 2 3

(c) This is a binomial probability question with $n = 40$ and $p = 0.2$. If X is the number of questions guessed correctly, then

$$P(X = 8) = \binom{40}{8}(0.2)^8(0.8)^{32} \approx 0.156,$$

so theoretically $500 \cdot 0.156$, or 78, simulations should have produced 8 correct guesses. Similar methods show 69.3 simulations should produce 9 correct guesses and 53.7 should produce 10 correct guesses.

(d) The chi-square test for goodness of fit is called for.

Step 1: Hypotheses

H_0: The populations of all simulation results follow the proportions predicted by the binomial distribution.

H_0: The populations of all simulation results do not follow the proportions predicted by the binomial distribution.

Step 2: Check requirements for the goodness of fit test.

(i) The results constitute a simple random sample of the population.

(ii) All expected cell counts are greater than 5.

Step 3: Calculations

$$\chi^2 = \sum \frac{(\text{obs-exp})^2}{\text{exp}} = \frac{(82-78)^2}{78} + \frac{(60-69.3)^2}{69.3} + \frac{(58-53.7)^2}{53.7} = 1.80.$$

Degrees of freedom = 3 – 1 = 2

p-value > 0.25 from the chi-square table

Step 4: Conclusion

The p-value is larger than any reasonable significance level (such as 0.05 or 0.01), so there is no evidence to reject the null hypothesis. The simulation results shown are consistent with the predictions made by the binomial distribution.

APPENDIX: FORMULAS AND TABLES

YOU MAY USE THE FOLLOWING FORMULAS AND CHARTS ON THE TEST.

(I) Descriptive Statistics

$$\bar{x} = \frac{\sum x_i}{n}$$

$$s_x = \sqrt{\frac{1}{n-1}\sum(x_i - \bar{x})^2}$$

$$s_p = \sqrt{\frac{(n_1 - 1)s_1^2 + (n_2 - 1)s_2^2}{(n_1 - 1) + (n_2 - 1)}}$$

$$\hat{y} = b_0 + b_1 x$$

$$b_1 = \frac{\sum(x_i - \bar{x})(y_i - \bar{y})}{\sum(x_i - \bar{x})^2}$$

$$b_0 = \bar{y} - b_1\bar{x}$$

$$r = \frac{1}{n-1}\sum\left(\frac{x_i - \bar{x}}{s_x}\right)\left(\frac{y_i - \bar{y}}{s_y}\right)$$

$$b_1 = r\frac{s_y}{s_x}$$

$$s_{b_1} = \frac{\sqrt{\frac{\sum(y_i - \hat{y}_i)^2}{n-2}}}{\sqrt{\sum(x_i - \bar{x})^2}}$$

(II) Probability

$$P(A \cup B) = P(A) + P(B) - P(A \cap B)$$

$$P(A \mid B) = \frac{P(A \cap B)}{P(B)}$$

$$E(X) = \mu_x = \sum x_i p_i$$

$$\text{Var}(X) = \sigma_x^2 = \sum(x_i - \mu_x)^2 p_i$$

If X has a binomial distribution with parameters n and p, then

$$P(X = k) = \binom{n}{k} p^k (1 - p)^{n-k}$$

$$\mu_x = np$$

$$\sigma_x = \sqrt{np(1-p)}$$

$$\mu_{\hat{p}} = p$$

$$\sigma_{\hat{p}} = \sqrt{\frac{p(1-p)}{n}}$$

If $\bar{x}$ is the mean of a random sample of size n from an infinite population with mean $\mu_{\hat{p}}$ and standard deviation σ, then

$$\mu_{\bar{x}} = \mu$$

$$\sigma_{\bar{x}} = \frac{\sigma}{\sqrt{n}}$$

(III) Inferential Statistics

Standardized test statistic: $\dfrac{\text{statistic} - \text{parameter}}{\text{standard deviation of statistic}}$

Confidence interval: statistic ± (critical value) · (standard deviation of statistic)

Single-Sample

Statistic	Standard Deviation of Statistic
Sample mean	$\dfrac{\sigma}{\sqrt{n}}$
Sample proportion	$\sqrt{\dfrac{p(1-p)}{n}}$

Two-Sample

Statistic	Standard Deviation of Statistic
Difference of sample means	$\sqrt{\dfrac{\sigma_1^2}{n_1} + \dfrac{\sigma_2^2}{n_2}}$ Special case when $\sigma_1 = \sigma_2$ $\sigma\sqrt{\dfrac{1}{n_1} + \dfrac{1}{n_2}}$
Difference of sample proportions	$\sqrt{\dfrac{p_1(1-p_1)}{n_1} + \dfrac{p_2(1-p_2)}{n_2}}$ Special case when $p_1 = p_2$ $\sqrt{p(1-p)}\sqrt{\dfrac{1}{n_1} + \dfrac{1}{n_2}}$

Chi-square test statistic $= \sum \dfrac{(\text{observed} - \text{expected})^2}{\text{expected}}$

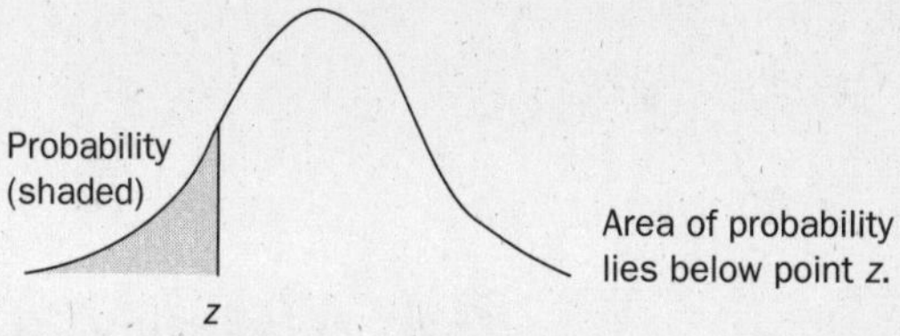

TABLE A STANDARD NORMAL PROBABILITIES

z	.00	.01	.02	.03	.04	.05	.06	.07	.08	.09
−3.4	.0003	.0003	.0003	.0003	.0003	.0003	.0003	.0003	.0003	.0002
−3.3	.0005	.0005	.0005	.0004	.0004	.0004	.0004	.0004	.0004	.0003
−3.2	.0007	.0007	.0006	.0006	.0006	.0006	.0006	.0005	.0005	.0005
−3.1	.0010	.0009	.0009	.0009	.0008	.0008	.0008	.0008	.0007	.0007
−3.0	.0013	.0013	.0013	.0012	.0012	.0011	.0011	.0011	.0010	.0010
−2.9	.0019	.0018	.0018	.0017	.0016	.0016	.0015	.0015	.0014	.0014
−2.8	.0026	.0025	.0024	.0023	.0023	.0022	.0021	.0021	.0020	.0019
−2.7	.0035	.0034	.0033	.0032	.0031	.0030	.0029	.0028	.0027	.0026
−2.6	.0047	.0045	.0044	.0043	.0041	.0040	.0039	.0038	.0037	.0036
−2.5	.0062	.0060	.0059	.0057	.0055	.0054	.0052	.0051	.0049	.0048
−2.4	.0082	.0080	.0078	.0075	.0073	.0071	.0069	.0068	.0066	.0064
−2.3	.0107	.0104	.0102	.0099	.0096	.0094	.0091	.0089	.0087	.0084
−2.2	.0139	.0136	.0132	.0129	.0125	.0122	.0119	.0116	.0113	.0110
−2.1	.0179	.0174	.0170	.0166	.0162	.0158	.0154	.0150	.0146	.0143
−2.0	.0228	.0222	.0217	.0212	.0207	.0202	.0197	.0192	.0188	.0183
−1.9	.0287	.0281	.0274	.0268	.0262	.0256	.0250	.0244	.0239	.0233
−1.8	.0359	.0351	.0344	.0336	.0329	.0322	.0314	.0307	.0301	.0294
−1.7	.0446	.0436	.0427	.0418	.0409	.0401	.0392	.0384	.0375	.0367
−1.6	.0548	.0537	.0526	.0516	.0505	.0495	.0485	.0475	.0465	.0455
−1.5	.0668	.0655	.0643	.0630	.0618	.0606	.0594	.0582	.0571	.0559
−1.4	.0808	.0793	.0778	.0764	.0749	.0735	.0721	.0708	.0694	.0681
−1.3	.0968	.0951	.0934	.0918	.0901	.0885	.0869	.0853	.0838	.0823
−1.2	.1151	.1131	.1112	.1093	.1075	.1056	.1038	.1020	.1003	.0985
−1.1	.1357	.1335	.1314	.1292	.1271	.1251	.1230	.1210	.1190	.1170
−1.0	.1587	.1562	.1539	.1515	.1492	.1469	.1446	.1423	.1401	.1379
−0.9	.1841	.1814	.1788	.1762	.1736	.1711	.1685	.1660	.1635	.1611
−0.8	.2119	.2090	.2061	.2033	.2005	.1977	.1949	.1922	.1894	.1867
−0.7	.2420	.2389	.2358	.2327	.2296	.2266	.2236	.2206	.2177	.2148
−0.6	.2743	.2709	.2676	.2643	.2611	.2578	.2546	.2514	.2483	.2451
−0.5	.3085	.3050	.3015	.2981	.2946	.2912	.2877	.2843	.2810	.2776
−0.4	.3446	.3409	.3372	.3336	.3300	.3264	.3228	.3192	.3156	.3121
−0.3	.3821	.3783	.3745	.3707	.3669	.3632	.3594	.3557	.3520	.3483
−0.2	.4207	.4168	.4129	.4090	.4052	.4013	.3974	.3936	.3897	.3859
−0.1	.4602	.4562	.4522	.4483	.4443	.4404	.4364	.4325	.4286	.4247
−0.0	.5000	.4960	.4920	.4880	.4840	.4801	.4761	.4721	.4681	.4641

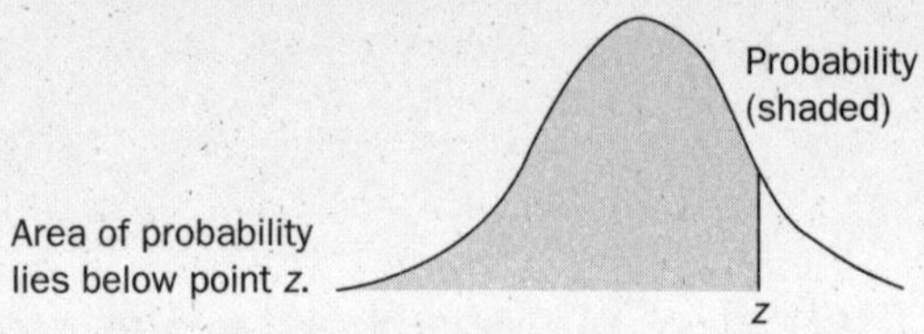

TABLE A (CONTINUED)

z	.00	.01	.02	.03	.04	.05	.06	.07	.08	.09
0.0	.5000	.5040	.5080	.5120	.5160	.5199	.5239	.5279	.5319	.5359
0.1	.5398	.5438	.5478	.5517	.5557	.5596	.5636	.5675	.5714	.5753
0.2	.5793	.5832	.5871	.5910	.5948	.5987	.6026	.6064	.6103	.6141
0.3	.6179	.6217	.6255	.6293	.6331	.6368	.6406	.6443	.6480	.6517
0.4	.6554	.6591	.6628	.6664	.6700	.6736	.6772	.6808	.6844	.6879
0.5	.6915	.6950	.6985	.7019	.7054	.7088	.7123	.7157	.7190	.7224
0.6	.7257	.7291	.7324	.7357	.7389	.7422	.7454	.7486	.7517	.7549
0.7	.7580	.7611	.7642	.7673	.7704	.7734	.7764	.7794	.7823	.7852
0.8	.7881	.7910	.7939	.7967	.7995	.8023	.8051	.8078	.8106	.8133
0.9	.8159	.8186	.8212	.8238	.8264	.8289	.8315	.8340	.8365	.8389
1.0	.8413	.8438	.8461	.8485	.8508	.8531	.8554	.8577	.8599	.8621
1.1	.8643	.8665	.8686	.8708	.8729	.8749	.8770	.8790	.8810	.8830
1.2	.8849	.8869	.8888	.8907	.8925	.8944	.8962	.8980	.8997	.9015
1.3	.9032	.9049	.9066	.9082	.9099	.9115	.9131	.9147	.9162	.9177
1.4	.9192	.9207	.9222	.9236	.9251	.9265	.9279	.9292	.9306	.9319
1.5	.9332	.9345	.9357	.9370	.9382	.9394	.9406	.9418	.9429	.9441
1.6	.9452	.9463	.9474	.9484	.9495	.9505	.9515	.9525	.9535	.9545
1.7	.9554	.9564	.9573	.9582	.9591	.9599	.9608	.9616	.9625	.9633
1.8	.9641	.9649	.9656	.9664	.9671	.9678	.9686	.9693	.9699	.9706
1.9	.9713	.9719	.9726	.9732	.9738	.9744	.9750	.9756	.9761	.9767
2.0	.9772	.9778	.9783	.9788	.9793	.9798	.9803	.9808	.9812	.9817
2.1	.9821	.9826	.9830	.9834	.9838	.9842	.9846	.9850	.9854	.9857
2.2	.9861	.9864	.9868	.9871	.9875	.9878	.9881	.9884	.9887	.9890
2.3	.9893	.9896	.9898	.9901	.9904	.9906	.9909	.9911	.9913	.9916
2.4	.9918	.9920	.9922	.9925	.9927	.9929	.9931	.9932	.9934	.9936
2.5	.9938	.9940	.9941	.9943	.9945	.9946	.9948	.9949	.9951	.9952
2.6	.9953	.9955	.9956	.9957	.9959	.9960	.9961	.9962	.9963	.9964
2.7	.9965	.9966	.9967	.9968	.9969	.9970	.9971	.9972	.9973	.9974
2.8	.9974	.9975	.9976	.9977	.9977	.9978	.9979	.9979	.9980	.9981
2.9	.9981	.9982	.9982	.9983	.9984	.9984	.9985	.9985	.9986	.9986
3.0	.9987	.9987	.9987	.9988	.9988	.9989	.9989	.9989	.9990	.9990
3.1	.9990	.9991	.9991	.9991	.9992	.9992	.9992	.9992	.9993	.9993
3.2	.9993	.9993	.9994	.9994	.9994	.9994	.9994	.9995	.9995	.9995
3.3	.9995	.9995	.9995	.9996	.9996	.9996	.9996	.9996	.9996	.9997
3.4	.9997	.9997	.9997	.9997	.9997	.9997	.9997	.9997	.9997	.9998

Area of probability p lies above point t^*. Confidence level C lies between $-t^*$ and t^*.

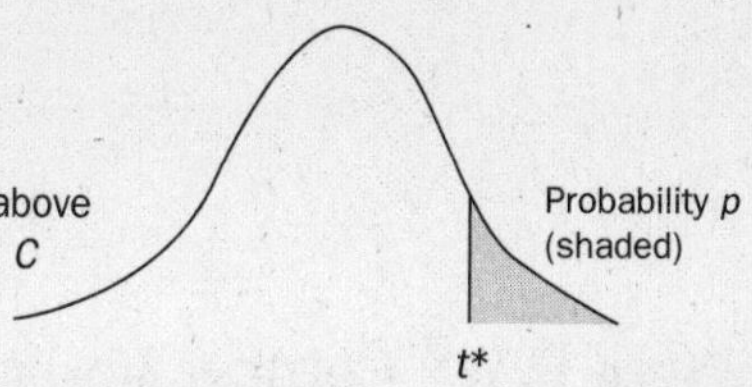

TABLE B *t*-DISTRIBUTION CRITICAL VALUES

	Tail Probability *p*											
df	.25	.20	.15	.10	.05	.025	.02	.01	.005	.0025	.001	.0005
1	1.000	1.376	1.963	3.078	6.314	12.71	15.89	31.82	63.66	127.3	318.3	636.6
2	.816	1.061	1.386	1.886	2.920	4.303	4.849	6.965	9.925	14.09	22.33	31.60
3	.765	.978	1.250	1.638	2.353	3.182	3.482	4.541	5.841	7.453	10.21	12.92
4	.741	.941	1.190	1.533	2.132	2.776	2.999	3.747	4.604	5.598	73173	8.610
5	.727	.920	1.156	1.476	2.015	2.571	2.757	3.365	4.032	4.773	5.893	6.869
6	.718	.906	1.134	1.440	1.943	2.447	2.612	3.143	3.707	4.317	5.208	5.959
7	.711	.896	1.119	1.415	1.895	2.365	2.517	2.998	3.499	4.029	4.785	5.408
8	.706	.889	1.108	1.397	1.860	2.306	2.449	2.896	3.355	3.833	4.501	5.041
9	.703	.883	1.100	1.383	1.833	2.262	2.398	2.821	3.250	3.690	4.297	4.781
10	.700	.879	1.093	1.372	1.812	2.228	2.359	2.764	3.169	3.581	4.144	4.587
11	.697	.876	1.088	1.363	1.796	2.201	2.328	2.718	3.106	3.497	4.025	4.437
12	.695	.873	1.083	1.356	1.782	2.179	2.303	2.681	3.055	3.428	3.930	4.318
13	.694	.870	1.079	1.350	1.771	2.160	2.282	2.650	3.012	3.372	3.852	4.221
14	.692	.868	1.076	1.345	1.761	2.145	2.264	2.624	2.977	3.326	3.787	4.140
15	.691	.866	1.074	1.341	1.753	2.131	2.249	2.602	2.947	3.286	3.733	4.073
16	.690	.865	1.071	1.337	1.746	2.120	2.235	2.583	2.921	3.252	3.686	4.015
17	.689	.863	1.069	1.333	1.740	2.110	2.224	2.567	2.898	3.222	3.646	3.965
18	.688	.862	1.067	1.330	1.734	2.101	2.214	2.552	2.878	3.197	3.611	3.922
19	.688	.861	1.066	1.328	1.729	2.093	2.205	2.539	2.861	3.174	3.579	3.883
20	.687	.860	1.064	1.325	1.725	2.086	2.197	2.528	2.845	3.153	3.552	3.850
21	.686	.859	1.063	1.323	1.721	2.080	2.189	2.518	2.831	3.135	3.527	3.819
22	.686	.858	1.061	1.321	1.717	2.074	2.183	2.508	2.819	3.119	3.505	3.792
23	.685	.858	1.060	1.319	1.714	2.069	2.177	2.500	2.807	3.104	3.485	3.768
24	.685	.857	1.059	1.318	1.711	2.064	2.172	2.492	2.797	3.091	3.467	3.745
25	.684	.856	1.058	1.316	1.708	2.060	2.167	2.485	2.787	3.078	3.450	3.725
26	.684	.856	1.058	1.315	1.706	2.056	2.162	2.479	2.779	3.067	3.435	3.707
27	.684	.855	1.057	1.314	1.703	2.052	2.158	2.473	2.771	3.057	3.421	3.690
28	.683	.855	1.056	1.313	1.701	2.048	2.154	2.467	2.763	3.047	3.408	3.674
29	.683	.854	1.055	1.311	1.699	2.045	2.150	2.462	2.756	3.038	3.396	3.659
30	.683	.854	1.055	1.310	1.697	2.042	2.147	2.457	2.750	3.030	3.385	3.646
40	.681	.851	1.050	1.303	1.684	2.021	2.123	2.423	2.704	2.971	3.307	3.551
50	.679	.849	1.047	1.299	1.676	2.009	2.109	2.403	2.678	2.937	3.261	3.496
60	.679	.848	1.045	1.296	1.671	2.000	2.099	2.390	2.660	2.915	3.232	3.460
80	.678	.846	1.043	1.292	1.664	1.990	2.088	2.374	2.639	2.887	3.195	3.416
100	.677	.845	1.042	1.290	1.660	1.984	2.081	2.364	2.626	2.871	3.174	3.390
1000	.675	.842	1.037	1.282	1.646	1.962	2.056	2.330	2.581	2.813	3.098	3.300
∞	.674	.841	1.036	1.282	1.645	1.960	2.054	2.326	2.576	2.807	3.091	3.291
	50%	60%	70%	80%	90%	95%	96%	98%	99%	99.5%	99.8%	99.9%
	Confidence Level *C*											

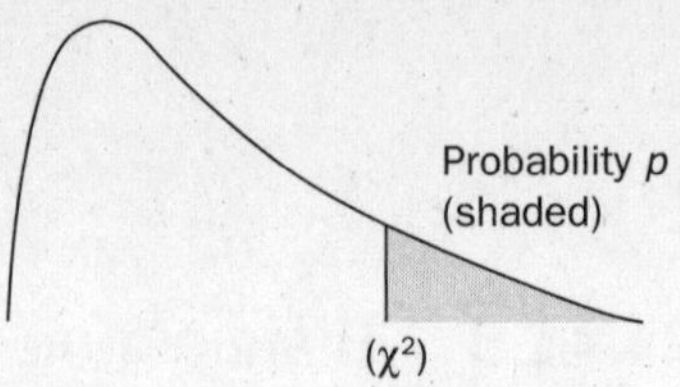

Area of probability lies above (χ^2).

TABLE C χ^2 CRITICAL VALUES

	Tail Probability p											
df	.25	.20	.15	.10	.05	.025	.02	.01	.005	.0025	.001	.0005
1	1.32	1.64	2.07	2.71	3.84	5.02	5.41	6.63	7.88	9.14	10.83	12.12
2	2.77	3.22	3.79	4.61	5.99	7.38	7.82	9.21	10.60	11.98	13.82	15.20
3	4.11	4.64	5.32	6.25	7.81	9.35	9.84	11.34	12.84	14.32	16.27	17.73
4	5.39	5.99	6.74	7.78	9.49	11.14	11.67	13.28	14.86	16.42	18.47	20.00
5	6.63	7.29	8.12	9.24	11.07	12.83	13.39	15.09	16.75	18.39	20.51	22.11
6	7.84	8.56	9.45	10.64	12.59	14.45	15.03	16.81	18.55	20.25	22.46	24.10
7	9.04	9.80	10.75	12.02	14.07	16.01	16.62	18.48	20.28	22.04	24.32	26.02
8	10.22	11.03	12.03	13.36	15.51	17.53	18.17	20.09	21.95	23.77	26.12	27.87
9	11.39	12.24	13.29	14.68	16.92	19.02	19.68	21.67	23.59	25.46	27.88	29.67
10	12.55	13.44	14.53	15.99	18.31	20.48	21.16	23.21	25.19	27.11	29.59	31.42
11	13.70	14.63	15.77	17.28	19.68	21.92	22.62	24.72	26.76	28.73	31.26	33.14
12	14.85	15.81	16.99	18.55	21.03	23.34	24.05	26.22	28.30	30.32	32.91	34.82
13	15.98	16.98	18.20	19.81	22.36	24.74	25.47	27.69	29.82	31.88	34.53	36.48
14	17.12	18.15	19.41	21.06	23.68	26.12	26.87	29.14	31.32	33.43	36.12	38.11
15	18.25	19.31	20.60	22.31	25.00	27.49	28.26	30.58	32.80	34.95	37.70	39.72
16	19.37	20.47	21.79	23.54	26.30	28.85	29.63	32.00	34.27	36.46	39.25	41.31
17	20.49	21.61	22.98	24.77	27.59	30.19	31.00	33.41	35.72	37.95	40.79	42.88
18	21.60	22.76	24.16	25.99	28.87	31.53	32.35	34.81	37.16	39.42	42.31	44.43
19	22.72	23.90	25.33	27.20	30.14	32.85	33.69	36.19	38.58	40.88	43.82	45.97
20	23.83	25.04	26.50	28.41	31.41	34.17	35.02	37.57	40.00	42.34	45.31	47.50
21	24.93	26.17	27.66	29.62	32.67	35.48	36.34	38.93	41.40	43.78	46.80	49.01
22	26.04	27.30	28.82	30.81	33.92	36.78	37.66	40.29	42.80	45.20	48.27	50.51
23	27.14	28.43	29.98	32.01	35.17	38.08	38.97	41.64	44.18	46.62	49.73	52.00
24	28.24	29.55	31.13	33.20	36.42	39.36	40.27	42.98	45.56	48.03	51.18	53.48
25	29.34	30.68	32.28	34.38	37.65	40.65	41.57	44.31	46.93	49.44	52.62	54.95
26	30.43	31.79	33.43	35.56	38.89	41.92	42.86	45.64	48.29	50.83	54.05	56.41
27	31.53	32.91	34.57	36.74	40.11	43.19	44.14	46.96	49.64	52.22	55.48	57.86
28	32.62	34.03	35.71	37.92	41.34	44.46	45.42	48.28	50.99	53.59	56.89	59.30
29	33.71	35.14	36.85	39.09	42.56	45.72	46.69	49.59	52.34	54.97	58.30	60.73
30	34.80	36.25	37.99	40.26	43.77	46.98	47.96	50.89	53.67	56.33	59.70	62.16
40	45.62	47.27	49.24	51.81	55.76	59.34	60.44	63.69	66.77	69.70	73.40	76.09
50	56.33	58.16	60.35	63.17	67.50	71.42	72.61	76.15	79.49	82.66	86.66	89.56
60	66.98	68.97	71.34	74.40	79.08	83.30	84.58	88.38	91.95	95.34	99.61	102.7
80	88.13	90.41	93.11	96.58	101.9	106.6	108.1	112.3	116.3	120.1	124.8	128.3
100	109.1	111.7	114.7	118.5	124.3	129.6	131.1	135.8	140.2	144.3	149.4	153.2

GLOSSARY

alternative hypothesis

An alternative statement to the null hypothesis that will be accepted if the null hypothesis is rejected. The alternative hypothesis is denoted H_a. An example is $H_a \neq \mu$. (chapter 17)

bad sampling frame

A general term that includes undercoverage, overcoverage, duplication of entries, etc., in a sampling frame. (chapter 9)

bias

Any systematic tendency to favor certain outcomes at the expense of others. (chapter 8)

binomial distribution

In probability, a distribution for an experiment with just two possible outcomes. (chapter 13)

bivariate data

Data that describes two characteristics of the subject or individual, such as height and weight. (chapter 3)

blinding

In an experiment, the act of keeping subjects from knowing which treatment each is receiving. This helps to keep the experiment free of bias. (chapter 10)

block

In an experiment, a group of randomly selected subjects who are separated based on a single characteristic (e.g., all freshmen). (chapter 10)

boxplot

A graphic display that represents the distribution of data by focusing on five key measures: the minimum value, the first quartile (25th percentile), the median (50th percentile), the third quartile (75th percentile), and the maximum value. (chapter 4)

categorical data

Data that is classified into one of several nonoverlapping groups or categories. Also referred to as qualitative data. (chapter 3)

census

A survey that gathers information from every member of a population. (chapter 8)

chi-squared distribution

In probability, a family of distributions that take only positive values and are skewed to the right. (chapter 15)

cluster sampling

A method of selecting entire groups at random. Each group, or cluster, should resemble the population and each other. (chapter 9)

complement

For an event *A*, the probability of not *A*, or $1 - P(A)$. (chapter 12)

conditional probability

In probability, an instance where the probability of a second event is dependent upon a first event having occurred. (chapter 12)

confidence level

The likelihood that a parameter will fall within a given margin of error or confidence interval. (chapter 16)

confounding variables

In an experiment, two or more variables whose effects on the response variable cannot be distinguished from one another. (chapter 10)

continuous

Said of quantitative data when there are an infinite number of data results possible, such as the exact weight of chickens in grams. (chapter 3)

control

The comparison of multiple treatments in an experiment. (chapter 10)

convenience sampling

Selecting a sample that is easy to collect rather than one that is representative of the population. (chapter 9)

correlation

A measure of the strength of a relationship between two variables. (chapter 6)

cumulative frequency plot

A histogram where any given bar represents the frequency or proportion of data values that fall into that value or anything less than that value. (chapter 3)

degrees of freedom

The number of independent observations. (chapter 15)

discrete

Said of quantitative data when every possible data value can be listed, such as scores on a statistics examination. (chapter 3)

dotplot

A number line representing the possible values of the data, with dots to show each data point. (chapter 3)

double blinding

In an experiment, the act of keeping both the subjects and the experimental staff from knowing which treatment each subject is receiving. Preferable to blinding. (chapter 10)

event

In probability, any outcome or set of outcomes of a random phenomenon. (chapter 12)

expected value

In probability, the mean of the probability distribution. (chapter 14)

experiment

In statistics, a study in which you are imposing a treatment. (chapter 8) In probability, any sort of activity whose results cannot be predicted with certainty. (chapter 12)

experimental unit

Any living or nonliving thing other than a person that is used as a participant in an experiment. (chapter 8)

explanatory variable

The cause of a measured effect in an experiment. (chapter 8)

geometric random variable

In probability, the number of trials required to get the first success in an experiment that has only two outcomes (failure or success). (chapter 13)

histogram

Similar to a bar chart; plots data in bars, where the height of the bar represents the count within the interval depicted on the *x*-axis. (chapter 3)

independent

In probability, the inability for any outcome or trial to affect the outcome of another trial. (chapter 12)

interquartile range

In a distribution, the difference between the third quartile value and the first quartile value. A quartile, or quarter, is one-fourth of the data. Thus, the first quartile, or Q_1, is the data value that has 25 percent of the data below it; the third quartile, or Q_3, is the data value that has 75 percent of the data below it. (chapter 4)

joint frequency

In a two-way contingency table, any frequency that occurs where a row category meets a column category. (chapter 7)

judgment bias

Bias derived from using samples selected through "expert judgment" rather than random sampling techniques. (chapter 9)

law of large numbers

In probability, law that states that the actual mean of many trials approaches the true mean of the distribution as the number of trials increases. (chapter 13)

least-squares regression

Process by which a line is fit through a set of data so that the sum of the squared deviations of all points on the line is minimized. (chapter 6)

linear correlation coefficient

Coefficient (r) is used to measure the strength of a linear relationship between two variables. The endpoints (−1 and +1) indicate a perfect correlation: −1 indicates a perfect negative correlation; +1 indicates a perfect positive correlation. A correlation of zero ($r = 0$) would indicate no correlation. The farther the measurement is from zero, the stronger the linear relationship between the two variables. (chapter 6)

lurking variable

In an experiment, a variable that might have an effect on the explanatory or response variable, through either common response or confounding. (chapter 10)

margin of error

A measurement that indicates the maximum amount the point estimate may deviate from the parameter. (chapter 16)

marginal frequency

In a two-way contingency table, the frequencies of the categories for the corresponding variable. (chapter 7)

matched-pairs design

In an experiment, the arrangement of subjects into pairs, where the subjects in each pair are as alike as possible. The members of each pair are randomly assigned to two treatments, and the response variable is measured for each. (chapter 10)

mean

The arithmetic average of a distribution. The mean divides an area of distribution into two equal halves. (chapter 4)

median

In a distribution, the middle value of the data set. Half of the values are above the median and half are below. (chapter 4)

multiplication rule

In probability, if A and B are independent events, P(A and B) = P(A)P(B). (chapter 12)

multistage sampling

Any random sampling scheme with at least two steps that incorporates elements of stratified sampling, cluster sampling, and/or simple random sampling. (chapter 9)

multivariate data

Data that describes multiple characteristics of the subject or individual, such as height, weight, percent body fat, blood pressure, temperature, resting pulse rate, or cholesterol level. (chapter 3)

mutually exclusive

In probability, two or more events that cannot occur together. For example, a flipped coin cannot yield both heads and tails on the same toss. (chapter 12)

normal distribution

In probability, a symmetric, unimodal density curve that is frequently used for statistical analysis. (chapter 14)

nonresponse

The lack of participation from certain individuals chosen as part of a sample. (chapter 9)

null hypothesis

A statement that the parameter equals a specific value. The null hypothesis is denoted H_0. An example is $H_0 = \mu$. (chapter 17)

observational study

A measurement of an attribute in a sample or population without intervening or altering conditions. Also referred to as a survey. (chapter 8)

ordered pair

In bivariate data, a pair of numbers (x, y) where x represents the independent variable, and y represents the response variable. (chapter 6)

outcome

In probability, one of the possible results of an experiment. (chapter 12)

outlier

A data point that lies outside of the general pattern of data. (chapter 3)

overcoverage

The inclusion of individuals in the sampling process who are not actually in the population. (chapter 9)

parameter

A measured characteristic about a population, using measurements calculated from a sample. (chapter 16)

placebo effect

In an experiment, an apparent change observed in the response variable simply because the subjects believe they are receiving a treatment, whether or not the treatment is having any genuine effect. (chapter 10)

point estimate

A single value that estimates a population parameter. (chapter 16)

population

The entire body of individuals to which statistics techniques are applied. (chapter 8)

probability

The long-term relative frequency of any outcome in a random experiment. (chapter 12)

probability density function

In probability, an expression giving the frequencies (and thus the probabilities) of continuous random variables. (chapter 14)

probability distribution

In probability, a listing or graphing of the probabilities associated with a random variable. (chapter 13)

probability sampling

The name given to any sampling scheme in which each member of the population has a given, fixed probability of being selected in a sample. (chapter 9)

qualitative data

See *categorical data*.

quantitative data

Data that represents a numerical measure or value. (chapter 3)

random

In probability, a phenomenon is random if individual outcomes are uncertain, but there is nonetheless a regular distribution to the outcomes over a large number of repetitions. (chapter 12)

random variable

In probability, a numerical measure of the outcomes of a random phenomenon. (chapter 13)

randomization

In an experiment, the random assignment of subjects to different treatment groups. (chapter 10)

randomized block design

In an experiment, the act of dividing subjects into blocks after the random selection process. (chapter 10)

range

In a distribution, the difference between the largest data value and the smallest data value. (chapter 4)

relative frequency histogram

A histogram that displays the percent or proportion of data points that fall into each bar. (chapter 3)

repeated measures design

In an experiment, the application of two treatments to each subject in random order. (chapter 10)

replication

In an experiment, the assignment of multiple subjects to each treatment group. (chapter 10)

residual

The difference between the *y*-value for an ordered pair and the predicted *y*-value for the *x*-value of that ordered pair. (chapter 6)

response bias

Any bias that results from the conduct of the interviewer in a survey, the circumstances in which the interview is conducted, or the interaction between the interviewer and the participant. (chapter 9)

response variable

A measured effect in an experiment. (chapter 8)

sample

A selection of individuals from a population. (chapter 8)

sample space

In probability, the set of all possible outcomes of an experiment. (chapter 12)

sample survey

An observational study that uses data gathered from a sample of a population. (chapter 8)

sampling distribution

In probability, the distribution of values of a statistic taken from all possible samples of a specific size. (chapter 15)

sampling frame

A list of all individuals from which a sample is drawn. (chapter 9)

sampling variability

The difference between two samples of the same population. (chapter 8)

scatterplot

A graph on the Cartesian coordinate system showing all the points that are represented by all ordered pairs in the data. (chapter 6)

simple random sample

A sample obtained by randomly selecting individuals from a population, with each individual equally likely to be selected. More precisely, for a simple random sample of *n* individuals from a population, every different sample of size *n* must have the same chance of being selected. (chapter 9)

size bias

Any tendency to favor either larger or smaller individuals in your sample. (chapter 9)

skewed distribution

A distribution where the majority of the values fall either to the left or the right when graphed, and the data is then spread out with a small number of values in the opposite direction. (chapter 3)

standard deviation

The square root of the variance of a distribution. (chapter 4)

statistic

An attribute of a sample. (chapter 4)

statistical inference

The process of drawing conclusions based on statistics. (chapter 8)

statistical significance

In an experiment, a characteristic achieved when the difference in values of the response variable from one treatment group to another is too large to be attributable to chance alone. (chapter 10)

stem and leaf plot

A type of graphic display of data where the first digit (the stem) represents one type of number (like the "tens" place, where "8" would represent all numbers from 80 through 89), and the second digit (the leaf) represents another type of number (such as the "units" place). A space separates the stem from the leaf, but numbers on the leaf for that stem are all grouped together. For example, a stem and leaf plot of 6 2233678 would represent the data 62, 62, 63, 63, 66, 67, and 68. (chapter 3)

stratified random sampling

The process of dividing a population into non-overlapping strata and then selecting an SRS from each stratum. (chapter 9)

subject

A person who participates (receives treatment) in an experiment. (chapter 8)

survey

See *observational study*.

symmetric distribution

A distribution that has two halves that are mirror images of each other. A vertical line can be drawn somewhere along the distribution, and when the graph is folded along this line, the two sides will match up. (chapter 3)

systematic sampling

Any regular, systematic way of selecting individuals from a sampling frame, such as taking every tenth name from a list of the population. (chapter 9)

***t*-distribution**

In probability, a family of distributions similar to normal distribution that is based on the number of degrees of freedom. (chapter 15)

treatment matrix

A representation of the set of all possible treatment options in an experiment. (chapter 8)

trial

In probability, the single running of a random phenomenon. (chapter 12)

two-stage sampling

Sampling design that combines cluster sampling with simple random sampling. (chapter 9)

two-way contingency table

A table used to express bivariate data where categorical data for each variable is expressed in columns and rows. (chapter 7)

undercoverage

The exclusion of a portion of a population during the sampling process. (chapter 9)

uniform distribution

A distribution where the values along the vertical axis all fall roughly on a horizontal line. (chapter 3)

univariate data

Data that describes a single characteristic of the subject or individual, such as height. (chapter 3)

variance

In a distribution, the amount of data spread found by examining the deviation of the data values from the mean. (chapter 4)

voluntary response bias

When participants self-select whether or not to participate as part of a sample. (chapter 9)